Agrometeorology of Plantation Crops

NIPA® GENX ELECTRONIC RESOURCES & SOLUTIONS P. LTD.
New Delhi-110 034

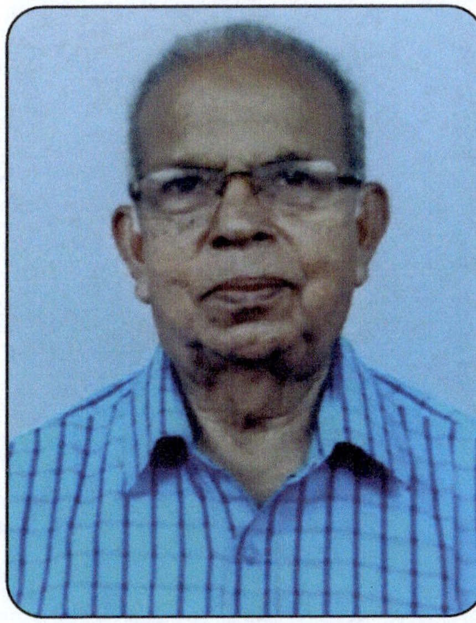

Dedicated to

Dr. R. Ravindran Nair 1936-2024

This book is dedicated to Late Dr. R. Ravindran Nair (Popularly known as Dr. R.R. Nair), Associate Director of Research (Rtd.), Kerala Agricultural University. He was a dedicated researcher, writer and strong administrator with whom my association was about a decade directly or indirectly. His liking towards the science of Agricultural Meteorology with his Agronomy background resulted to take up several research projects in the field of Agrometeorology of Plantation Crops and conducted a national seminar in 1987 and brought out the Proceedings. It was unique at that point of time in the field of Plantation Crops. It was a base for me and continued further in similar lines. The research work done in various projects finally led to this publication for the benefit of students and scientific community.

I am grateful to you, Sir. May His Soul Rest in Peace!

Agrometeorology of Plantation Crops

GSLHV Prasada Rao
Professor (Rtd.)
Kerala Agricultural University
Kerala

NIPA® GENX ELECTRONIC RESOURCES & SOLUTIONS P. LTD.
New Delhi-110 034

NIPA® GENX ELECTRONIC RESOURCES & SOLUTIONS P. LTD.

101,103, Vikas Surya Plaza, CU Block
L.S.C.Market, Pitam Pura, New Delhi-110 034
Ph. +91 11 27341616, 27341717, 27341718
E-mail: newindiapublishingagency@gmail.com
www: www.nipabooks.com

For customer assistance, please contact
Phone: + 91-11-27 34 17 17
Fax: + 91-11-27 34 16 16
E-Mail: feedbacks@nipabooks.com

Print ISBN: 978-93-5887-732-8

ebook ISBN: 978-93-5887-664-2

Composed and Designed by NIPA®.

Prof. B.V. Ramana Rao
M.Sc Tech., FAAM
Project Co-ordinator (Rtd.)
All India Coordinated Project on Agrometeorology
Central Research Institute for Dryland Agriculture
Indian Council of Agricultural Research
Santosh Nagar, Hyderabad

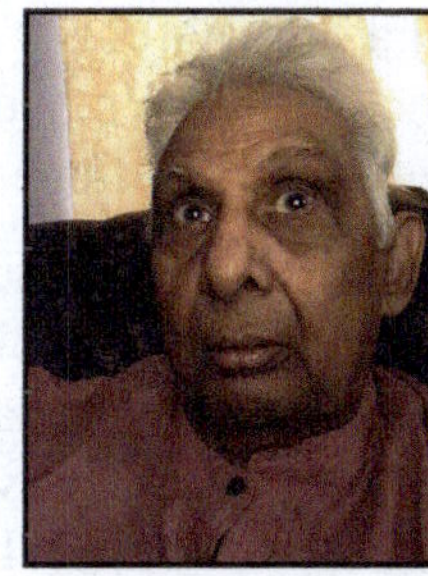

Foreword

The major plantation crops in India include coconut, cashew nut, cocoa, coffee, tea, rubber and spices like cardamom and pepper. The area under plantation crops in India is significant covering around 4271 thousand hectares. In terms of production, the total yield of plantation crops in India was approximately 16.05 million tons during the year 2022-23. Year to year variation in production can be attributed to several factors including climate change, weather variability and market fluctuations. The production of plantation crops has shown steady increase at the rate of 2.99 per cent per annum. The plantation crops contribute 10 to 12 per cent to India's agricultural GDP. India is a significant exporter of plantation crops, generating substantial foreign exchange in addition to providing livelihood for millions of rural households. The key producing states are Kerala, Karnataka, Andhra Pradesh, Tamil Nadu, Orissa, West Bengal and Assam. Most of these plantation crops are grown from sub humid to per humid climates in Peninsular and eastern parts of the Country. Coffee and tea are grown on mountain slopes in per humid climates.

The year to year variability in production can be attributed to changes in rainfall patterns, thermal regime, extreme weather events and pests and diseases which are manifested by weather conditions during critical period of productivity phase. Increased frequency of droughts can significantly impact productivity of coconut. Overall understanding of the year to year variation in production of plantation crops is crucial for policy makers, farmers and industries to adapt to changing climate conditions, market demands and pest management strategies. The Government initiatives for maintaining sustainable production of plantation crops include setting up of National Horticulture Mission, Coffee Board, Coconut Development Board, Tea Board and Spices Board etc. Despite the importance of plantation crops in National Economy, there was no concerted effort to consolidate information on the overall effects of weather and climate change on management and productivity of plantation crops to evolve future research strategies for increasing the production in the overall interest of National Economy.

In this context, the Initiative taken by Dr G.S.L.H.V. Prasada Rao to bring out a very valuable collection of consolidated information on Agricultural Meteorology of Plantation Crops is very much laudable and need of the hour. Dr G.S.L.H.V. Prasada Rao Rao is perhaps only Indian Scientist who has made significant research contribution in this particular area of research for more than four decades.

Undoubtedly, the present book will be very useful to the students in providing learning opportunities, researchers to redefine their research priorities and policy makers in refining targets for infrastructure development required to step up production. I take this opportunity to record my utmost satisfaction and appreciation to Dr G.S.L.H.V. Prasada Rao and the authors who have contributed different chapters.

Prof. B.V. Ramana Rao
M.Sc Tech., FAAM
Project Co-ordinator (Rtd.)
All India Coordinated Project on Agrometeorology
Central Research Institute for Dryland Agriculture
Indian Council of Agricultural Research
Santosh Nagar, Hyderabad

Preface

The plantation crops are predominant in Kerala which falls under the humid tropics and the homestead farming is the unique system of cultivation with various plantation and seasonal crop combinations. The mixed plantation with five-six crops and one or two seasonal crops also can be seen in multitier cropping system. The State's economy is predominantly a plantation economy and it is known as the Plantation State though the name of the State is originated from coconut. Coconut, cashew, cocoa, cardamom, rubber, tea, coffee, black pepper and areca nut are the important plantation crops in the State though rice is the main food crop. Among the plantation crops, coconut is the predominant one in terms of area and production in the State of Kerala. Cardamom dominates in Idukki District while coffee in Wayanad District across the high ranges. The crops that are grown across the high land and high ranges viz., tea, coffee, cardamom, cocoa and black pepper are thermo-sensitive and prone to rise in temperature and decline in rainfall in ensuing decades under warming Kerala. Crops such as coconut, cashew, areca nut and rubber are relatively tolerant in the humid tropics under profuse irrigated conditions. Though several research and education institutes are involved in crop improvement, crop management and crop protection in relation to plantation crops, research and education in the field of Agrometeorology of plantation crops are still in the infant stages. Realizing its importance the Association of Agrometeorologists prepared a course curriculum to improve the quality of education in the field of Agrometeorology at the postgraduate level and it is well documented in which a separate subject is incorporated on "Agrometeorology of Plantation crops" for the benefit of agromet students and this task is given to me for preparing a textbook in this field.

This book contains 12 chapters altogether. The first chapter deals with the Plantation Crops' Climate. It highlights the altitudinal distribution of crops, indicating that the crops that are grown across the highranges viz., tea, coffee and cardamom are not seen in the low and midland regions. Similarly the crops that are grown across the mid and highlands viz., cashew, rubber and cocoa are not seen across the highranges. Coconut is predominantly grown in the lowlands and midlands, beyond 600m it is not economic for commercial cultivation. Areca nut is seen cultivated from the lowlands to highlands and it is predominantly grown in valleys wherever water is plenty for summer irrigation while black pepper is seen everywhere mostly as intercrop with other plantation crops like coconut. With altitude as the surface air temperature decreases the thermo-sensitive crops like coffee, cardamom and tea are predominant. It reveals that the temperature distribution plays a predominant role in selection of plantation crops. Rainfall and temperature influences are summarised in plantation crops in the form of negative and positive effects. Influence of weather on quality aspects is also focussed in this chapter. The second chapter highlights on influence of weather and geographical coordinates on crop phenology of cashew. The cashew phenology at a given location shifts from year to year depending upon rainfall distribution during the monsoon season along the West Coast is well brought out in this chapter. Environmental constraints are focussed for improvement of cashew production across the cashew growing tracts. Incidence of tea mosquito is a constraint along the West Coast while cyclones along the East Coast. The threshold values of night temperature and hotspot areas of tea mosquito incidence across the cashew growing areas are presented in this chapter. The third chapter focuses on crop phenology of coconut and the influence of summer drought on monthly coconut yield and its price hike. The second phase of nut development in coconut is sensitive to temperature, deciding the final nut size. Crop-weather relationships are focussed in case of cocoa and difference in monthly cocoa yield is highlighted between bad and good yield years in chapter four. The quality of cocoa beans is superior once it is matured and harvested in October and November. The chapter five projects on agroecological zones of cardamom to assess cardamom yields across the Western Ghats. The influence of blossom and backing showers on coffee flowering and final yield are highlighted in chapter six. A combination of mixed crop with black pepper is focussed since one crop fails another crop performs under the similar weather conditions with examples. Summer rainfall with blossom and backing showers is beneficial to coffee while detrimental to black pepper if summer rains followed by dry spell is noticed since black pepper needs continuous rains till berry formation and development. Chapter seven deals with black pepper in relation to weather factors. Chapter eight deals with tea while rubber in chapter nine. The quality aspects are dealt in chapter ten. The quality of coconut, coffee and cashew nut is dealt in this chapter with reference to temperature, wind and monsoon rainfall. The quality of coffee can be well maintained if it is processed under continuous monsoon breeze along the West Coast with good amount of rainfall from July to September. Weather risk management is discussed in chapter eleven with various examples and local practices while the Chapter 12 deals with the overall initiatives taken in rubber research at RRII, Kottayam. Reprints of research papers presented in relation to Agrometeorology of plantation crops at the national seminar organized in Kerala Agricultural University at the Regional Agricultural Research Station, Pilicode during March 1987 are appended for the benefit of agromet students and the faculty.

The author wishes that the textbook will be informative and useful to the students and faculty and finally, but not the least, the author acknowledges the authorities of Kerala Agricultural University for extending technical support in research, teaching and extension during my carrier as Professor of Agricultural Meteorology. Retired in July 2012 as the Associate Director of Research and Special Officer, Academy of Climate Change Education and Research (renamed as the College of Climate Change and Environmental Science) in July 2012. On retirement, served as Consultant Professor in Kerala Veterinary and Animal Sciences University till September 2016 and took up short term international assignments in Govt. of Nepal, Kathmandu in 2018 and 2019.

Author

Contents

1

Plantation Crops' Climate

Coconut, cocoa (cacao), coffee, cashew, cardamom, black pepper, tea, rubber, oil palm and areca nut are the major plantation crops that are perennially grown along the coastal belt in southern States of the Country and to some extent in high rainfall zones of the northeastern States of India. These crops are also known as perennial crops. The life span of these crops varies from fifteen to sixty years or more depending upon the crop and its management. Coconut and cashew are also popularly known as the coastal trees. Cardamom and black pepper are also known as the spice crops. Once these plantation crops are well established, they mature for fruiting in three to eight years depending upon the nature of plantation crops and the crop management. The reproductive phase is seen round-the-year in case of coconut and cocoa. Cashew is one of the plantation crops with seasonality in fruiting from September to June. Cocoa is mostly grown as an intercrop in coconut gardens and in some pockets as a pure crop. Black pepper is trained to coconut and areca nut in many locations and seen as mono crop as well. All these crops are grown in laterite, sandy and alike in well drained soils. The hill slopes with terrace cultivation are also suitable for all the plantation crops. Tea, coffee and cardamom are seen on hillocks extending up to the base. Some of the plantation crops are seen in multitier cropping system as well as mixed/intercrops/homestead gardens. They are grown mostly under the rain fed and summer irrigation is beneficial for better harvest. The plantation crop distribution, production and productivity are influenced by air temperature, rainfall and the other meteorological variables viz. sunshine, cloud cover, solar radiation, vapour pressure, evapotranspiration, humidity and wind. Interestingly, coffee spread in Wayanad District while cardamom in Idukki District. Coffee experiences unimodel rainfall while bimodal in the case of cardamom. It is clear that every crop has its own cardinal points in terms of climatic requirements though they are classified under the umbrella of plantation crops. It is one of the reasons for altitudinal spread of these crops to avoid high temperatures and waterlogging due to heavy rains. Farmers prefer the northern slopes of hills for cultivation of almost all the thermo-sensitive crops to avoid exposure to direct sun light.

1.1 Altitudinal Sequence of Crop Distribution

The altitudinal sequence of crops in the Humid Tropics (Kerala), which may be related to temperature distribution. The altitudinal crop distribution is a classic example of the influence of temperature within the Humid Tropics and classified arbitrarily with the universally accepted thermal nomenclature **(Table1.1)**. The distinct feature is that the plantation crops which are grown in the high ranges like cardamom, coffee and tea are not seen in the lowlands and midlands while coconut is absent across the highranges on commercial scale. Coconut, cocoa, cashew, rubber, black pepper, oil palm and areca nut are seen mostly in the mid-high land regions. Of course, rice and banana are mostly in valleys and homestead gardens. It may be due to temperature distribution which is reflected by altitude. Areca nut also comes up in high altitudes under rain fed condition though it is an irrigated crop mostly seen in valleys and low lands **(Plate 1.1)**. Interestingly farmers are selective to choose mountain slopes facing north to cultivate some of the plantation crops like coffee and black pepper to avoid direct sunlight. All the plantation crops are susceptible to waterlogging and insect pest and diseases in high rainfall zones.

Plate 1.1: Altitudinal sequence of crops in Kerala (Influence of surface air temperature on crop distribution)

Table 1.1: Altitudinal sequence of crops in Kerala

Class	Region	Temperature conditions	Altitude (AMSL)	Crops
Megatherms	Lowland	Moderate temperature throughout the year	0–7.5 m	Coconut, rice and banana*and cashew
Megatherms	Midland	Moderate temperature throughout the year	7.5–75 m	Coconut, cocoa, areca nut, rubber, cashew and black pepper
Mesotherms	Highland	Moderate temperature throughout the year, winter temperature is low with altitude	75–700 m	Tea, coffee (robusta), rubber, black pepper and coconut
Mocrotherms	High ranges	Low temperature throughout the year	700–1500m	Cardamom, coffee (arabica) and tea

Rice and banana are cultivated everywhere in valleys irrespective of altitudinal difference

1.2 Coconut

Coconut productivity in Kerala is relatively low (5890 nuts/ha as against 7572 nuts/ha of the national average) though the total coconut production of the State is high due to area coverage. It appears that the name of Kerala is derived from the word KERA, meaning coconut. Coconut is mostly grown in laterite and sandy soils and on bunds below the mean sea level in Kuttanad belt. With increase in area under coconut, Kerala contributes around 45.2 per cent of the total coconut production of the country while there is no breakthrough in nut productivity of coconut. This could be attributed to several factors *viz.*, high rainfall during monsoon period, insignificant rainfall in several parts of Kerala during summer, nutrient status of soil, lack of agronomic practices and prevalence of root wilt disease in southern districts while stem bleeding in northern districts of Kerala. Coconut palms in Kerala not only experience severe soil moisture stress during summer but also subjected to waterlogging due to heavy rains during monsoon periods. The above situation is more predominant in the northern districts of Kerala where rainfall is very high during the monsoon months (June-September), followed by no significant rainfall from November to May if rainfall during the northeast monsoon and summer showers fail. The northern districts experience uni-model rainfall while bi-model rainfall in southern districts of Kerala. The coconut production of Kerala suffered to a great extent during 1983-84 due to unprecedented drought that occurred during summer 1983, followed by 2004 and 2013. The effect of high rainfall from June to September and prolonged dry spell during summer adversely affect the coconut production in the following year though the phenology of coconut from the primordium initiation to nut harvest takes place in 44 months. It takes 10-12 months from the spathe opening/ inflorescence to nut harvest. In all other coconut growing areas outside Kerala, coconut comes up well with well assured irrigation conditions only even if coconut is seen under warm conditions/heat waves during summer.

1.3 Cocoa

The biotic events such as flowering, cherelle production and pod set in cocoa are seen throughout the year with significant variation seasonally. However, the flowering appears to be very low during the rainy season (June to September). It normally initiates in October/November depending upon rainfall distribution from June to September and reaches to its peak by May. Similar trend is seen in the case of cherelle production and pod set. The above biotic phenomena is true only under rain fed conditions in the Humid Tropics, where heavy rainfall is noticed from June to September, followed by prolonged dry spell. Profuse flowering is also noticed during summer months if one or two good showers are received 15 days prior to commencement of flowering in cocoa. Interestingly, the number of flowers appears to be more under open conditions when compared to shade due to better availability of sunlight under good crop management practices. The number of cocoa (cacao) pods is high during summer, but pod weight is low. A harvest of five pods during October/November may be equal to eight/nine pods harvested in summer, showing superiority in cocoa pods harvested in October/November. There is a lag period of 4 to 6 months between weather effects and cocoa yield. The maximum temperature from January to March influences cocoa yield to a considerable extent. A prolonged dry spell from November to May with high maximum temperature adversely affects the pod yield to the tune of 40 per cent depending upon the crop management. High rainfall versus cocoa yield showed inverse relationship, indicating that high rainfall during rainy season may not be conducive in the case of cocoa due to waterlogging and lack of soil aeration. Overall, it reveals that high maximum temperature during summer, followed by heavy rains during the southwest monsoon may not be conducive for obtaining good yield in cocoa under the Humid Tropics. High rainfall has a malevolent effect in places where the crop is grown under waterlogged conditions, while summer rainfall has a benevolent effect on pod yield. On augmenting the dry season's rainfall by the use of irrigation, the results have not always been improved because the water lost from the leaves is so high with low humidity that the roots cannot match the loss. Crop yields are very low with heavy shade and increases with increasing light up to the 50% level. Cocoa yield is affected by the presence or absence of fertilizer if the level of light is above 50%. With added fertilizer, yields increase almost up to full light, whereas in the absence of fertilizer the yields fall off. The theory has therefore been advanced that the light regime for optimum yield of cocoa is a function of its mineral nutrition. The loss of apical dominance in cocoa is seen constantly at a temperature of above 32°C. The loss of apical dominance implies the development of side shoots of the plant. High maximum temperatures (> 36°C) during February and March, 2004 resulted in low pod yield in cocoa over the central region of Kerala. Cocoa is grown as an intercrop in coconut or part of homestead garden. Both the crops cocoa and coconut respond identically with reference to weather and climate under the Humid Tropics. Of course, the phenology of both the crops is totally different. In case of cocoa, the difference in monthly yield is high during the monsoon season between good and bad yield years when compared to other seasons **(Fig.1.1)**.

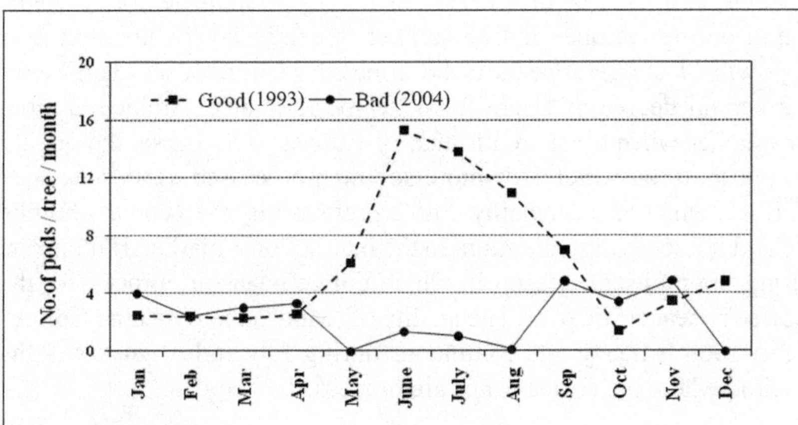

Fig. 1.1: Monthly cocoa yield of good and bad yield years (1993 and 2004)

1.4 Cardamom

A sharp decline in area under cardamom was noticed across the cardamom tract of the Western Ghats in recent decades. However, the production and productivity of small cardamom were increasing. The important physiological stages like panicle initiation and subsequent growth (forwarding) depend on receipt of showers from January to May. The failure of showers during this period results in poor growth and crop yield. The unprecedented drought in summer 1983 adversely affected the growth and yield of cardamom in Coorg district, and the same trend prevailed in other cardamom-growing tracts of South India. The drought that prevailed during 1982–83 resulted in as high as 50% mortality in some of the cardamom estates. In the Palney hills of Tamil Nadu, both panicle and flower production are seen in October and November, during which northeast monsoon rains are received. It indicates that both panicle and flower production in cardamom are commensurate with the rainfall pattern of the cardamom tract. There is a close relationship between cardamom production and rainfall distribution during summer. Summer rains during March, 2008 have benefited the cardamom crop of 2008-09 up to 20–30%. Therefore, summer rains have significant positive influence on cardamom.

1.5 Coffee

Arabica prefers high altitudes when compared to that of Robusta. Due to increase in temperature, maturity of crop appears to be early and weight of beans is also less. The major factors that affect production of the arabica and robusta varieties of coffee are related to the blossom and backing showers during summer. Coffee yield is very poor in the absence of either of them. That is why coffee growers invariably prefer sprinklers or drip irrigation facility to manage coffee yields better. Rainfall during October and November is beneficial for the next year crop. The growth of coffee appears to be affected when exposed, and hence it requires a certain degree of shade for better performance. Arabica is more sensitive to weather when compared to that of Robusta. Of course, the quality of Arabica appears to be better. It is more so when processed as "Monsooned Malabar Coffee", which is a speciality coffee preferred by the people who like black coffee and it fetches high premium in the international market. If monsoon is weak during the monsoon season in which coffee beans are processed, the quality of coffee appears to be poor. The quality of "Monsooned Malabar Coffee" is better if monsoon is heavy and continuous during July and August over the Malabar Region, where the coffee beans are processed.

1.6 Black Pepper

Black Pepper requires a warm and humid climate. An annual rainfall of 2500 mm is ideal for proper growth of the crop. Rainfall is believed to be the chief pollen vector in pepper. The flowering process in pepper is initiated by the application of water equivalent of 70 mm or more rainfall within the period of three weeks, followed by a dry spell. Rainfall influences the flower bud differentiation process. Spike elongation and berry development in pepper may be ceased if there is prolonged dry spell immediately after good summer showers. Hence, good summer showers followed by a dry spell may be detrimental to black pepper production. Moderate and continuous rainfall till berry initiation may be favourable to black pepper. It may not come up well in high ranges, where low temperature prevails throughout the year. The green berry weight varied significantly with soil moisture depletion. In contrast to coconut, cardamom and cocoa, black pepper responds relatively better under prolonged dry spells during summer. Of course, the mortality rate is high under poor management conditions during prolonged dry spell conditions in the case of young black pepper vines. It was the reason, why, black pepper gardens were wiped out in Wayanad District during summer 2004. Another important aspect is that whenever black pepper yield is low the coffee yield is better. October and November rains augment black pepper yields in the next crop season. It is understood that crop in recent years matures early and berry weight declined. It is attributed to global warming and climate change.

1.7 Cashew

Continuous rain during rainy season without critical dry spell may delay the bud break in cashew under the Humid Tropics like in Kerala. A cashew plant undergoing a first dry spell may wait for the second dry spell to occur for bud break. It appears that both the dry spells, at the time of bud break and thirty days prior to bud break, with bright sunshine are critical for bud break in cashew. However, the influence of dry spell and bright sunshine on bud break of cashew is yet to be understood. Cashew flowering is highly erratic and depends on many environmental factors. The soil moisture stress has no relevance in bud break of cashew, as it commences much before soil moisture stress starts across the West Coast. Unusual rain between November and December may lead to delay in the reproductive phase of cashew. The delay in crop duration could be explained with geographical coordinates, once the bud break in cashew is initiated immediately after the rain spell is over. Relatively dry atmosphere with mild winter may be conducive for better flowering in cashew. Mild winter may be defined as "low minimum surface air temperature ranging between 15 and 20^{0}C coupled with more dew nights having moderate dew". Better availability of soil moisture during flowering period (December and January) may not influence flowering in cashew. Tea mosquito is a menace in cashew along the West Coast and it is not so along the East Coast. Cyclone is a threat along the East Coast while not so along the West Coast. Cashew environment along the West Coast and East Coasts and within Kerala is well explained in 'Chapter 1' under cashew.

1.8 Tea

Two growth periods are seen in tea (mid-March to mid-June and midAugust to mid-November). Night temperature and relative humidity are the most important factors which determine the productivity of tea in addition to rainfall. In Annamalai, night temperature occasionally drops below 5^{0}C and humidity is below 10% when the dry atmosphere is prevalent. This situation leads to tea production at lower levels. In tea, there is a time lag between the physiological function of tea bush and the vegetative harvest. It is about five weeks in the pruning year and four weeks in other years. As such the weather conditions of the previous month are generally correlated to the current month's harvest. Irrigation during cool dry season (mid-January to mid- March) will not produce the desired effect on production as temperature is the limiting factor during the period. Irrigation during the hot and dry period (March to September) that occurs frequently in Nilgiris will be beneficial in crop production. The occurrence of heat wave during March 2004 adversely affected tea crop up to 50 percent over Himachal Pradesh. Decline in tea productivity across the Country is attributed to global warming and climate change. It is more so in the northeast parts of the country, where tea plantations are predominant.

1.9 Rubber

The area, production and productivity of rubber have been increasing over a period of last four decades. Rubber plants are traditionally grown under rainfed conditions. Irrigation is practiced only in nurseries. Under rainfed conditions, monthly rainfall should be sufficient to meet the water requirement of the crop. In the tropical monsoon climate, the potential evapotranspiration rate is about 4-5 mm per day and therefore, an amount of 120-150 mm per month is considered essential to maintain optimum growth of plants. In traditional rubber growing areas like Kerala, annual rainfall ranges from 1000 to 4000 mm. The distribution pattern of rainfall in Kerala is such that it is having two peaks one in southwest monsoon season and another one in post monsoon period. However, the bimodal pattern of distribution of rainfall is restricted from central and southern districts where the influences of both the monsoons are felt. Rubber being a crop requires well distributed rainfall during its yielding phase may not perform well under a unimodal rainfall regime. That is why, rubber production is comparatively low in areas where unimodel rainfall is seen such as northern districts of Kerala. Heavy rainfall from June to September followed by weak rainfall in October - November is the rainfall scenario of Kannur and Kasaragod districts. If the pre-monsoon showers fail most of the plantation crops are subjected to soil moisture stress under rainfed conditions. However, such long dry spells are not conducive to rubber as it yields poor under dryspell / drought situation. This is more so in northern Kerala consisting of Kannur and Kasaragod districts where the rubber area is comparatively poor. Rubber, being a crop adapted to moderate temperatures, gets affected by extremes in temperatures. Higher temperature results in higher rates of evapotranspiration and thereby to severe soil moisture stress in the absence of rainfall. Rubber is grown extensively across the midlands of Kerala. It can thrive well and produce better in the midlands of Kerala where the thermal environment is conducive to its growth.

Rubber yield is affected due to short term monsoon uncertainties like prolonged rain during monsoon, followed by extended rain during post monsoon as seen in 2007 and 2010. Similarly, prolonged dry spells during summer may adversely affect rubber yield under rainfed conditions.

1.10 Areca Nut

The areca nut palm is capable of growing under a variety of climatic and soil conditions. It grows well from almost sea level up to an altitude of 1000 m in areas of abundant and well distributed rainfall or under irrigated conditions. Areca nut is mostly grown in valleys under profuse irrigation in summer. Areca nut may not come up well under waterlogged conditions. The climatic requirements of arecanut appear to be similar to that of coconut, but it is relatively more sensitive to soil moisture stress and waterlogging. Due to waterlogging, areca nut is susceptible to yellowing, which is a dreaded disease. Most of the areca nut palms were affected with yellowing after heavy rains in 2014 in Wayanad District. The area under areca nut was decreasing very fast due to various reasons. The quality of areca nut has also come down. It is attributed to global warming and climate change. In Wayanad District, rainfall decline and increase in temperature are noticed in tune with global warming and climate change. Rate of increase in temperature across the Wayanad Disrict is alarming. Warm nights are noticed. Hailstorms are also not uncommon during summer in some parts of Wayanad District. Fruit drop is common during hailstorms in fruit crops like mango and flowering in coffee is also adversely affected. Areca nut or banana or combination of both require large amount of water for irrigation to grow successfully. Paddy fields/wet lands were replaced by the above crops in some locations, leading to prolonged drought during summer since the water resources are dried up. Wayanad District experiences frequent droughts due to replacement of wet lands with areca nut and banana. The wet lands act as water sink during rainy season and water source during summer/lean season. Replacing such wet lands with other heavy water consuming crops is detrimental. Whenever high intensity rainfall is noticed within short span of time, hill cultivation is in threat due to soil erosion and land slides.

1.11 Weather and Plantation Crops

The effect of weather elements on plantation crops is summarised in the form of positive and negative signs based on the above discussion and presented in Table 1.2. The influence of seasonal rainfall on plantation crops is the only variable discussed in literature, while there is scant mention of other variables. Kerala is predominantly a plantation state, falls under the Tropical

Humid Climate (B_4-B_3 as per the Thorthwaite's climate classification). The State as a whole was moving from wetness to dryness within the Humid Climate. Monsoon rainfall across the State was declining since last 60 years (cyclic trend of 40-60 years is also noticed with annual rainfall) while rise in temperature is evident. Of course, rate of increase in temperature was alarming across the High ranges (where cardamom, coffee and tea are grown) due to deforestation. It is also true to some extent along the Coast (low land) due to increase in sea surface air temperature. The decade 1981-90 was the driest and warmest decade and warming Kerala is true since last four decades. About 70 per cent of annual rainfall (3000 mm) is received during summer monsoon (southwest monsoon) with heavy cloudiness, followed by prolonged summer (November to May) is the climatic characteristic of the State. In majority of the locations/regions within the State experiences uni-model rainfall except few pockets in southern parts of Kerala, where bi-model rainfall is received due to influence of both southwest and northeast monsoons. Therefore, the plantation crops' production is relatively better towards south of Kerala except in the case of cashew, black pepper and coffee since the length of crop growing season is better under rain fed conditions due to the influence of northeast monsoon. Cashew, coffee and black pepper appear to be better across the mid-high lands of northern region of Kerala since the prolonged dry spell from November to May may not be detrimental for better yield in terms of quality and quantity depending upon the nature of the crop under rain fed conditions. Of course, coffee requires blossom and backing showers during summer for better performance of crop. Sprinkler irrigation is practiced in the case of coffee to meet the blossom and backing showers in the absence of rains. The coffee growers are also of the opinion that drip irrigation during summer can be a substitute to sprinkler irrigation and beneficial to the crop. In the case of black pepper, the mortality rate is high in young pepper vines due to prolonged summer drought as noticed in summer 1983 and 2004. Many black pepper vines were wiped out in summer 2004 across the Wayanad District due to hydrological drought and hot summer. Heavy rainfall during monsoon season may not be detrimental to black pepper, cashew, coffee, tea and rubber (number of tapings is adversely affected-this effect is minimized if rain guards are practiced). While prolonged high rainfall may be detrimental to other crops such as coconut, areca nut, cocoa, rubber and cardamom due to waterlogging. In the case of mixed cropping of black pepper and coffee, satisfactory yields from both the crops are in doubt in any given year since one crop yields better other crop fails. Therefore, mixed cropping is better to sustain crop income against adverse weather. In general, growth is restricted during rainy season under water-logged conditions, while crop growth is better during summer if irrigation is provided and other environmental factors are not limiting.

Table 1. 2: Effect of weather extremes on plantation crops

Weather elements	Coconut	Areca nut	Cocoa	Pepper	Cashew	Rubber	Cardamom	Coffee	Tea
Rainfall(monsoon)	-	-	-	+	+	-	-	+	+
Rainfall(summer)	+	+	+	-	-	+	+	+	+
Rise in maximum Temperature.	-/+	-/+	-	-	+/-	+/-	-	-	-
Rise in minimum Temperature	Not known	Not known	Not known	-	-	Not known	-	-	-
RH (%)	Leads to insect pest and diseases. Hence, negative effect on all the crops. Tea requires high relative humidity for better yield.								
Solar radiation	Not adequate during monsoon period and hence it affects crop production. High rainfall with low solar radiation during the monsoon while low rainfall with high solar radiation during summer adversely affect most of the plantation crops production in the Humid Tropics.								
Wind	Moderate in Kerala except in the Palakkad Gap region. Cashew flowering and nut set may be affected in Thrissur and Palghat regions due to strong wind from November to February. Beneficial to crops where winds are moderate.								
Evaporation	Evapotranspiration is restricted during the monsoon due to heavy rains. Atmosphere is relatively saturated under the above conditions. Therefore, uptake of nutrients may be restricted in all the plantation crops.								
Drought	All plantation crops do suffer due to prolonged summer drought except cashew, coffee and black pepper. Of course, mortality rate is high in young pepper vines during disastrous summer drought.								

1.12 Climate Change and Plantations

The State of Kerala, being a coastal state located in the Humid Tropics, is prone to frequent weather and climate related disasters due to heavy rainfall during monsoon season, followed by insignificant rains from November to May if pre-monsoon showers fail. It is more so in the event of climate change since the frequency of weather related disasters is likely to increase in ensuing decades. The issues related to climate change / variability over the Humid Tropics (Kerala) appears to be decline in rainfall and rise in temperature. The monsoon rainfall is declining while increasing during the post monsoon season (Fig.1.2). The maximum, minimum and the mean surface air temperature were increasing at the rate of 0.99°C, 0.3°C and 0.65°C, respectively over a period of time from 1956 to 2014 (Fig. 1.3). The last two decades (2001-10 and 2011-2020) were warmer and the recent summers 2023 and 2024 also recorded warmer. It indicates that rise in temperature may be a continuous phenomenon year after year due to global warming. The rate of increase in temperature is also not uniform within the State of Kerala. It is relatively high across the high ranges, followed by low lands along the Coast. Deforestation and rise in sea surface temperature appear to be the reasons in variations of rate of increase in temperature. As a whole, the State of Kerala was moving from wetness to dryness within the Humid Climates. Of course, the UV-B radiation is filtered to the ground irrespective of seasons due to ozone depletion, which is harmful to living organisms. Sunburns are also reported now and then since 2010 onwards in central part of Kerala (Palghat and Thrissur Districts). Vegetation burning was also noticed along the Coast in 2015. The State Action Plan on Climate Change (SAPCC) indicated that the temperature rise is likely around 2°C by 2050 and the rate of increase is likely to be more across the high ranges 2–4.5°C. If sea level rises by one metre, 169sq.km surrounded Kochi to be inundated. Salinity intrusion along the Coast due to sea level rise is also alarming. Alappuzha, Palakkad, Wayanad and Idukki are the 'Climate Change Hot Spots' in Kerala while Alappuzha and Palakkad are the districts most vulnerable to climate change. Idukki, Wayanad, Thiruvananthapuram and Kannur are highly vulnerable to Climate Change. Climate change in the form of climate variability is a major threat to plantation crops' productivity. It is also observed that crop mix/integrated farming may withstand relatively better against the ill effects of climate variability/change as seen in the case of coffee and black pepper. When black pepper performs better coffee suffers and vice-versa. Recent findings indicate that the climate change effects not only crop output but also commodity quality and price. Therefore, it is high time that the government agencies and policy makers to be proactive with short term and long term strategies to mitigate the ill effects of weather related disasters with the people's participation and minimize the losses against the weather related disasters. Need of the hour is expect the unexpected guests in the form of weather related disasters under the projected climate change scenario and tune to the system to minimize the losses with well preparedness at various levels. Keeping the above in view, the National Action Plan on Climate Change (NAPCC) identified 'Sustainable Agriculture' is one of the thrust areas to tackle the climate change related issues. The mission intends making Agriculture more resistant to climate change by identifying and developing new varieties of crops that are thermal resistance and capable of withstanding extreme weather. These issues are being tackled as a part of National Initiative on Climate Resilient Agriculture (NICRA). When it comes to the State of Kerala such initiatives need to be taken up on priority basis in developing technologies against the weather extremes such as saline/flood resistance and drought resistance/tolerance rather than thermal resistance, for which a definite State Action Plan on Climate Change (SAPCC) in Plantation Sector need to be highlighted and tackled with concerted efforts. There is need to have an action plan similar to that of NICRA at the State level with technical people in the form of 'State Initiative on Climate Resilient Agriculture' (SICRA). It may be related to plantation crops or horticulture or medicinal plants or seasonal or field crops or as a whole integrated/homestead faming since it dominates the whole Agriculture Sector in Kerala. Such system should be made operational and in vogue at the earliest possible time since climate change/variability is a big threat in Plantations and Agriculture sector as a whole under the Humid Tropics. The altitudinal sequence of crops within the Humid Tropics is quite interesting since the plantation crops like coffee, cardamom and tea that are grown across the highranges are not at all seen in mid and lowlands, indicating that they cannot withstand to high temperature and the crop distribution is a reflection of prevailing temperature as influenced by altitude and lapse rate.

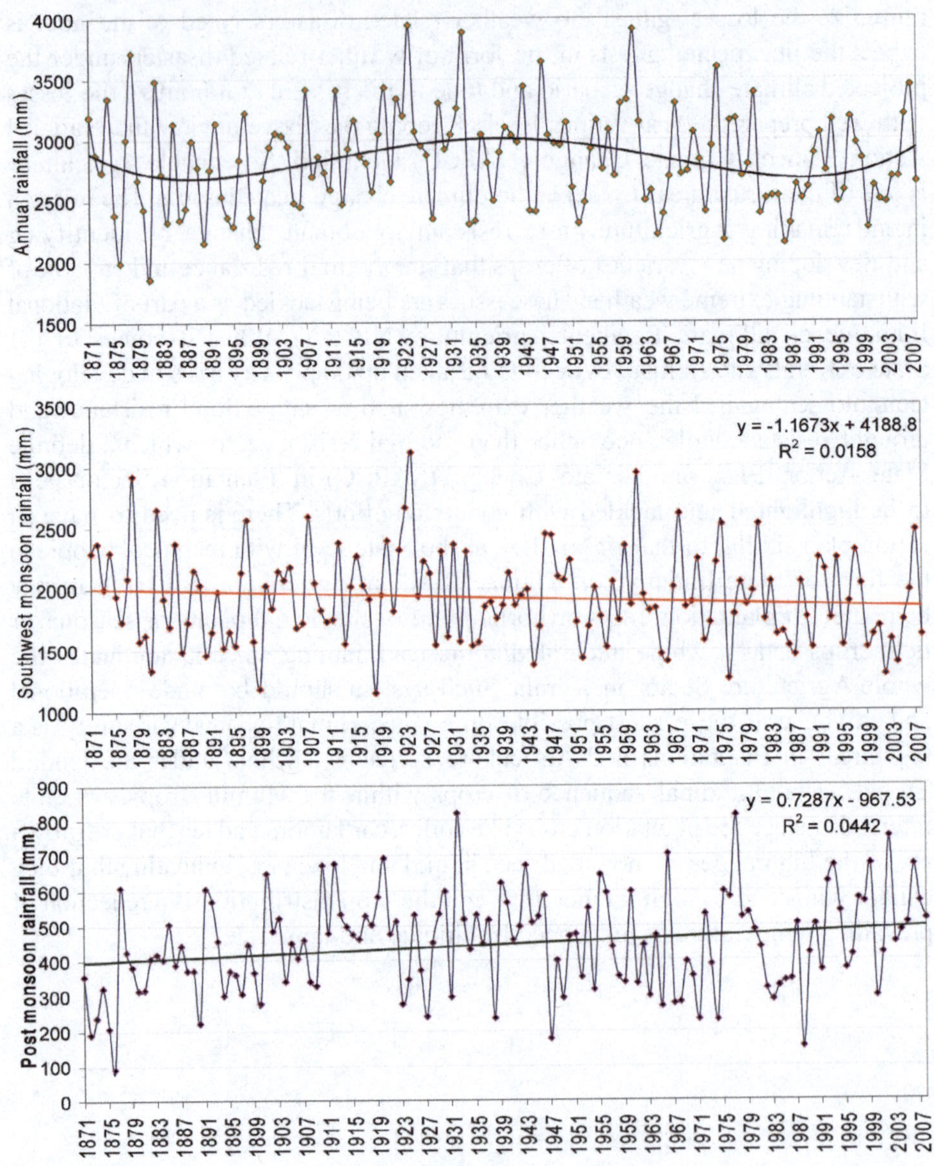

Fig. 1.2: Trend in annual, southwest and post monsoon rainfall over Kerala

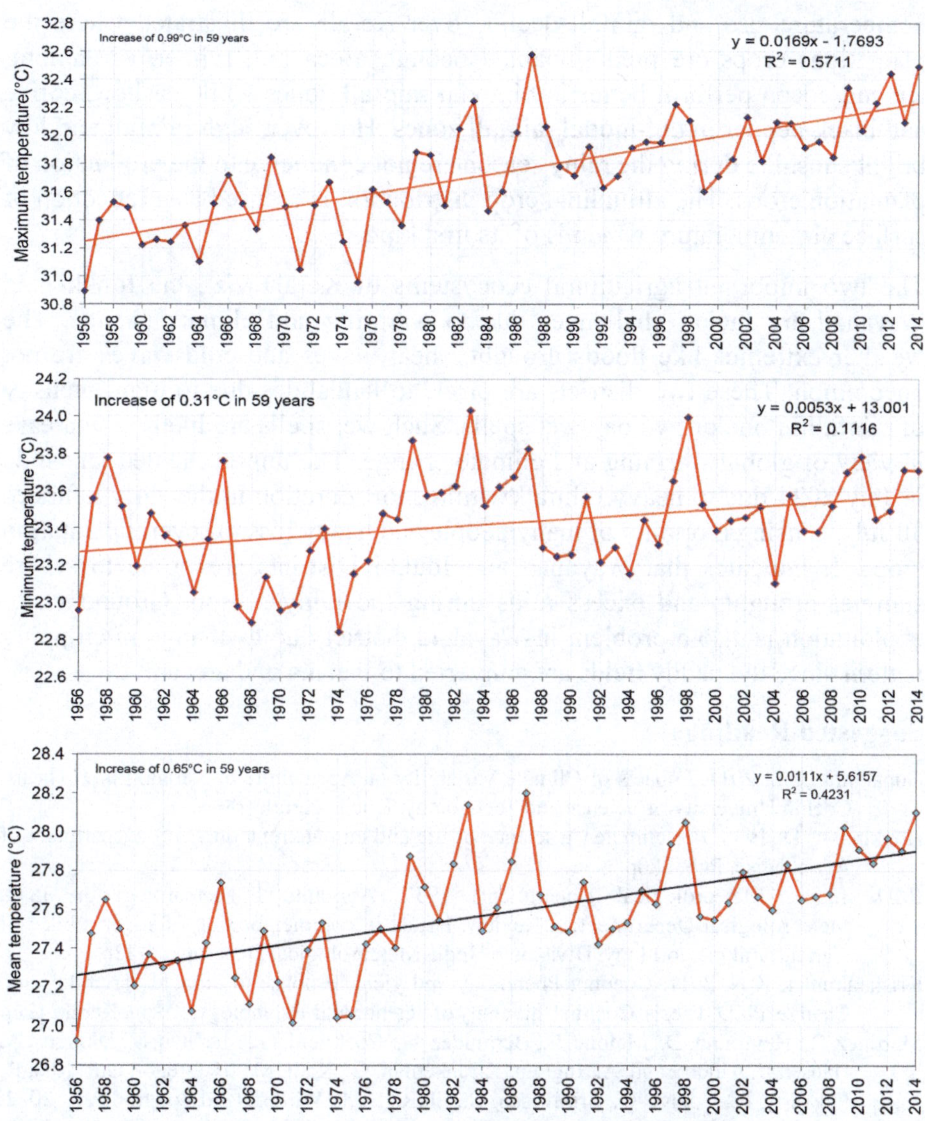

Fig. 1. 3: Maximum, minimum and mean surface air temperatures over Kerala

Summary

The plantation crops like coconut, cashew, cardamom, coffee, cocoa, black pepper, tea, areca nut and rubber play a major role in export economy. In fact, the quality produce is the one which plays a pivotal role in plantation crops in deciding the marketability outside. A brief summary of all the plantation crops together in response to weather aberrations is provided in the first Chapter.

Temperature rise and rainfall decline over Kerala are illustrated where the plantations crops are predominant. Coconut, areca nut, rubber, cardamom, tea and cocoa perform better in bimodal rainfall zones while cashew, coffee and black pepper in uni-modal rainfall zones. However, high rainfall and low bright sunshine during the rainy season are not conducive to the production of plantation crops. The altitudinal crop distribution is focused as a reflection of surface air temperature which is of its first kind.

The two important agricultural ecocystems of Kerala viz., the Idukki and Wayanad are facing challenges due to warming and climate change. The weather extremes like floods droughts, heat waves and cold waves are not uncommon. These two districts are prone to landslides due to high intensity of rainfall in one or two day wet spells. Such wet spells are likely to increase in view of global warming and climate change. The unprecedented landslide in July 2024 due to heavy rainfall within short duration in the early hours of 30 July, causing mortality of many people and heavy loss to several plantation crops. It indicates that Wayanad and Idukki Districts are prone to severe summer droughts and excess rains during the rainy season. Ground water exploitation is also a problem in Wayanad district due to change in cropping pattern since the paddy fields are converted to banana and areca nut.

Suggested Readings

Gopakumar,C.S. 2011. Impacts of Climate Variability on Agriculture in Kerala. Ph.D. Thesis. Cochin University of Science and Technology. Kochi, Kerala.286p

Hopkins, A. D. 1938. Bio climates, a science of life and climate relations. U.S. Department of Agric. Misc. Pub, 280p

IMD (Indian Meteorological Department) 1957. Agricultural Meteorology in India Meteorological Department: A Review. Issued in commemoration of silver jubilee of Agricultural meteorology, Division of India Meteorological Department, 42p

Krishnakumar, K.N. 2011. Coconut Phenology and yield response to climate variability and Change. Ph.D. Thesis. Cochin University of Science and Technology. Kochi, Kerala.196p

Martinez, R., Hemming, D., Malone, L., Bermudez, N., Cockfield, G., Mushtaq, S., Diongue, A. Hansen, J., Hildebrand, A., Ingram, K., Jakeman, G., Kadi, M., McGregor, G.R. Rao, P., Pulwarty, R., Ndiaye, O., Srinivasan, G., Eh.R., Seck, White, Nand Zougmore, R. 2012. Improving Climate Risk Management at Local Level – Techniques, Case Studies, Good Practices and Guidelines for World Meteorological Organization Members-Chapter 21 (477-532): In Risk Management-Current Issues and Challenges (Ed: Nerija Banaitiena). In Tech Publishers. Rijeka, Croatia, 584p

Nybe, E.V. and Rao, G.S.L.H.V.P. 2005. Drought management in plantation crops. Kerala Agricultural University. Thrissur, Kerala, India. 176p.

Rao, B.V.R. and Rao, V.U.M. 2024. Agricultural Meteorology. Jain Brothers, New Delhi. 154p

Rao, G.S.L.H.V.P. 2021. Agricultural Meteorology (5th Reprint). PHI Learning Private Limited, New Delhi, India. 364p.

Rao, G.S.L.H.V.P. 2015. Climate Change and Horticulture-Chapter-5. Basics of Horticulture (Ed. K.V.Peter). 2nd Revised and Enlarged Edition. New India Publishing Agency. New Delhi, India. 704p.

Rao, G.S.L.H.V.P. 2015. Climate Change and Agriculture - Chapter-1. Climate Resilient Crops for the Future. New India Publishing Agency. New Delhi, India. 450p.

Rao, G.S.L.H.V.P. 2002. Climate and cashew. Kerala Agricultural University. Thrissur, Kerala, India. 100p.

Rao, G.S.L.H.V.P. 2011. Climate change adaptation strategies in agriculture and allied sectors. Scientific Publishers (Jodhpur), India, 336p.

Rao, G.S.L.H.V.P. 2016. Weather Extremes and Plantation Crops in the humid tropics. Mausam, 67(1):251-258

Rao, G.S.L.H.V.P. and Beevi, P. 2010. Tea mosquito forewarning in cashew. Kerala Agricultural University. Thrissur, Kerala, India. 74p

Rao, G.S.L.H.V.P. and Gopakumar, C.S. 2011. Climate change and plantation crops in Kerala. Kerala Agricultural University. Thrissur, Kerala, India. 255p.

Rao, G.S.L.H.V.P. and Krishnakumar, K.N. 2011. Climate change and coconut productivity in Kerala. Kerala Agricultural University. Thrissur, Kerala, India. 219p.

Rao, G.S.L.H.V.P. and Manikandan, N. 2009. Climate Variability and small cardamom across the Western Ghats. Kerala Agricultural University. Thrissur, Kerala, India. 117p.

Rao, G.S.L.H.V.P. and Nair, R.R. 1988. Agrometeorology of plantation crops. Proceedings of National Seminar, held at RARS, Pilicode. Kerala Agricultural University. Thrissur, Kerala, India. 187p.

Rao, G.S.L.H.V.P. and Pushphalatha, P.B. 2011. Monsoon Malabar Coffee. Kerala Agricultural University. Thrissur, Kerala, India. 134p.

Rao, G.S.L.H.V.P. 2002. Climate variability and coconut production in the Humid Tropics. Kerala Agricultural University. Thrissur, Kerala, India. 38p.

Rao, G.S.L.H.V.P., Rao, A.V.R.K., Krishnakumar, K.N. and Gopakumar, C.S. 2009. Climate variability and food security. Kerala Agricultural University. Thrissur, Kerala, India. 73p.

Rao, G.S.L.H.V.P., Rao, G.G.S.N. and Rao. V.U.M. 2010. Climate change and agriculture over India. PHI Learning Private Limited, New Delhi, India. 320p.

Rao, G.S.L.H.V.P, Gopakumar, C.S and Krishnakumar, K.N. 2013. Impacts of Climate Change on Horticulture Across India. In: Climate-Resilient Horticulture: Adaptation and Mitigation strategies. H.P.Singh et al (eds). Springer Publication. DOI 10.1007/978-81-322-0974-4_2, pp 13-20.

Rao, G.S.L.H.V.P. and Krishnakumar, K.N. 2024. Impact of weather and climate on plantation crops in the Humid Tropics. Chapter 21 (417-432pp). Agrometeorological Applications for Climate Resilient Agriculture (Eds.): Chattopadhyay, N., Stefanski, R., Attiri, S.D. and Rathore, L.S. Springer International Publishing AG. Singapore.445p

Sharangi, A.B., Rao, G.S.L.H.V.P., Das, S., Krishnamurthy, K.S.Upadhyay, T.K., Gopakumar, C.S. and Acharya. S.K. 2024. Brunt of Climate Change and Spice Crops: Scenario, Response, and Resilience. 755-812pp. Eds: Ravindran, P.V., Sivaraman, K.,Devasahayam,S., Nirmal Babu, K. Handbook of Spices in India: 75 Years of Research and Development. Springer Nature Singapore Pte. Ltd.535p

Questions

I Elucidate on

1. Weather and plantation crops
2. Weather extremes and plantation crops
3. Plantation crops' climate

II Write short notes on

1. Altitudinal sequence of crops
2. Hill crops
3. Climate and coconut
4. Climate and cashew
5. Climate and cocoa
6. Climate and cardamom
7. Climate and coffee
8. Climate and black pepper
9. Climate and tea
10. Climate and rubber
11. Climate and areca nut

III Fill in the blanks

1. ----, ------, -----, --------, ------,--------, ------,------ and ------ are known as the plantation crops
2. Plantation crops are also known as -------------------- crops
3. Life span of plantation crops varies from ----------- to ----------- years
4. ----------- plantation crops are known as -------------- trees
5. -------- and ---------- crops are known as spice crops
6. First fruiting is seen in coconut between -------- and ----------- years
7. Black pepper is trained to ------ and ----------- crops
8. --------, ------ and ------ crops are cultivated on hillocks
9. -------------- crop is predominant in Wayanad District
10. --------------- crop is predominant in Idukki District
11. ------- and ------ crops are predominant in valleys
12. -------------- decides crop distribution
13. -------------- decides crop productivity
14. ------------ mountain slope is preferred for cultivation to avoid direct sunlight
15. ------------ and --------------- rainfall models are noticed in Kerala

16. ------------ model rainfall is predominant in southern districts of Kerala
17. ----------- model rainfall is predominant in northern districts of Kerala

IV State True or False

1. 44 months is the duration from primordium initiation to harvest in coconut
2. 12 months is the duration from female flower fertilization to harvest in coconut
3. Coconut and cocoa produce harvest in every month
4. Rise in summer temperature and high rainfall during the rainy season are not conducive for better performance in plantation crops
5. The difference in cocoa yield is large between good and bad yield years during the Southwest monsoon season
6. Robusta is more sensitive to weather in coffee
7. The weight of green pepper berries goes down with depletion of rainfall
8. Dry and wet spells are critical for bud break initiation in cashew
9. 150 cm is the optimum monthly evapotranspiration for rubber plantations
10. The girth of rubber plant initially better in the same age group towards southern parts of Kerala
11. January and May are the two growth periods in tea

V Match the following

Coconut	Root diseases 7
Cocoa	Foot Rot 9
Cashew	Yellowing 6
Cardamom	Stem bleeding 1
Coffee	Leaf fall 8
Areca nut	Katte4
Tea	Rodents2
Rubber	Tea mosquito 3
Black pepper	Berry borer5
	Leaf spot

VI Expand

NAPCC SAPCC NICRA SICRA CPCRI

VII Distinguish between

1. Microtherms and Macrotherms
2. Climates of coffee and cardamom
3. Climates of cashew and rubber
4. Low land and mid land crops
5. High land and highrange crops
6. Southwest and Northeast monssons

2

Cashew

Cashew (*Anacardium occidentale* L.) is a native of Brazil and was introduced in India by the Portuguese during the16th Century. It is grown in India, Brazil, Vietnam, Tanzania, Mozambique, Indonesia, Sri Lanka and other tropical Asian and African countries. The crop is grown within the latitudinal belt of 27°N and 28°S of the equator experiencing the tropical climates for its growth and development. It is categorised under one of the plantations along the West and East Coasts. To a limited extent cashew is cultivated in Gujarat, Chattisgarh, Manipur, Meghalaya, Tripura, Andaman and Nicobar Islands. The crop is mostly grown in low and midlands along the West and East Coasts of India in sandy and laterite soils. Since it is mostly grown along the Coast across the Country it is also known as the Coastal Tree. Cashew grows along the West and East Coasts of India from South to North, lying between 09°N and 23°N latitudes. Though it grows beyond 23°N latitude, its production is relatively low as the reproductive phase coincides with cold injury/low temperature during winter (December-February). Being a tropical coastal tree, it cannot thrive under low winter temperature. There is significant yield variation in cashew among the different agroclimatic zones of the country. Year-to-year yield variation is also significant as the crop is weather sensitive. This crop may come up even in waste lands with specific crop management practices along the West Coast and as a shelter belt and wind break along the East Coast.

2.1 Cashew Production and Productivity

India is one of the largest producers of cashew nuts in the world, with several states contributing significantly to its production. Cashew cultivation is a vital part of India's agricultural economy, providing livelihoods to millions of farmers. Cashew cultivation in India spans 0.7 million hectares with annual production exceeding 0. 8 million tonnes. Recent increase since last one decade highlighting a significant growth in cashew production. The total raw cashew nut production in India depends upon the performance of the crop along the West Coast as it accounts relatively more area with high productivity when compared to that of the East Coast under rain fed conditions. Among the cashew grown States, Maharashtra is ranked first in area, production and productivity of cashew in the Country (Fig. 2.1). Maharashtra contributes 32.4 per cent to total cashew production from an area of 19 per of the total. It has the credit of having the highest productivity of 1500 kg/ha. It is the largest cashew production State in India with an annual production of 199.7 thousand tonnes. The State's favourable climate, particularly in the Konkan Region, supports extensive cashew cultivation. This significant production not only boosts Maharastra's economy but also provides livelihoods for numerous farmers, making a key player in India's cashew Industry. Andhra Pradesh, Orissa, Karnataka, Tamil Nadu and Kerala stands in second, third, fourth, fifth and sixth positions, respectively in terms of cashew production. Interestingly, Kerala was the leading State in cashew area, production and productivity in the Country in 1990s and now stands in sixth position. Kerala's share to cashew production is only 11 per cent from an area equivalent to that of 8 per cent of the total area under cashew in the Country. Its productivity is comparable to that of national productivity (900 kg/ha). More than 50 per cent of the State's cashew production is from northern Districts viz., Kannur and Kasaragod and superior in size and quality, hence it fetches more value in the international market. This could be attributed to the prevailing weather to which the crop is exposed during the reproductive phase, commencing from flushing to harvest of nuts. In non-traditional areas cashew is grown under irrigated conditions and the productivity is much higher. It is true in majority of cashew plantations along the East Coast and act as shelterbelt and wind breaks too in the State like Orissa.

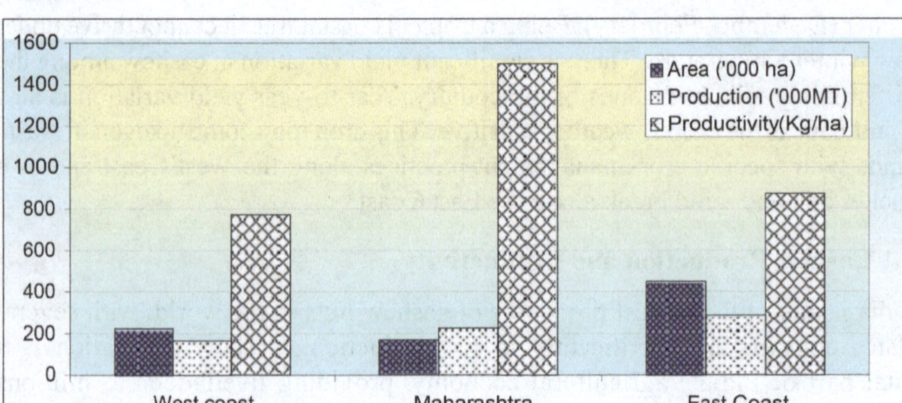

Fig. 2.1: Area, production and productivity of cashew across the West Coast, East Coast and Maharashtra

2.2 Influence of Geographical Co-ordinates on Cashew Flowering

There is a change in flowering time of cashew depending upon the genotype and geographical co-ordinates viz., latitude, longitude and altitude. For example, peak flowering is seen between August and September in Brazil and Tanzania while in Mosambique, maximum flowering occurs in October. Interestingly, the peak flowering is during March in Philippines. The peak flowering in cashew is in early January and time of peak harvest is early April in Mangalore.(13°N) and Anakkayam (11°N). At Bapatla, 17°N on the East Coast, peak flowering occurs between mid January and February and the harvest time is April end. Further North in Berhampur, harvest is prolonged until the end of May. Studies revealed that there exists a gradual delay in flowering as one moves North or South of the equator across the cashew tract. The relationship between latitude and flowering reflects the effect of latitude on temperature in regions away from the equator. Hopkins (1938) attempted to express the importance of latitude, longitude and altitude (Geographical Co-ordinates) in the distribution and rate of development of plants by means of a "Bioclimatic law". It may be stated as " A biotic event in north America will, in general, show a lag of four days for each degree of latitude, five degree of longitude and 400 feet of altitude, northwards, eastward and upward in spring and early summer". In mango, the India Meteorological Department also established that there is a delay in flowering from South to North of India, which generally followed Hopkins' bio-climatic law (IMD, 1957). Across the West and East Coasts of India also, flowering of cashew shows a delay in time from South to North depending upon the latitude and altitude (Figs. 2.2 and 2.3). The law stated by Hopkins appears to be sound in biotic events of cashew also in the tropical monsoon climates under better crop management in rain fed situations, provided the genotype and rainfall distribution are uniform. However, there was a difference in number of days delayed in cashew flowering at each degree of North latitude while the effect of altitude on time of cashew flowering is similar to that of Hopkins (1938). In the tropical monsoon climates, there appeared to be a delay of 6 days in cashew flowering at every 1° of North latitude and for every 100 m of altitude, the delay in cashew flowering was 3 days. However, the flowering in cashew depends on several biotic and abiotic factors viz., soil type, climate, exposure to sunlight, crop management, variety, age of the cashew plantations and incidence of pest and diseases in the previous crop season (Fig. 2.4).

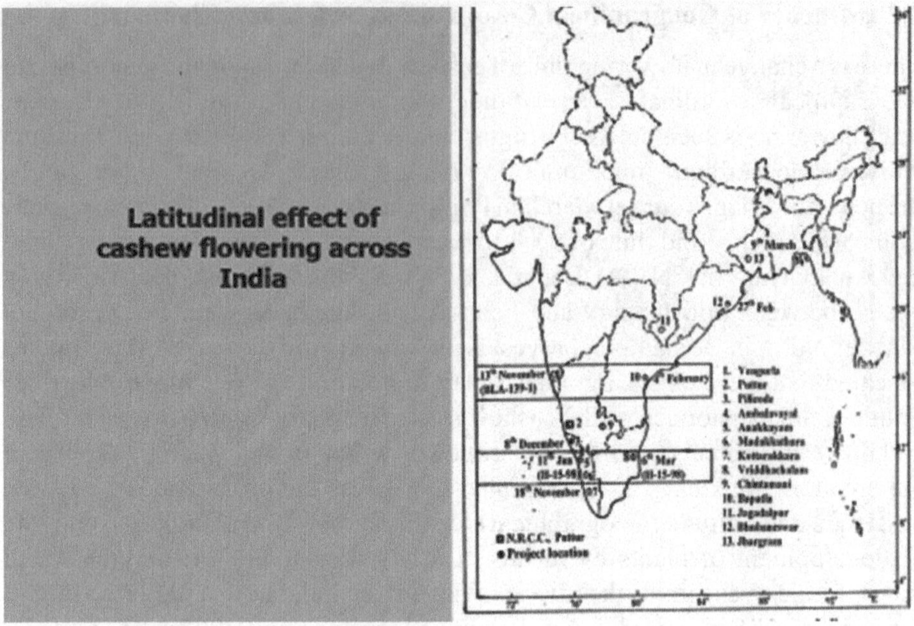

Fig. 2. 2: Effect of latitude on cashew flowering

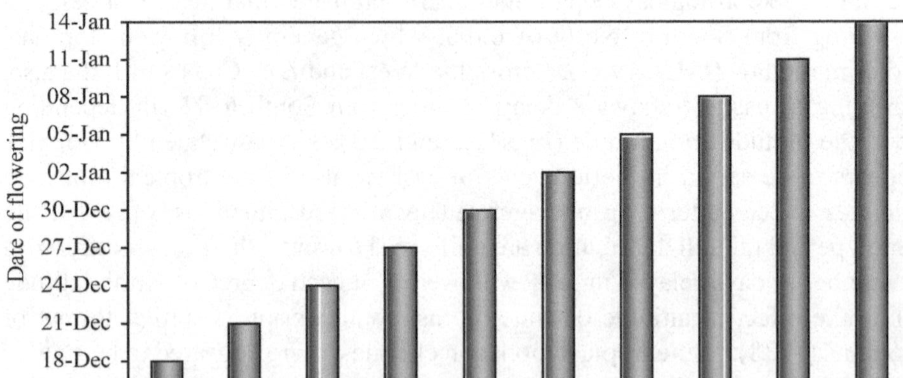

Fig. 2.3: Influence of altitude on cashew flowering

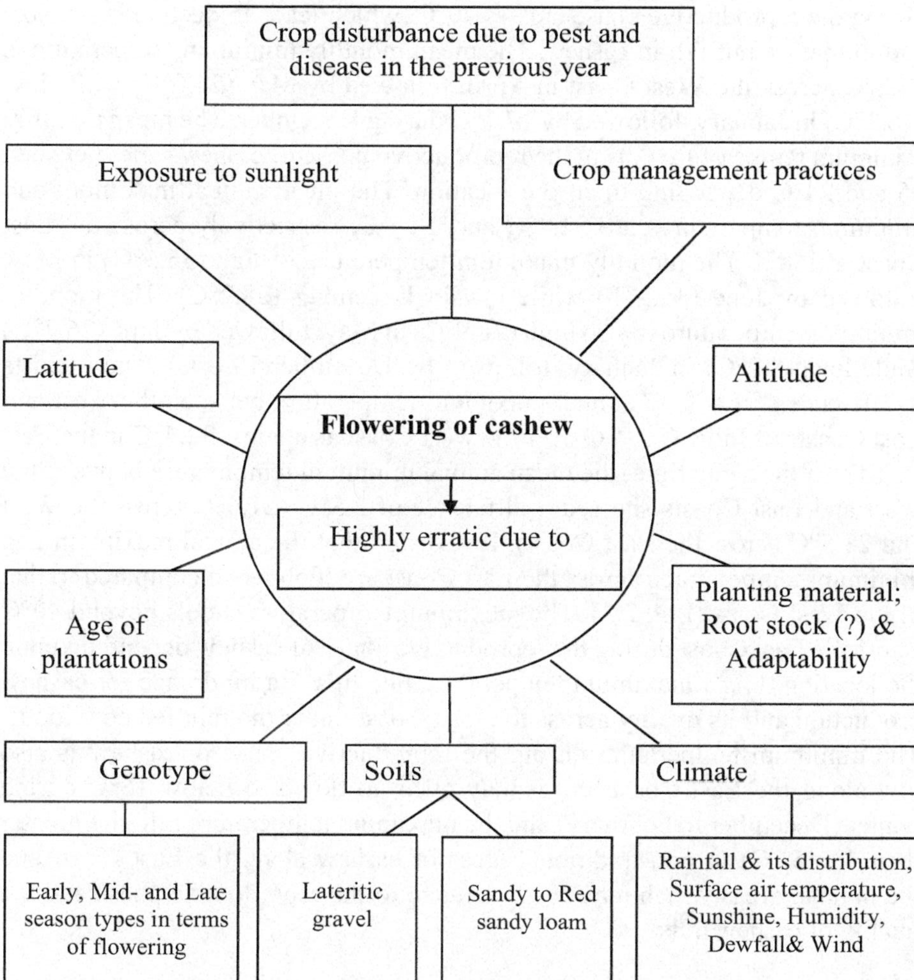

Fig. 2.4: Effect of abiotic and biotic factors on flowering of cashew

2.3 Thermal Regime of Cashew

The mean annual maximum and minimum temperature across the West Coast of India is 30.9 °C and 21.0°C, respectively. The monthly maximum temperature is high (34.1°C) across the West Coast in April, followed by May (33.8°C) while low (28.7°C) in July. The mean maximum temperature during the reproductive phase of cashew (December–March) never crossed 35°C across the West Coast of India. This may be one of the reasons for obtaining large sized nuts when compared to that of the East Coast of India where the maximum temperature during the reproductive phase crosses 40°C, which leads to decline in nut size and immature nut fall in cashew. The mean monthly minimum temperature is 24.3°C across the West Coast in April, followed by May (24.0°C) while low (16.3°C) in January, followed by 17.1°C during December. The mean monthly minimum temperature during the reproductive phase of cashew varies between 15 and 24°C depending upon the location. The mean annual maximum and minimum temperatures are 32.4°C and 23.5°C, respectively across the East Coast of India. The monthly maximum temperature is high (36.5°C) in May, followed by June (35.5°C) while low in December (28.2°C). The monthly minimum temperature is also high (26.9°C) in May, followed by June (26.7°C) while low(18.7°C) in January, followed by December(19.0°C). There exists a difference of 1.5°C in annual maximum temperature between the West and East Coasts of India (30.9°C over the west Coast as against 32.4°C in the East Coast). At the same time, the mean annual minimum temperature between the West and East Coasts showed a difference of 2.5°C (21.0°C across the West and 23.5°C across the East Coast). It revealed that the annual maximum and minimum temperatures across the East Coast are high when compared to that of the West Coast (Fig. 2.5). The maximum temperature shoots beyond 40°C across the East Coast during the reproductive phase of cashew depending upon the location. High maximum temperature like this is a hindrance for cashew production and its quality across the East Coast under the rain fed conditions. The minimum temperature during the reproductive phase of cashew is also low along the East Coast and it may even go down to below 15°C during winter (December to February) and the maximum temperature is high (greater than 40-45^{0} C) in non-traditional areas of cashew along the East Coast and the maidan areas which may cause damage to the reproductive phase and the quality of cashew nuts.

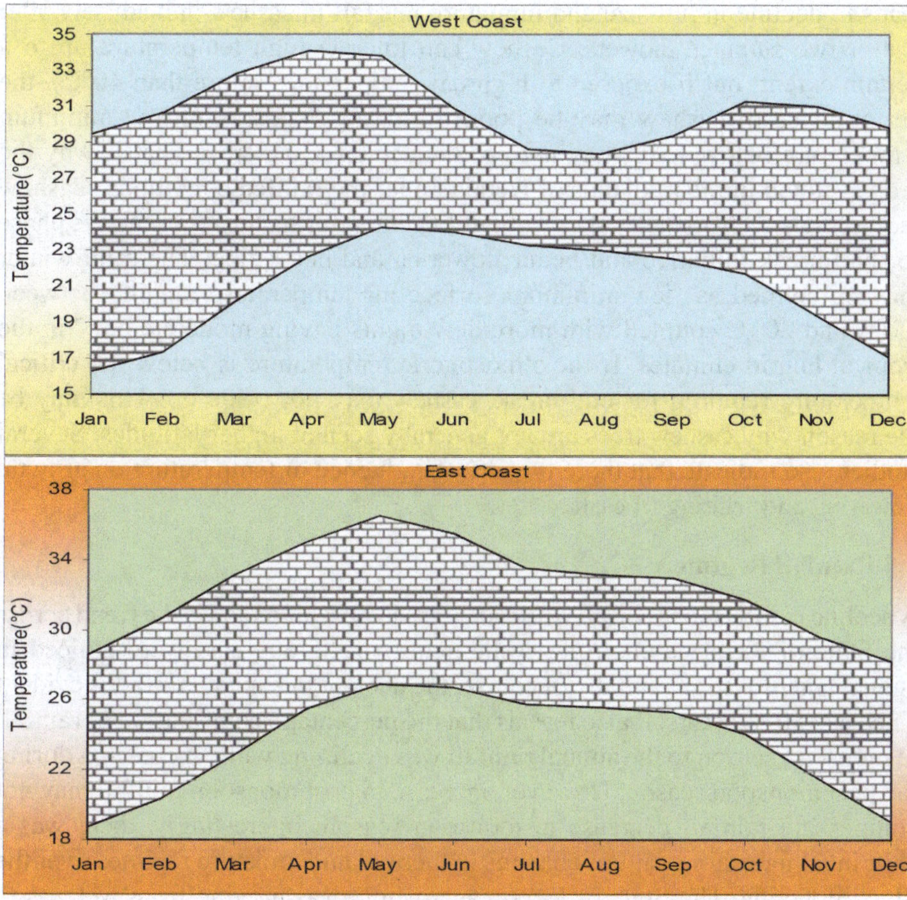

Fig. 2.5: Thermal regime of cashew across the West and East Coasts of India

2.3.1 Thermal Regime of the mid-lands over Kerala

The mean annual maximum temperatures across the midlands of Kerala is 32.6°C while 22.8°C in the case of mean annual minimum temperature where cashew is extensively grown. The mean monthly maximum temperature varies between 29.2°C in July and 36.7°C in March. In the case of minimum temperature it varies between 21.3°C in January and 24.4°C in April. The thermal environment in the midlands of Kerala is conducive for the growth and development of cashew. The mean maximum temperature during the reproductive phase of cashew (December– March) never crossed 35°C across the mid-lands of Kerala. This may be one of the reasons for obtaining large cashew nuts when compared to that of the East Coast of India where the maximum temperature during the reproductive phase crosses 40°C, which leads to decline in nut size and immature nut fall in cashew immediately after one or two summer showers. Cashew can tolerate high temperature up to a certain extent, but if exposed to high day temperature (more than 40°C), the performance of cashew may be poor. Similarly, exposure to low minimum temperature below 15°C may not be conducive for the crop. That is why the crop is suited in the midlands of Kerala under the rainfed conditions. Cashew requires relatively long dry spell with mild winter during the reproductive phase for bud break initiative and better flowering and nut setting. The mild winter may be defined as "low minimum surface air temperature ranging between 15^0 C and 20^0 C coupled with more dew nights having moderate dew" in the tropical humid climates. If the atmospheric temperature is below the critical temperature required for bud break, cashew may not come up. This may be the reason why cashew trees are not generally seen at higher latitudes. Several studies also indicate that there is a strong influence of temperature on growth, flowering and fruiting of cashew nut.

2.4 Rainfall Regime

A decline in monsoon rainfall while increase in post monsoon is the trend across the State of Kerala though the cyclic trends of 40–60 years were noticed in annual and monsoon rainfall. Of course, the decline in annual rainfall is evident since last 50–60 years. It also reveals that the percentage contribution of rainfall in monsoon season to the annual rainfall was declining while increasing during the post monsoon season. However, increase in post monsoon rainfall may not compensate rainfall decrease in monsoon season. Interestingly, there was a shift in climate as a result of changes in thermal and moisture regimes over the State of Kerala. The State of Kerala as a whole was moving from wetness to dryness within the Humid Climate from B_4 to B_3 since last 100 to 150 years. One of the major factors could be due to alarming deforestation and forest fires that took place during the above said period across the Western Ghats within and outside the State of Kerala. The distribution pattern of rainfall in Kerala is such that it is having two peaks one in southwest monsoon season and another one in post monsoon period. However, the bimodel pattern of distribution of rainfall is restricted from south to the central districts of Kerala where the influence of both the monsoons is felt. Cashew being a crop requires long dry spell during the reproductive phase, may not perform well under a bimodel rainfall regime. If the reproductive phase (November to April) of the crop coincides with wet spell, cashew production is adversely affected. That is the reason why, performance of cashew is better in northern Kerala, where unimodel rainfall is felt. Heavy rainfall from June to September followed by weak rainfall, in October–November is the rainfall scenario of Kannur and

Kasaragod Districts, where cashew is predominantly grown **(Fig. 2.6)**. The dormant phase of cashew coincides with high rainfall from June to September, followed by cashew budding, flushing and flowering are subjected to moderate soil moisture while the nut set and development experience severe soil moisture stress. If the pre-monsoon showers fail most of the plantation crops are subjected to soil moisture stress under rain fed conditions. Such periods may not be harmful to cashew as it performs better under dry spell/drought to a certain extent in rain fed conditions since the crop is almost in mature stage with flower drops. This is more so in northern Kerala consisting of Kannur and Kasaragod Districts where the cashew area is more. More than 50 per cent of the cashew production is from these districts in Kerala. The cashew nuts produced from these districts are also superior in quality and therefore fetches more money in the international market.

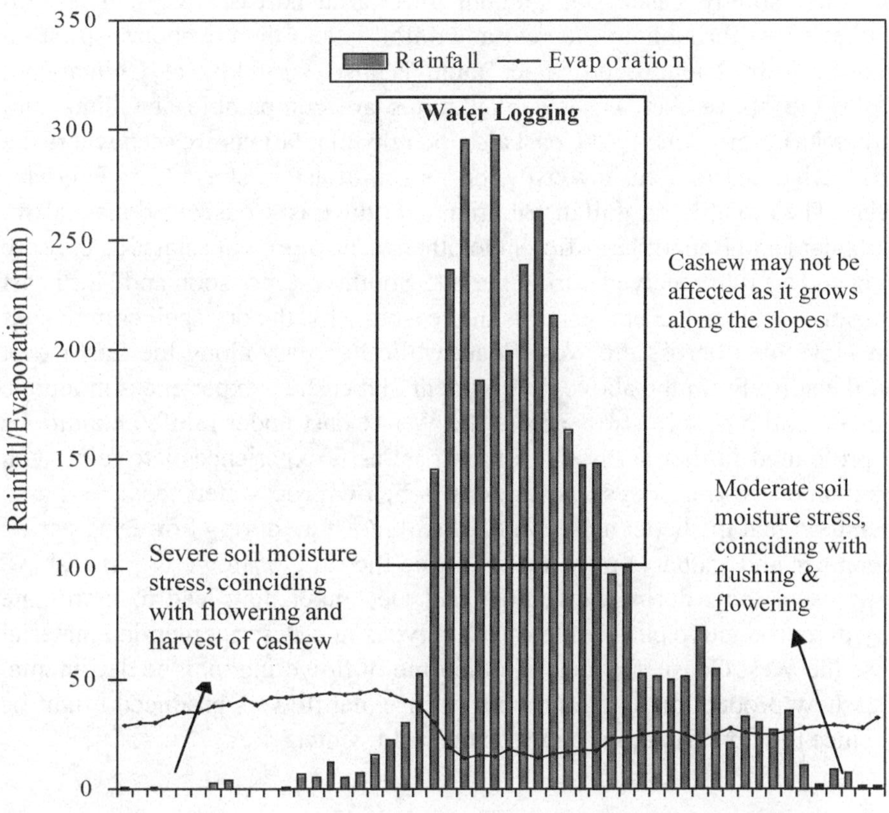

Fig. 2.6: Mean weekly rainfall versus evaporation at RARS, Pilicode, Kasaragod District, Kerala

2.4.1 Rainfall and Pan Evaporation

In spite of the fact that the reproductive phase of cashew at all locations along the West Coast (Kottarakkara, Madakkathara, Anakkayam, Pilicode, Puttur and Vengurla) is under soil moisture stress, the cashew plantations are healthy under rainfed conditions if the cashew is cultivated scientifically. The annual rainfall is relatively less over the cashew tract of Orissa and West Bengal and the distribution of rainfall is alike to some extent as in the case of West Coast (Fig.2.7a) due to the influence of south west monsoon. Across the cashew tract of Andhra Pradesh (Bapatla), rainfall is relatively better from July/August onwards and sometimes, rains extend up to November due to cyclones, which are not uncommon. In contrast, Tamil Nadu receives more rains from September to November due to influence of Northeast monsoon and it may extend up to December where cashew is grown (Figs.2.7b&c). Quite interestingly, cashew at Chintamani (Karnataka) is always under soil moisture stress throughout the year as rainfall is less than evapotranspiration round-the-year. The ground water table is also very low at Chintamani. Despite the above fact, cashew plantations are comparable at Chintamani (Karnataka). Across the East Coast also, healthy plantations are seen where the reproductive phase of cashew experiences soil moisture stress from February to May. The monthly rainfall distribution over the East Coast (Vriddhachalam, Chintamani and Bapatla) is different to other cashew growing states as they are influenced by rain received during the late Southwest monsoon and Northeast monsoon (October-December). It is the reason, why, the dry spell commences from November across the West Coast while February along the East Coast (Tamil Nadu). From the above, it is evident that cashew experiences moderate to severe soil moisture stress across the West Coast under rainfed conditions and prolonged further across the East Coast as it experiences late rains with less number of rainy days and rainfall is below crop water need. It is also understood that the better availability of soil moisture during flowering period (December and January) may not influence the time of flowering in cashew. The unusual rains during November and December may lead to inordinate delay in reproductive phase of late season types/non-descript planting material across the West Coast. The delay in the time of flowering may be detrimental for cashew production as the number of bisexual flowers produced might be less since the flowering phase misses the mild winter.

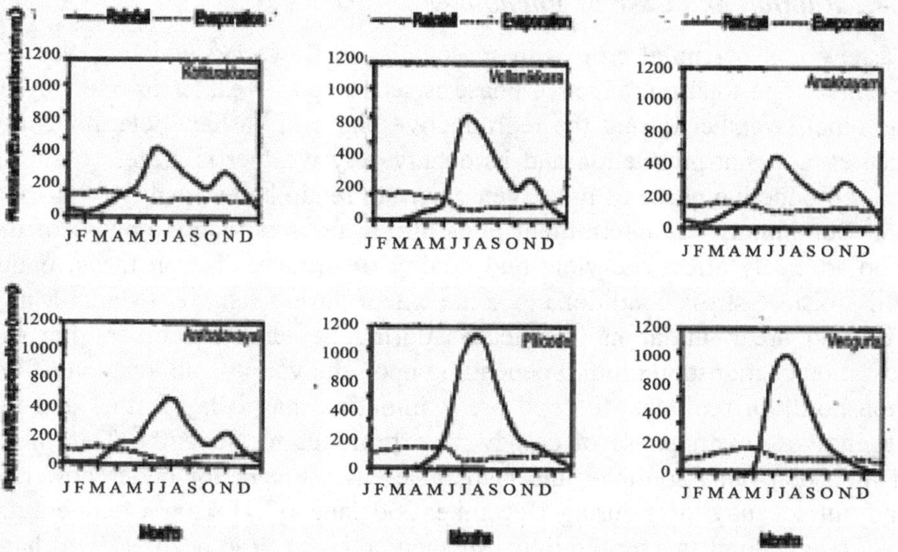

Fig. 2.7a: Rainfall and pan evaporation (mm) at differest focatiie: across the west coast of India

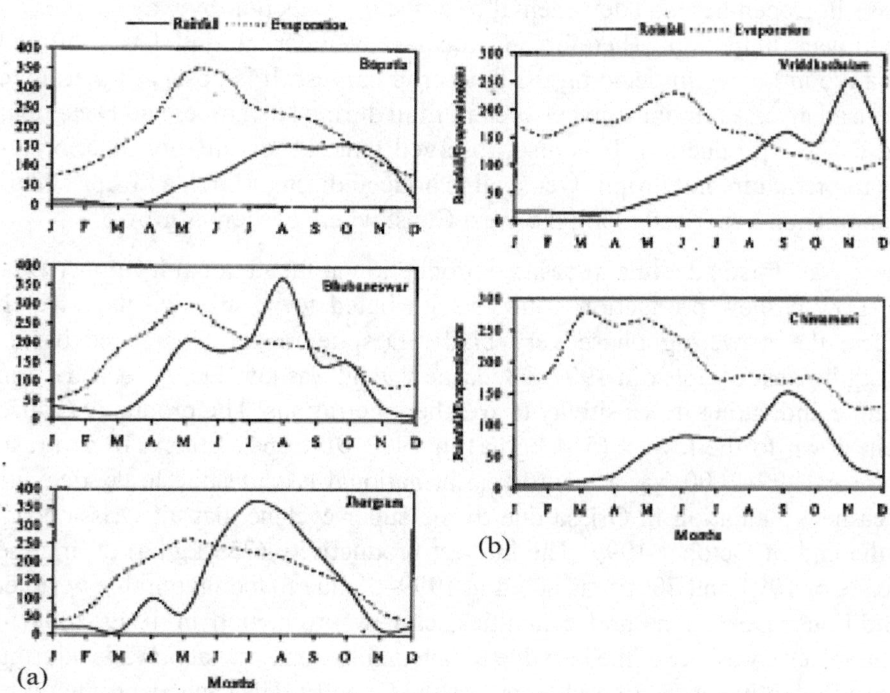

Figs. 2.7b, a and **b** Rainfall and Evaporation (mm) at cashew locations across the West and East Coasts of India

2.4.2 Rainfall and Cashew Phenology

Cashew is a perennial crop with seasonality in flowering which is weather sensitive. The total reproductive phase is sensitive to weather in one way or the other. Weather during the reproductive phase of cashew determines the final cashew nut production and its quality. Dry weather is preferred during the reproductive phase of nut development to retain better quality of cashew nut. Continuous and intermittent rains during the reproductive phase of the crop adversely affect the yield and quality of produce. It can thrive under soil moisture stress conditions to some extent during summer (March-May). The bud break initiation commences during September/October after the cessation of monsoon rainfall depending upon the variety and location. If the monsoon rain prolongs, the bud break initiation may delay further and the vegetative growth phase of cashew also prolongs to November/December. If monsoon rains continue, the favourable environment for better flowering and nut set may miss during December and January. The night temperature may rise beyond favourable limits in such an event. Rao *et al.*, (1999) have observed that untimely rains during November and December and rise in night temperature (>20°C) with less dew nights which coincide with flowering phase in December was detrimental to cashew production by way of altering the flower induction. Therefore, October–November rainfall has profound negative influence in deciding the final crop harvest. It is observed that there exists an inverse relationship between rainfall during November and December and cashew production. It is also observed that cashew quality deteriorates due to premature nut drop if wet spell is noticed during March and April. This phenomenon is noticed along the West Coast where cashew is grown.

There was drastic decline in cashew production at the national level in 1997–98. Low cashew production could be attributed to adverse weather, which affects the flowering phase very badly. Despite meagre incidence of tea mosquito pest complex in 1997–98, cashew yield was low due to wet and warm weather, indicating its sensitivity to weather aberrations. The productivity also went down to the lowest (514 kg/ha) in 1997–98 to pest attack. In contrast, the year 1999–2000 was a record year at national level despite heavy damage to cashew plantation in Orissa due to the super cyclone that hit Orissa coast in the end of October 1999. The highest productivity (758 kg/ha) during the decade of 1991 and 2000 was noted in 1999–00 due to the favourable weather conditions. Despite natural calamities, cashew production in 1999–2000 at national level was one of the best due to favourable weather that prevailed during the reproductive phase of cashew across the Country. The cashew productivity has been dwindling between 600–700 kg/ha since last two-and-a half decades, except in 1999–2000 and has reached 778 kg/ha in the recent years. As a whole, at the national level the cashew production and productivity of cashew is in increasing trend though inter annual variations are significant through better crop improvement, management and protection measures.

2.5 Wind

In the Central region of Kerala, viz., in Thrissur and Palakkad Districts a gradual increase in wind speed is noticed from the second fortnight of November to mid February or till the end of February during which the process of cahew nut set takes place . The wind as it passes through the Palakkad Gap of the Western Ghats, gains momentum and blows as easterly wind towards the West. The influence of wind is felt up to coastal areas of Thrissur district. The average daily wind speed quite often crosses 10 km/hr whereas the mean annual wind speed is 5 km/hr. Instantaneous wind speed is much higher than the average wind speed during these periods. Studies reveal that the wind speed from November to February which coincides with the flowering period of cashew is increasing since last one-and-a-half decade over the Central region of Kerala. The strong wind prevails during the peak flowering of cashew during November to February may be able to wither away the flowers/ flower fall, and thereby causes yield loss. Strong dry wind also causes flower infertility and flower scorching. However, effect of increased wind speed on yield loss of cashew is yet to be quantified. The wind never caused mechanical damages to cashew plantations as noticed in the East Coast during cyclonic storms. Hence, as far as Kerala is concerned, mild wind speed is having a favorable effect on cashew production. Whereas strong wind during the flowering phase of cashew is a constraint for cashew production in Palakkad and Thrissur Districts. Rao (2002) reported that strong wind during the flowering phase of cashew prevailing in the Central region of Kerala may be an environmental constraint in obtaining potential cashew yield in the central belt of Kerala.

2.6 Phototropism

Cashew requires enough sunlight for reproductive phase, commencing from September with the bud break and flushing. When we closely examine the cashew plantations at individual trees, the flushing is noticed early on top of the canopy, followed by East and West in South direction (Figs. 2.8a&b). Cashew shows a tendency to grow more towards South in response to sunlight in the Northern Hemisphere. It is more evident in higher latitudes towards North in search of sunlight. The direction-wise occurrence of abiotic events indicated that there is a clear-cut difference in the number of flushes produced, number of panicles and number of fruits set/m2 within the canopy of a tree.

The branches of cashew in south produced maximum number of fruits, followed by West and East while the lowest towards North. Cashew responds well to sunlight, indicating that it has a predominant phototropism character. Interestingly, harvest is estimated to be more in South (60-80%) compared to that of North (20-40%) of cashew trees. A delay of one week to 10 days in appearance of all the biotic events of cashew towards north within the cashew tree can also be noticed. It also appears to be photosensitive as the reproductive phase is confined to season bound, varying between March and May across the West and East coasts of India. However, harvesting falls in June/July at higher altitudes (Ambalavayal and Chintamani). Of course, decolourisation of nuts happens and their weight and quality are very poor as regular rains start by that time. Though delay in bud break and flowering is seen towards North (Bapatla, Bhubaneswar and Jhargram), the crop matures early and completes by April-end or first week of May. This could be attributed to high maximum temperature (35-40^{0}C) that prevails during summer (March-May). It appears that the cashew is not only photosensitive but also thermosensitive as the biotic events respond to low minimum and high maximum temperatures. It seems that a temperature range between 13 and 35^{0}C may be optimum for better growth and production of cashew nuts. However, the optimum temperature range for cashew is yet to be worked out since it is only observational.

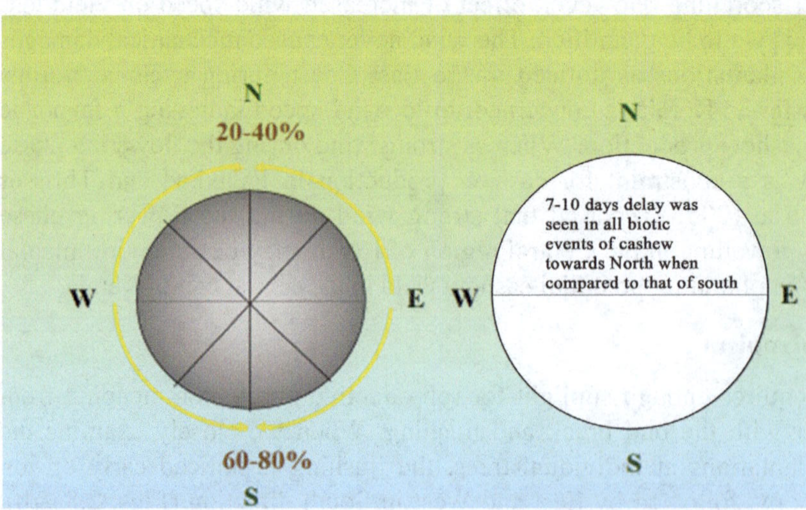

Direction-wise yield pattern in cahsew tree

Delay in biotic events between North & South of a cashew tree

2.8a **2.8b**

Fig. 2.8 a and b Direction wise yield pattern and biotic events

2.7 Drought versus cashew productivity

The rainfall distribution across the cashew tract of West Coast is uni-model, followed by prolonged dryspell from November and can extend up to May, if pre monsoon showers fail. The dryspell from October/November to April/May coincides with the reproductive phase of cashew, resulting moderate to severe soil moisture stress across the West Coast. This pattern is predominant in coastal districts of northern Kerala. The cashew tract in Maharashtra experiences prolonged dryspell of five to seven months as rain ceases by the end of October and the crop experiences severe soil moisture stress. However, in the West Coast during the dormant phase (June to September) of cashew no soil moisture stress is seen. Though the reproductive phase of cashew is under soil moisture stress, cashew plantations at different locations (Kottarakkara, Madakkathara, Anakkayam, Pilicode, Puttur and Vengurlae) across theWest Coast are healthy. Similar was the situation in Orissa and West Bengal where cashew is cultivated across the East Coast. Though the annual rainfall in Orissa and West Bengal may be different, the monthly pattern of rainfall is similar to that of the West Coast to some extent. In contrast, Tamil Nadu receives more rains from September to November due to the influence of Northeast monsoon. It may, sometimes extend up to December over Tamil Nadu where cashew is grown. Across the cashew tract of Andhra Pradesh, the monthly rainfall may exceed sometimes evapo-transpiration from August to October depending upon the location. Cyclones are not uncommon across the East Coast (Tamil Nadu, Andhra Pradesh and Orissa) during October and November, which bring copious rains, leading to floods and uproot cashew plantations due to heavy wind speed. Also, rains may extend up to mid December due to cyclones, which are not uncommon over the coastal belt of Andhra Pradesh and Tamil Nadu. The reproductive phase of cashew from February to May in these areas is under soil moisture stress as the crop is grown under sandy to sandy loam with poor water holding capacity. The cashew at Chintamani (Karnataka) is always under soil moisture stress throughout the year as rainfall was less than evapotranspiration round the year except a few weeks between September and November. Despite of the above fact, cashew plantations are better and comparable at Chintamani (Karnataka) with better crop management and crop protection measures. The cashew farmers practice summer irrigation wherever it is possible for obtaining better yields. The cashew yield at Vengurlae is the highest (18.5 kg/tree), even though the crops is under moderate to severe soil moisture stress for five to seven months during its reproductive phase **(Fig. 2.9)**. The cashew yield of the superior varieties was 10.5 kg/ tree at Madakkathara, where a prolonged dryspell of three to five months is seen during the reproductive phase of cashew. Bhubaneswar and Vriddhachalam recorded 12.8 kg/tree and 8.9 kg/

tree, respectively where the maximum temperature during the nut setting and development phase (March to May) is high and considered to be a constraint occasionally under severe soil moisture stress. At Chintamani, the cashew yield was 8.1 kg/tree where crop is under soil moisture stress for a prolonged period unlike at other cashew growing locations. It is evident that cashew productivity was relatively low where the crop is under soil moisture stress for eight to ten months as the case at Vriddhachalam and Chintamani. However, the yield was much better at cashew research stations where cashew is grown under rainfed conditions in better crop management situations when compared to that of country's average (less than 4 kg). Also, the cashew production is relatively high in Kannur and Kasaragod districts in Kerala, where prolonged dryspell is noticed from November to March, coinciding with the reproductive phase of cashew. At the same time, the yield was low whenever intermittent rain spells received from November to May across the West and East Coasts of India. Immature fruit drop was noticed if wetspells are seen in March/ April, which lead to poor yield and quality. From the above, it is understood that cashew experiences moderate to severe soil moisture stress across the west coast under rainfed conditions and prolonged further across the east coast as rain received is late and it is below crop water need. At Chintamani, cashew almost suffers due to soil moisture stress round the year except a few weeks between September and November.

With all the supporting data on the performance of cashew under dryspell/ drought, it can be inferred that cashew thrives relatively better under drought situation. Infact, cashew is exposed to dryspell/drought during its reproductive phase and yields better. However, the abnormal drought situation affects cashew yield very much under rainfed conditions. This is possible if the duration of dryspell is prolonged from the normal as the varieties/cultivars are susceptible to abnormal severe soil moisture stress. The varieties/cultivars are tolerant to soil moisture stress under normal rainfed conditions due to their adaptability over a period of long time, beyond which they may fail to yield at the desired levels. Also, the cashew yield and its quality are affected due to prevalence of high maximum temperature (350 C and beyond) during the reproductive phase of cashew (flowering/nut setting to harvest) in a situation where prolonged abnormal drought is noticed. The above situation is noticed across the East Coast at Vriddhachalam, Bapatla, Bhubaneswar and Jhargram. Sometimes, the maximum temperature shoots up to 42°C during the nut development period across the east coast. However, it is not the case across the West Coast. The moderate (30–33°C) maximum temperature that persists across West Coast during the reproductive phase may be ideal for better nut development and its quality. However, at Madakkathara in Kerala along the

West Coast, high wind speed during November and December and maximum surface air temperature beyond 35 degree Celsius during summer (February–April) may be limiting factors for obtaining potential yields in cashew. It needs further examination. Also, the physiology of crop is to be totally understood to find out the mechanisms, which are working internally in favour of drought tolerance in cashew.

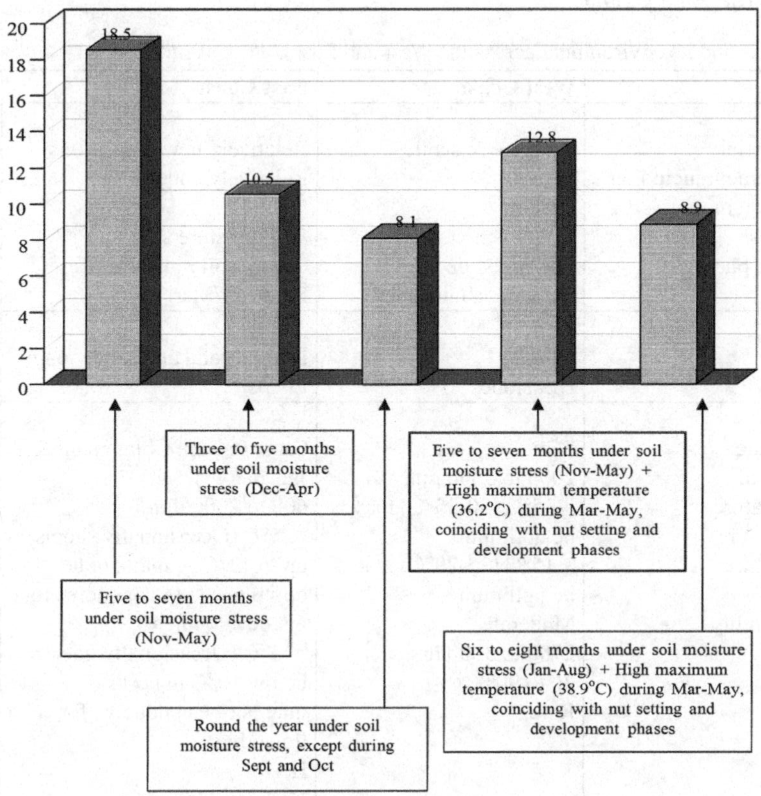

Fig. 2.9: Drought versus cashew productivity

2.8 Environmental Constraints

The climatic requirements during the reproductive phase of cashew are relatively favourable across the West Coast expect at few pockets where the maximum surface air temperature goes beyond 35 degree Celsius. As a whole, the major threat to cashew production across the West Coast is the incidence of tea mosquito bug complex. It is, again, triggered by favourable weather. Suitable genotypes and crop management under the field conditions along with scientific control measures against the pest complex will go a long way to sustain cashew production across the cashew tract of the West Coast. The meteorological factors like night temperature (<10° C) and day maximum temperature (>35° C) during the reproductive phase of cashew may not be conducive for better flowering along the East Coast. In addition, the occurrence of cyclones is not uncommon. It is a major threat to cashew plantation across the East Coast (Table 2. 1). Unless the cashew plantation is well protected with wind breaks/shelter belts, it will not be a viable economic proportion to grow cashew across the East Coast.

Table 2.1: Cashew and its environment across the West and East Coasts of India

Environment	West Coast	East Coast
Plant		
1. Planting material	Pedigree & senile	Relatively new plantations
2. Crop management including crop protection measures	Poor	Relatively better
3. Dormant phase	No moisture stress	Soil moisture stress
4. Reproductive phase	Soil moisture stress	Severe soil moisture stress
5. Flowering	December/January	February/March
Soil		
1. Soil	Laterite	Littoral sand and sandy loam
2. Topography	Hill slopes	Plains
Climate		
1. Coastal climate	Same	Same (except Chintamani & Jagadalpur)
2. Annual rainfall	2000 to 3600 mm	600 to 2000 mm
3. Max. temperature	< 35°C (30-35°C) may be optimum	>35°C (Occasionally shoots up to 42°C) – may not be optimum - not conducive for nut development
4. Min. temperature	> 15°C (15-20°C) may be optimum	<15°C (Occasionally goes below 10°C in north – may not be conducive for flowering)
5. Soil temperature	Moderate	High
6. Sunshine	2700 to 3000 hrs	2700 to 3000 hrs
7. Relative humidity	70 to 90%	50 to 70%
8. Cyclones	Rare	Not uncommon
Irrigation	No summer irrigation	Irrigation given in some areas
Pest Complex	Incidence is common	Incidence is not common
Tea mosquito bug	Moderate to severe	Low

2.8.1 Environmental Constraints Over Kerala

The benevolent and malevolent effects of cashew environment across Kerala are highlighted separately as it plays a major role in cashew industry. As a whole, the incidence of pest complex is the threat to cashew production over Kerala. In few pockets like Palghat and Thrissur, the adverse effects of high wind speed and the maximum surface air temperature beyond 35 degree Celsius during the reproductive phase on nut yield and its characteristics are yet to be understood. Similarly, lack of mild winter and occasional wet spells towards the South of Kerala may also hinder cashew production (Table 2.2).

Table 2.2: Cashew and its environment over Kerala

Environment	Existing condition	Effect
Plant		
Planting Material	Pedigree & Senile	Malevolent
Crop management including crop protection measures	Very poor	Malevolent
Dormant phase	No soil moisture stress, exposed to low solar radiation due to heavy rainfall	Benevolent/Malevolent
Reproductive phase	Soil moisture stress	Malevolent
Flowering	December/January	Early season type may escape pest complex-Benevolent
Soil	Very deep and well drained laterites OC-Marginal N- low to medium K_2O- low	Benevolent/Malevolent
Topography	Sloppy	Benevolent
Climate		
Climate	Coastal climate	Benevolent
Annual rainfall	2000 to 3600 m	Malevolent
Max. Temperature	< 35°C (30-35°C) –may be optimum	Benevolent
Min. Temperature	>15°C (15-20°C) –may be optimum	Benevolent
Soil temperature	Moderate	Benevolent
Sunshine	2700 to 3000 hrs	Benevolent
Relative humidity	70 to 90 %	Malevolent
Wind	Low to moderate	Benevolent- It is doubtful in case of Palghat and Thrissur districts where wind speed is high
IRRIGATION	No summer irrigation	Malevolent
PEST & DISEASES Tea mosquito	Moderate to severe	Malevolent-threat to cashew production

2.9 Forewarning of Tea Mosquito Bug

Tea mosquito bug, *Helopeltis antonii* Signoret (Heteroptera: Miridae), is an important pest of cashew. It sucks the sap of tender parts of cashew including inflorescence, panicles and young cashew nuts. The insect injects its saliva into the plant during its feeding which causes drying of the affected parts. It is a secure menace across the cashew tract of the West Coast and the East Coast, to some extent. The pest menace is also seen in “Maidan areas” of Karnataka located over highranges where cashew is predominantly grown. The pest population commences in October and November depending upon the location and synchronizes with flushing and flowering stages and reaches its peak during the blossom period. The flower scorching is normally noticed during the reproductive phase of cashew irrespective of tea mosquito bug incidence is prevalent or not. It may lead to suspicion of pest incidence as the flower scorching is a common symptom in cashew. Nevertheless, the flower scorching can also be due to cloudy weather as noticed on many occasions.

2.9.1 Seasonal Pest Incidence

The most favourable period for rapid multiplication and population build up of pest are noticed between December and February synchronizing with flushing and flowering of cashew **(Fig. 2.10).** The pest population of tea mosquito bug is low and insignificant round-the-year except during the reproductive phase from December to February. The tea mosquito bug incidence during the reproductive phase is detrimental in the case of cashew productivity. By and large, the seasonal tea mosquito incidence is predominant across the West Coast when compared to that of any other cashew tract of India. The pest population is relatively high towards the West Coast.

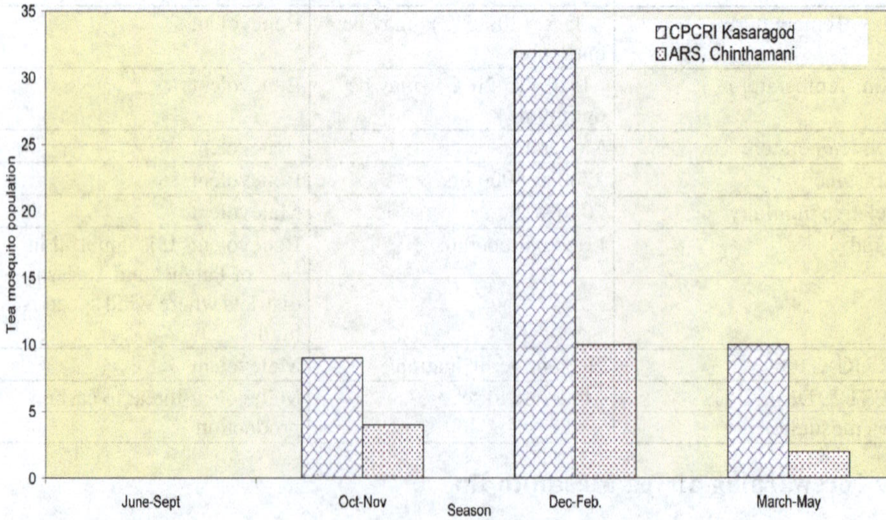

Fig. 2.10: Season-wise tea mosquito population in cashew (after Pillai et al., 1984 and after Rao, 2002)

2.9.2 *Influence of Weather on Pest Incidence*

There is zero pest population from June to September during which the relative humidity and rainfall are quite high and the period of bright sunshine is very low across the West Coast and build-up of pest population from a very low level commences in October and reaches its peak in December/January, when the minimum temperature is very low with low relative humidity. During this period, no rainfall is seen and the duration of bright sunshine is quite high with clear sky. Several studies also reveal that the most favourable period for rapid multiplication and pest population build-up of tea mosquito is during December, January and February, when the mean duration of bright sunshine is noticed around nine hours per day. Even though tea mosquito bug is known to be as insect which normally shuns bright sunlight, it was observed to continue its destructive activity hiding on the ventral surface of leaves and feeding on the midrib, petiole, tender shoot or panicle. Moreover, it is during the period the host plant also provided an abundant supply of succulent plant parts. The succulent plant parts may not be normally available on grown up trees, and as such, the tea mosquito population was also absent. The population build-up of the pest was found to be negatively correlated with the meteorological factors like minimum temperature, low relative humidity with no rainfall and the sunshine is maximum. A comparative study at Pilicode revealed that the night temperature and relative humidity in the afternoon were less during the flushing/flowering phase of cashew in 1998–99 during which the pest population was high when compared to that of 1997–98. In a period of two-and- a-half months, the pest might have completed three to four overlapping generations, thus increasing in number by six to seven fold, causing serious damage during 1998–99 while the pest population was totally absent/meagre (during 1997–1998). The minimum night temperature below 20oC along with relatively dry weather in the afternoon during majority of the days from December to January might have triggered the pest population during 1998-99 (Rao, *et al* 2002). The build-up of pest population was relatively late and coincided with mid-and-late season types as early season types escaped from the incidence of tea mosquito and yielded better in 1998–99. At Chintamani also, the pest population was high during flushing and flowering initiation, coinciding with winter season (December to February) during which the maximum and minimum temperatures were relatively low when compared to that of the other seasons. High surface air temperature with afternoon relative humidity reflects negatively on pest population. Rainfall is also negatively correlated with the pest incidence and development while the bright sunshine had positive effect on pest population (Table 2.3). Obviously, it is seen that the bright sunshine is more during flushing and flower initiation

Table 2.3: Relationship between weather and TMB incidence at ARS, Chinthamani from 1988 to 1999

Season	Surface air temperature (°C)		Relative humidity (%)		Rainfall (mm)	No. of Rainy days	Bright sunshine hours (h/day)
	Max	Min	Forenoon	Afternoon			
June-Sept	29.7	20.5	78.4	59.2	413.0	24.3	5.7
Oct-Nov	27.8	18.3	78.9	63.3	180.2	11.9	5.8
Dec-Feb	27.8	14.7	74.6	49.8	25.2	1.9	8.6
Mar-May	34.0	20.7	68.3	39.7	103.3	7.0	9.4
Correlation coefficients with pest population	-0.385**	-0.664**	0.095	-0.173**	-0.305**	-0.397**	0.339**
Direct effect through path coefficient analysis	-0.364	-0.351	0.077	0.016	0.016	-0.110	0.262

when compared to that of rainy season during which the pest population is meagre. The negative indirect effect of maximum surface air temperature could be explained on pest population as the pest population was declining rapidly during summer months though the reproductive phase is active through flowering and fruiting phases. At the same time, the number of rainy days had direct negative effect, indicating that the pest population was negligibly small during the southwest monsoon season (Rao, *et al.*, 2002). When weather parameters were related with weekly mean TMB population of absolute and sample estimates, only minimum temperature has shown consistently negative relationship. Among weather factors, monitoring minimum temperature alone may be an useful tool for early warning of TMB population incidence and explosion. It is also mentioned that continuous low minimum temperature in the current week may cause increase in abundance of TMB population in the succeeding second-third week period. In some of the years, marked variations in minimum temperature during December–January was seen particularly in two outbreak years (1998–99 and 2000–01) and it was between 16 and 20°C. Similar results were obtained at Vengurle also as the minimum temperature was found to be negatively correlated with pest incidence. The pest incidence increased with decrease in minimum temperature. The pest population was high at Vengurle during 1982–83 and 1983–84, followed by 1985–86 and the least in 1984–85, indicating that the pest population is dependent on favourable weather conditions under the same crop environment. The study at Pilicode also indicated that the pest population was high as the night temperature range varied between 17 and 20°C for majority of the days from December to February when compared to that of Madakkathara, where wind speed is high and relative humidity in the afternoon is too low in addition to high night temperature of 19–24°C. On an average, the mean night temperature during the reproductive phase (December–February) was 20°C at Pilicode while 22.1°C at Madakkathara In the peak pest year 2005–06, night temperature was relatively low (21.7°) when compared to that of 2003–04 (22.3°C) and 2004–05 (22.3°C) at Madakkathara also. The number of nights recorded below 20.0°C at the above stations was also more (23 days) in 2005–06 when compared to that of 2003–04 (18 days) and 2004–05 (7 days). It was 39 days, 51 days and 46 days during 2003–04, 2004–05 and 2005–06, respectively at Pilicode where night temperature was on an average below 20.0°C (Table 2.4). Probably, it was one of the reasons for high pest incidence at Pilicode and it is a hot spot for the tea mosquito incidence when compared to that of Madakkathara **(Fig. 2.11)**. In a peak pest year, such trend was noticed at Madakkathara also though the pest population was relatively less when compared to that of Pilicode. Thus, the night temperature was found to be more critical in pest build up. A high significant negative relationship was found between TMB population and weather factors such as minimum temperature, relative humidity, rainfall and the number of rainy days. Due to the above factors, the pest population is meagre during the rainy season and more during winter (December to February), which coincides with the reproductive phase of cashew. It reveals that the pest population is moderate to severe (8–20/q/tree) if the minimum temperature varies between 15 and 22°C. It is low to moderate (4–8/q/tree) if the minimum temperature is between 12 and 15° C and 22 and 24° C. It is less than four per quadrant per tree if the minimum temperature is between 10 and 12° C and greater than 24° C (Table 2.5).

Table 2.4: Night temperature during the reproductive phase of cashew at Madakkathara and Pilicode from 2003-04 to 2005-06

Year	Night temperature (°C) during reproductive phase (December -February)	
	Madakkathara	Pilicode
2003–04	22.3 (7 days)	20.4 (39 days)
2004–05	22.3 (18 days)	19.5 (51 days)
2005–06	21.7 (23 days)	20.2 (46 days)
Mean	22.1 (16 days)	20.0(45 days)

Table 2. 5: Minimum temperature versus intensity of tea mosquito bug

Minimum temperature (°C)	TMB population (No.q/tree)	Intensity of TMB
>24	<4	Low
22 and 24	4 to 8	Low to moderate
15 and 22	8 to 20	Moderate to Severe
12 and 15	4 to 8	Low to moderate
10 and 12	<4	Low
<10	0	Nil

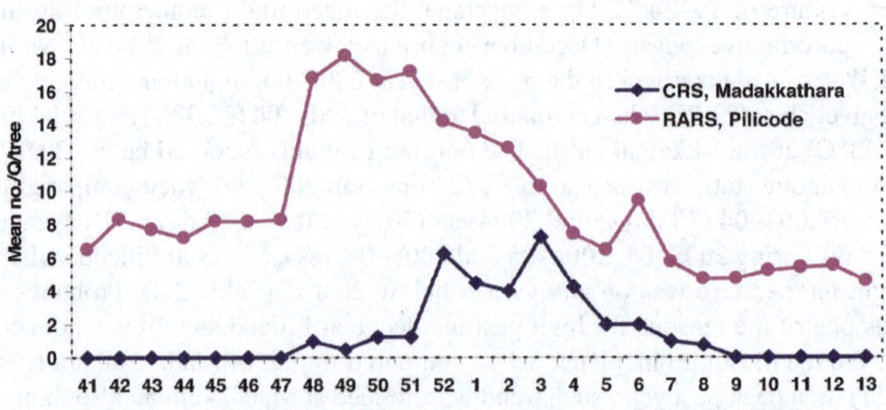

Fig. 2.11: TMB population at Madakkathara and Pilicode (2005-06)

2.9.3 Hotspot Areas of Tea Mosquito Bug

The hot spot areas of tea mosquito across the cashew tract of the whole country have been demarcated taking into account the optimum night temperature during flushing and flowering stages. The pest population may be moderate to severe across the West Coast, commencing from northern districts of Kerala to southern districts of Maharashtra, where cashew is predominantly grown (Fig. 2.12). The pest population may be low to moderate across the East Coast and high ranges (Ambalavayal in Kerala and Chintamani in Karnataka) and relatively free in West Bengal and the Northern States as the pest may not survive in cold weather. The potential areas of cashew production across the West Coast may experience the menace of the pest. The hotspot areas of tea mosquito population provides a bird's eye view of the area that need intervention through integrated pest management as well, for expansion of cashew cultivation across India.

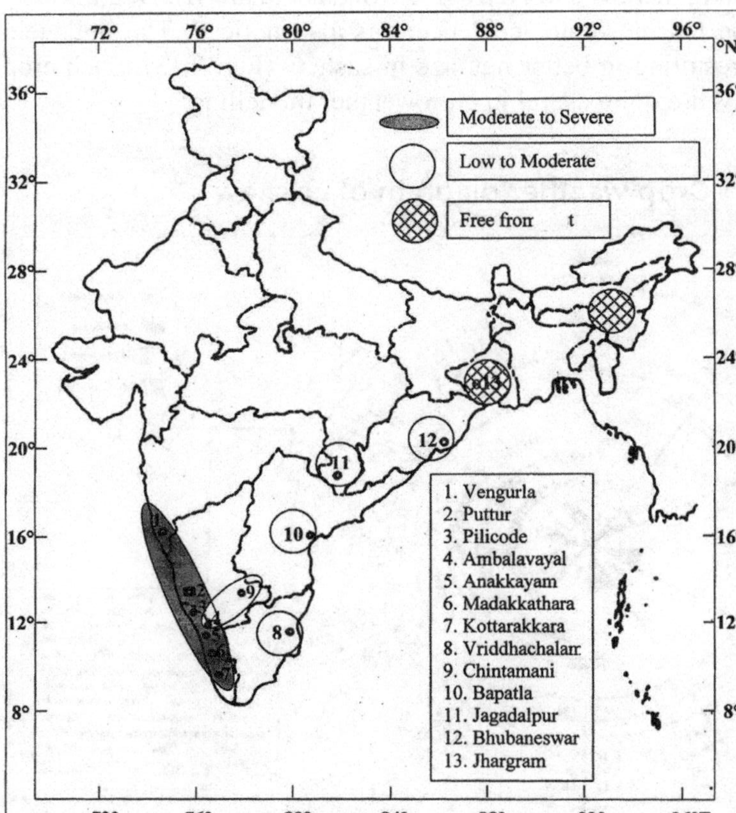

Fig. 2.12: Hot sport areas of Tea Mosquito Bug across India

2.10 Crop Weather Diagrams

It is handy to understand the crop performance in relation to weather. The biotic events such as bud break, flushing, nut set and nut harvest corresponding to rainfall and temperature are monitored in bad, moderate and good yield years. It gives an idea that the type of weather conditions are conducive for better or poor crop set including the incidence of pest and diseases. The crop season 1998-99 was an example to understand the effect of tea mosquito bug together with circospore, during which the crop damage was up to 100% in majority of cashew plantations in Kannur and Kasaragod Districts. The crop season1997-98 was another example to understand the poor yielding year due to unfavourable weather conditions prevailed during the flowering phase of cashew in the absence tea mosquito incidence. The moist weather during November and December due to rains in 1996 delayed the reproductive phase of cashew in all the three selected varieties viz., BLA 139-1 (Anakkayam), BLA 39-4(Madakkathara I) and NDR -2-1 (Madakkathara II). Due to delay in flowering, the tea mosquito incidence was also noticed. The nut yield was moderate, resulting in better nut size in cashew (Fig. 2.13). Such crop weather diagrams are quite useful in crop weather modelling.

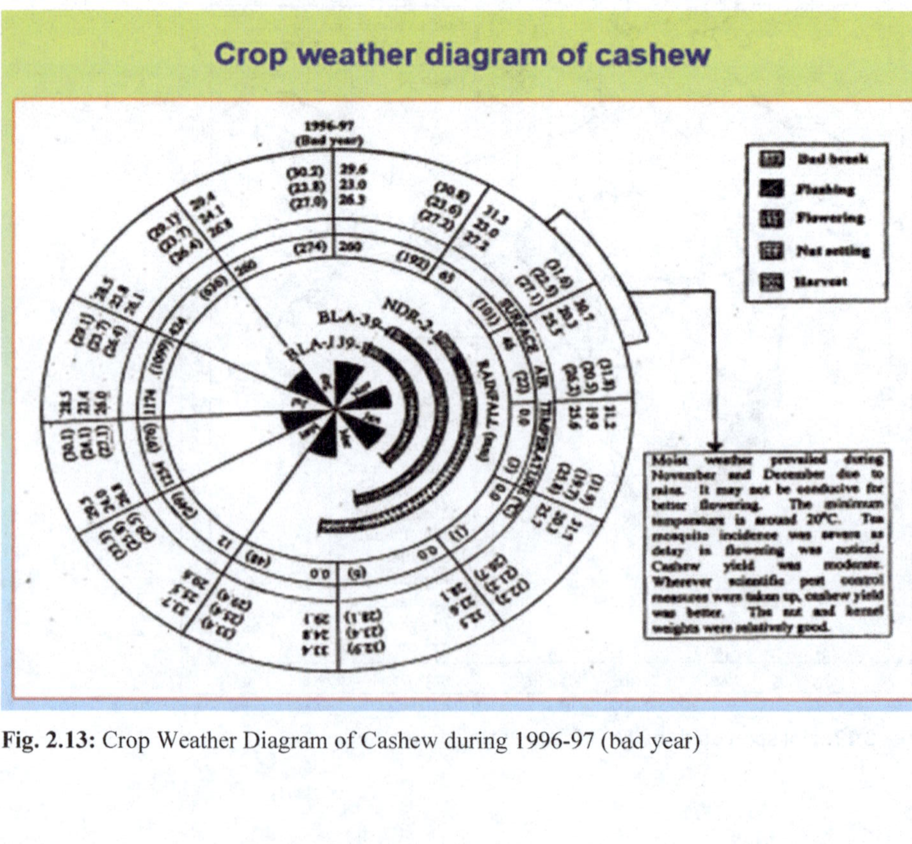

Fig. 2.13: Crop Weather Diagram of Cashew during 1996-97 (bad year)

2.11 Agroclimatic Zones of Cashew

Bhaskara Rao et al. (2002) made an attempt to indicate agroecological sub-regions and climatic conditions of cashew-growing areas in different states. Taking into consideration the variations in fruiting phase, temperature variations along the West and East Coasts, rainfall behaviour including summer showers, different agroecological zones are demarcated and suggested suitable varieties. The length of crop growing period for cashew varies from 90 to 270 days depending on agroclimatic / ecological regions and varieties. Kerala is the major cashew growing region falls under the two Agro-ecological Zones, numbered as 19.2 and 19.3 and described as follows:

The Agro ecological Zone 19.2: Kerala, Central and south Sahayadris, hot moist sub humid transitional ESR with deep, loamy to clayey red and lateritic soils, low to minimum AWC and LGP 210-270 days. The areas with medium AWC and LGP fallunder this zone.

Varieties recommended under 19.2 : VRI 3, Kanaka, Dhana, VRI 2, NDR-2-1, Amrita, Ullal 1,3,4

The Agro ecological Zone 19.3: Konkan, Karanataka and Kerala: Coastal plain, hot humid to per humid transitional ESR with deep, clayey to loamy, acidic, coastal alluvium-derived soils, low AWC and LGP 240-270 days. Coastal areas of Kozhikode, Kannur, Kasaragod, Malappuram, part of Thrissur and part of Trivandrum, fall in this humid transitional zone. Coastal areas of Maharashtra Kalahandi, Phulbani, Balangir, Sambalpur, Sundargarh and Koraput fall in this zone including the interior areas of this region.

Varieties recommended under 19.3: VI, V3, V6, V7, Dhana, Kanaka, Goal, NDR-2-1, Ullal, 3, 4, VRI 3,2 and BPP8

Based on the information available on various aspects of cashew production in relation to weather, soil and yield potential, an attempt has been made to demarcate “Agroclimatic zones for cashew” across the West and East Coasts of India by Prasada Rao. Cashew productivity can be improved towards North across both the Coasts of West and East while it may decline towards inland plateau of Peninsular India (Fig. 2.14). The Northeast region may also be not conducive due to occurrence of frequent cold waves during fruiting season.

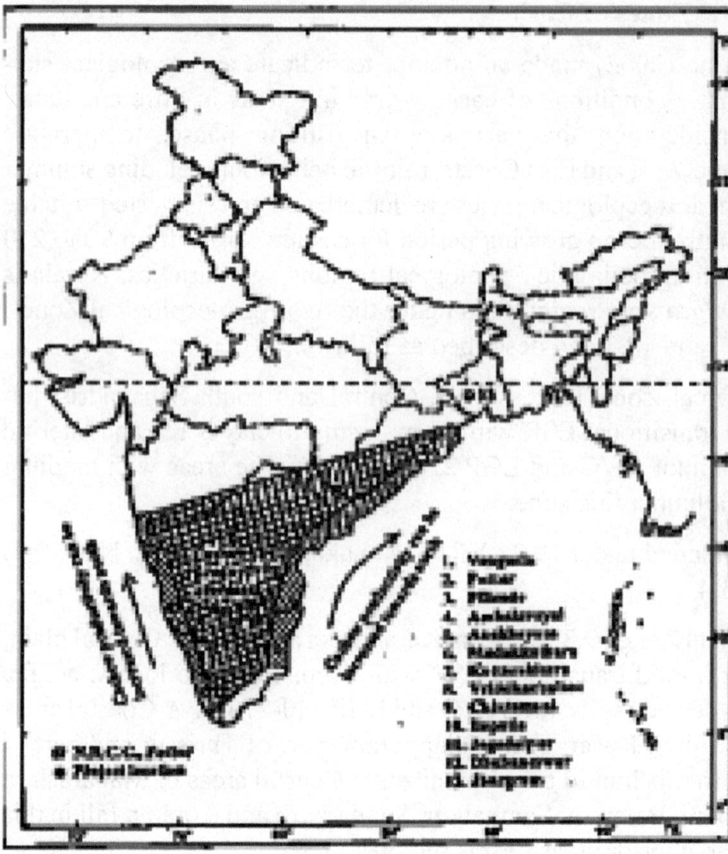

Fig. 2. 14: Agroclimatic zones of cashew

2.12 Yield forecasting models for cashew

It is evident that the reproductive phase of cashew is very sensitive to weather aberrations and final crop output depends upon weather conditions if no pest incidence is noticed. The variations in crop output could be known only after the crop is over or in due course of reproductive phase of cashew. It may not be much useful for the cashew industry since information on cashew production is not known before the crop season commences. It means, a sort of crop forecasting mechanism is to be developed well ahead of crop season commences. Such a mechanism is evolved for crop output of cashew for the State of Kerala based on rainfall from September to December (Fig. 2.15). Similar attempt can be made for the country as a whole so that the information can be used in EXIM policy. In view of drastic decline in cashew area across Kerala in recent years the model needs to be re-validated for forecasting cashew production.

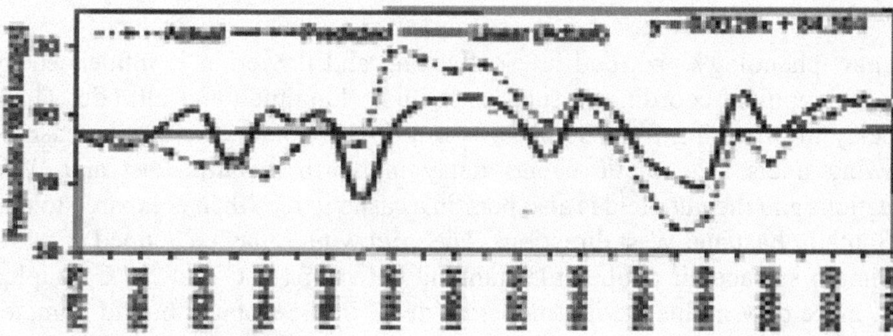

Fig. 2.15: Actual and estimated production of cashew over Kerala from1978-79 to 2004-05

2.13 Heat Units and Cashew Yield

A significant variation in heat unit requirement of cashew is noticed depending upon genotype. Early variety like Anakkayam (ANK-1) requires only 1953 day°C for reproductive phase (25% bud break to 75% harvest) while 2483 day°C in case of late season types like Madakkathara (MDK - 2). The heat unit requirement of mid-season type like MDK-1 is 2245 day°C. It reveals that early season types require less heat units while late season types require more heat units and the heat unit requirement of mid-season types is intermediary. The heat units versus cashew yields of test varieties showed that there is no definite relationship exists between them as negative relationship is seen in 50% of the cases and positive in other 50% of cases. However, the mean nut yield follow the number of heat units except in M D K-2. It is mainly attributed to incidence of the pest complex, which is not uncommon during the flowering season of cashew across the West Coast. Interestingly, the kernel weight of all the test varieties followed the heat units. It can be inferred that the cashew yield and kernel weight may follow the availability of heat units if other environmental conditions are homogenous. It reveals that the biotic events and cashew yield attributes not only respond to rainfall regimes but also thermal regimes too.

Summary

Cashew phenology viz., bud break, flushing and flowering is influenced by the geographical coordinates such as latitude, longitude and altitude. There is delay in flowering from South to North of the Country across the cashew growing tracts. The biotic events delay in North towards East and West directions and the nut yield is also poor in a cashew tree when compared to that of South in East and West directions. The mild winter may be defined as " low minimum surface air temperature ranging between 15^0 C and 20^0 C coupled with more dew nights having moderate dew" in the tropical humid climates (Rao et al., 2001). Extended rainfall during November and December is not conducive for reproductive phase of cashew viz., bud break initiation, flushing and flowering. The reproductive phase is delayed due to extended rains during November and December and misses mild winter along with moderate dew, resulting in low production of cashew along the West Coast. It is the reason, why, bi-model rainfall due to influence of southwest and northeast monsoon is not favourable for cashew production. Along the East Coast, a few intermittent showers during summer favours for better nut size in cashew though nut drop is noticed. Wet spells during nut development phase (summer) are detrimental to quality of cashew nuts. High temperature (40-45^0C) during summer along the East Coast may be detrimental to nut size. It is one of the reasons for relatively small size nuts along the East Coast when compared to the West Coast under rain fed conditions. Tea mosquito is a threat along the West Coast while cyclones are threat along the East Coast in case of cashew production. Hotspot areas of tea mosquito are delineated based on night temperature. Tea mosquito incidence is severe along the West Coast while moderate along the East Coast. It appears to be negligible in Northeast region where low day and night temperatures prevail during the reproductive phase of cashew.

Suggested Readings

Bhaskara Rao, E.V.V., Yadukumar, N. and Bhat, N.G 2002. Agroecological regions and varietal classification of cashew in developmental strategies. Proc. of the S'11 National Seminar on Indian cashew in the New Millennium, Vishakhapatnam, 12-13 April 2002, pp.43- 51

Haldankar, P.M., Deshpande, S.B., Chavan, V.G., Dalvi, M B., Patil, B P., Joshi, G.D., and Bhaskara Rao, E V V 2003. Weather associated yield variability in cashew nut. J. of Agrometeorology 5(2): 73–76.

Haldankar, P.M., Deshpande, S.B., Chavan, V.G., Nadkarni, H.R., Patil, B.P., Joshi, G.D., and Rao, E V V.B 2002. Crop weather association in mixed population of cashew hybrids. J. of Plantation Crops 2002, 30(3):37–40.

Hopkins, A. D. 1938. Bio climates, a science of life and climate relations. U.S. Department of Agric. Misc. Pub, 280p

IMD (Indian Meteorological Department) 1957. Agricultural Meteorology in India Meteorological Department: A Review. Issued in commemoration of silver jubilee of Agricultural meteorology, Division of India Meteorological Department, 42p

Naik, B.J., Gopakumar, C.S., Subash, N and Rao, G.S.L.H.V.P. 1997. Anther and anther dehiscence in cashew (*Anacardium occidentale* L.). In: Proceedings of the Tenth Kerala Science Congress, January 1-3, 1998. Kozhikode, pp. 193-195

Nybe, E.V. and Rao, G.S.L.H.V.P. 2005. Drought management in plantation crops. Kerala Agricultural University. Thrissur, Kerala, India. 176p.

Rajagopalan, A., Rao, G.S.L.H. V. P. and Devi, T.S. 1992, Leaf area index - A measure of vigour in softwood grafts of cashew. The Cashew. 6(1) :10

Rao, B.V.R. and Rao, V.U.M. 2024. Agricultural Meteorology. Jain Brothers, New Delhi.154p.

Rao, G.S.L.H.V.P, Giridharan, .P, Naik, B.J.P., Gopakumar, C.S., Krishnakumar., K.N and Tony, X . 2001. Influence of weather factors on bud break, flowering and nut quality of cashew (*Anacardium occidentale* L.). Indian J. of Agric. Sciences 71(6): 399-402.

Rao, G.S.L.H.V.P., Giridharan. M.P., Naik, B.J. and Gopakumar, C.S. 1998. Weather inflicted damage on cashew production- A remedy. The Cashew. 12 (4):36-43

Rao, G.S.L.H.V.P., Krishnakumar, K.N. and Tony, X. 2001. Agroclimatic zones for cashew across India. Proceedings of TROPMET 2000, February 1-4, 2000. Cochin, pp. 556-559

Rao, G.S.L.H.V.P., Raju, G.T.T., Puttaswamy., Krishnakumar, K.N. and Tony, X. 2002. Forecasting tea mosquito bug (Helopeltis antonii) of cashew (*Anacardium occidentale* L.). Journal of Agrometeorology. 4(1), 45-52pp.

Rao, G.S.L.H.V.P. 2002. Climate and cashew. Kerala Agricultural University. Thrissur, Kerala, India. 100p.

Rao, G.S.L.H.V.P. 2021 (5th Reprint). Agricultural Meteorology. PHI Learning Private Limited, New Delhi, India. 364p.

Rao, G.S.L.H.V.P. and Beevi, P. 2010. Tea mosquito forewarning in cashew. Kerala Agricultural University. Thrissur, Kerala, India. 74p

Rao, G.S.L.H.V.P. and Gopakumar, C.S. 1999. Latitudinal and altitudinal influence on cashew flowering over Kerala. Proceedings of XI Kerala Science Congress, February 27-March 1, 1999, Central Plantation Crops Research Institute, Kasaragod, pp. 242-246

Rao, G.S.L.H.V.P. and Gopakumar, C.S. 2009. Agroclimatology of Cashew. In: Cashew Research and Development with emphasis on cashew apple processing (Eds: Jose Mathew, Mini, C. and Mareen Abraham). Directorate of Extension, KAU, Mannuthy, Thrissur, Kerala, India.160p

Rao, G.S.L.H.V.P. and Krishnakumar ,K.N. 2004. Yield forecasting of cashew. www. CommodityIndia.Com. April 4(4): 51-52

Rao, G.S.L.H.V.P. and Krishnakumar, K.N. 2024. Impact of weather and climate on plantation crops in the Humid Tropics (417-439pp). In: Agrometeorological Applications for Climate Resilient Agriculture (445p). Eds. Chattopadhyay, N. Robert, S., Attiri, S.D. and Rathore, L.S. Springer International Publishing AG. Singapore.

Rao, GS.L.H.V.P., Krishnakumar, K.N. and XavierTony, 2000 Crop weather diagrams of cashew- an agroclimatic tool for crop forecasting. Proc. of Twelfth Kerala Science Congress, January 2000, Kumily, pp.520-26

Questions

I. Elucidate on

1. Agroclimatic zones of cashew
2. Flowering in cashew
3. Forewarning of tea mosquito incidence
4. Environmental constraints in cashew
5. Thermal and Moisture Regimes of cashew

II. Define

1. Hopkins Bioclimatic law
2. Mild winter in cashew

III. Write Short Notes on

1. Geographical Co-ordinates and cashew flowering
2. Phototropism in cashew
3. Night temperature and tea mosquito incidence
4. Hotspot areas of tea mosquito incidence
5. Seasonal tea mosquito incidence
6. Thermal regime of cashew
7. Rainfall regime of cashew
8. Rainfall and bud break in cashew
9. Drought and cashew across India
10. Heat units and size in cashew nut

IV. Fill in the Blanks

1. Bud break in cashew depends on -------------
2. ------------- model rainfall is good for cashew
3. ------------- and ---------- crops compete each other
4. Cashew along the East Coast is prone to -------------
5. Cashew along the West Coast is prone to -----------
6. Sudden decrease in maximum temperature is noticed during -------------season
7. ----------------- soil moisture stress coincides with flushing and flowering in cashew

8. ---------------- soil moisture stress coincides with flowering, nut set and harvest in cashew
9. Waterlogging may not affect cashew as it grows along --------------
10. ----------------------- phase in cashew is sensitive to weather aberrations
11. For every 100 m altitude there is a delay of ---------- days in flowering of cashew
12. Cashew almost suffers for want of soil moisture round-the-year at ----------
13. -------------- is a constraint for cashew flowering in Thrissur and Palghat Districts
14. TMB incidence is more at ----------- when compared to that of Madakkathara
15. Intensity of TMB is moderate to severe if night temperature is between ----- and ----^{0}C

V. State True or False

1. Cashew is an inland crop
2. Cashew is a coastal tree
3. Coefficient of Variation is more in cashew productivity at Pilicode and Puthur
4. Cashew productivity is high in Kerala
5. Cashew area is more in Maharashtra
6. There is a positive relationship between NEM rainfall and cashew yield
7. The quality of cashew nuts deteriorates if nut development matches with rainy season
8. The uni-model monthly distribution of rainfall is conducive for better cashew production
9. The super cyclone hit Orissa Coast and destroyed cashew plantations in 2010
10. The quality of cashew nuts is superior in Kannur and Kasaragod Districts
11. The reproductive phase of cashew delays if NEM rainfall is prolonged
12. Dry spells are required for bud break in cashew
13. Biotic events are seen earlier towards the North of a tree
14. In a cashew tree, the yield is uniform in all directions
15. The area under cashew is more along the West Coast

16. High temperature beyond 40^{0}C during summer leads to small size in cashew nuts
17. Sudden summer rainfall for an hour or so along the East Coast leads to fruit drop in cashew
18. Coastal sandy soils are predominantly seen with cashew cultivation along the East Coast
19. Cashew nut size is lesser and lesser from the first harvest to final harvest
20. Cashew productivity and nut size are superior with summer irrigation

VI. Match the following

Temperature	Delay in cashew flowering 2
Rainfall	Early biotic events 4
Wind	Nut size 1
Sunshine	TMB incidence 5
Dew fall	Nut set 3
Bud break	

VII. Distinguish between

1. Temperature profiles across the East and West Coasts of India
2. Bi-model and Uni-model rainfall

3

Coconut

The Coconut palm, also known as the *Cocos Nucifera* L., is one of the most fascinating and beautiful palms in the world. The coconut is believed to have originated in southeast Asia (Indonesia, Malaysia, Philippines) or Micronesia. Coconut has been cultivated in India since ages and plays an important role in social, economic and cultural activities of the people. The palm is amenable to both plantation and homestead management and it can be either a major crop or a minor one in a homestead garden of mixed crops. While responding favourably to scientific management, the palm also tolerates negligent farming to a certain extent. Thus, it can adapt to the divergent farming situations and management practices that are prevalent in different agro-climatic regions. The coconut palm is usually found near the sea on sandy beaches where it can tolerate salt spray and brackish soils. It is also grown under different soil types such as loamy, laterite, coastal sandy, alluvial, clayey and reclaimed soils of the marshy low lands. The ideal soil conditions for better growth and performance of the palm are proper drainage, good water holding capacity, and presence of water table within three metres and absence of rock or any hard substratum within two meters of the surface. The coconut palm can also grow well inland, within altitudinal limit of 600 m to 1000 m in the Tropics. It requires an equatorial climate with high humidity. A year-round warm and humid climate favours the growth of coconut. The palm does not withstand prolonged spells of extreme weather/climate variations under rain fed conditions. A well-distributed rainfall of 1300–2300 mm per annum is preferred. It is known as a coastal tree since it is grown along

the Coasts and Islands and does not tolerate high diurnal temperature range, that is the difference between the maximum and minimum temperatures.

3.1 Coconut Position in India

India occupies the second position in the world with an annual production of 15 billion nuts. Our country with 15.6 % of coconut cultivation land contributes to 24.1 % of coconut production at the global level. There are about 5 million coconut holdings in the country with 98 per cent of such holdings occupying below two hectares. The coconut palm is grown in many diverse agro climatic zones of India, except sub tropic and temperate regions. However, they are favourably adapted to the coastal agro ecosystem coastline of 8129 km and its Peninsular Region bounded by the Arabian Sea on the West, the Bay of Bengal on the East and the Indian ocean on the South. Andaman and Nicobar Islands in the Bay of Bengal and Lakshadweep Islands in the Arabian Sea are unique inland eco-systems where coconut plantation is widely grown. The four southern States put together account for 90 per cent of the total coconut production in the country (Kerala 36.89%, Tamil Nadu 34.11%, Karnataka,13.83%, Andhra Pradesh 6.17% and other States 9.01%). Among the major coconut growing States, Tamil Nadu has the highest productivity (13771 nuts/ha). Tamil Nadu ranks first with regards to the expansion of area and production. Higher productivity in Tamil Nadu can be attributed to large size of coconut farms under irrigated conditions and better management practices on commercial scale. While Kerala accounts for the largest production of coconut in the country, productivity of coconut in Kerala is 7365 nuts/ha, even below the All India average yield of 8303 nuts/ha. The other coconut producing States in the country include Assam ,Goa, Gujarat, Maharashtra, Orissa,Tripura, West Bengal, Andaman & Nicobar Islands, Lakshadweep, Puducherry and Nagaland. Productivity-wise, some of these States, viz., Maharashtra (8338 nuts/ha), West Bengal (12430 nuts/ha), Lakshadweep (19630 nuts/ha) and Puducherry (14619 nuts/ha) performed better than some of the major producing States.

3.2 Coconut Position in Kerala

Kerala State is located between 8°15` N and 12°50`N latitudes and between74°50`E and 77°30`E longitudes. The State of Kerala is popularly known as the “Gateway of monsoon” over India. It is a strip of land running almost in North - South direction and is situated between the West Arabian Sea on the West and the ranges of Western Ghats and Nilgiri hills on the East both running parallel to each other. From the Western Ghats, the State undulates to the West and presents a series of hills and valleys intersected by numerous rivers. On extreme West, the State is more or less flat. These characteristics demarcate the State into three natural regions viz., the eastern high lands, the hilly midlands and western low lands. The changes in the geographical and topographical features due to man - made interventions are likely to influence atmospheric circulation altitudinally to a large extent. It may be one of the reasons in recent times for uncertainties in monsoon variability and rainfall distribution over Kerala. Kerala is blessed with tropical rain forests, rich biodiversity and has many rivers and streams and is spread in 13 natural agro-ecological zones. In Kerala State alone, there are about 2.5 million holdings. Among the 14 districts in Kerala, Kozhikode ranked first (851 million nuts), followed by Malappuram (832 million nuts) and Thrissur (575 million nuts) in terms of coconut production. In the case of productivity, Malappuram (7944 nuts/ha) stood first followed by Thiruvananthapuram (7755 nuts/ha) and Thrissur (7038 nuts/ha). Kozhikode, Malappuram and Kannur districts share 38 per cent of the coconut area. But it is interesting to observe that despite the incidence of the dreaded disease (root wilt) in the southern districts of Kollam, Alappuzha, Kottayam, Ernakulam, Idukki and Thrissur, the per palm productivity in these districts has been consistently on par with that of the disease free northern districts of Malappuram, Kozhikode, Kannur and Palakkad. It could be attributed to uniform distribution of rainfall towards the South when compared to that of northern districts of Kerala. Low productivity in Kerala is attributed to several factors, viz., high rainfall during monsoon. insignificant rainfall in several parts of Kerala during summer months, lack of irrigation during summer, poor nutrient status of soils, lack of agronomic practices, prevalence of dreaded root wilt disease in the problem zone of Kerala, incidence of stem bleeding in northern zone, of late attack of coconut mite and bud rot and the farmers` reluctance in application of fertilizers due to price fluctuations. In addition, it is mostly grown under rain fed conditions. The seasonal and annual variations in coconut production is mainly attributed due to dryspells within the monsoon and soil moisture deficiency from December to May, if pre - monsoon showers fail, which is not uncommon in the Humid Tropics under rain fed conditions. The severity of soil moisture deficiency is more towards northern districts of Kerala due to uni- model high rainfall (June–September), followed by prolonged dryspell for four to six months (December to May) if pre-monsoon showers fail during summer (March–May). Of course, the soil moisture status depends not only on soil type but also on topography, which is a significant feature in Kerala as coconut cultivation is seen in low land, mid land and high lands.

3.3 Onset of Monsoon

The monsoon directory of Kerala indicates that the earliest onset of monsoon was on 11th May in 1918 while belated monsoon on 18th June in 1972. The normal date of onset of monsoon was worked out and it is on 1st June with a standard deviation of seven days, varying between 25th May and 8th June (Fig. 3.1). The occurrence of maximum frequency (32.4% of years) of the onset of monsoon was between 30th May and 3rd June, followed by 4-8th June and 25-29th May (20.1 % of years in each interval) and 9-13th June (11.5 % of years). The chances of early onset of monsoon before 25th May were relatively less (5-5.8% of years). It was true in the case of belated monsoon beyond 8th June as 3.6 % of years only fell under the category, indicating that there was no change in normal date of onset of monsoon since last 150 years if the period from 1870 to 2020 was considered as a whole though inter-annual uncertainties were noticed within the limits of +/- one standard deviation. It was 30th May in 2024, 8th June in 2023, 29th May in 2022 and 3rd June in 2021. The onset of monsoon in recent years also indicated that it is seen within the normal onset of monsoon that is 1st June with +/-one standard deviation. It is very clear that there is no change in onset of monsoon due to climate change and inter-annual variations are not uncommon within the boundaries of one standard variation.

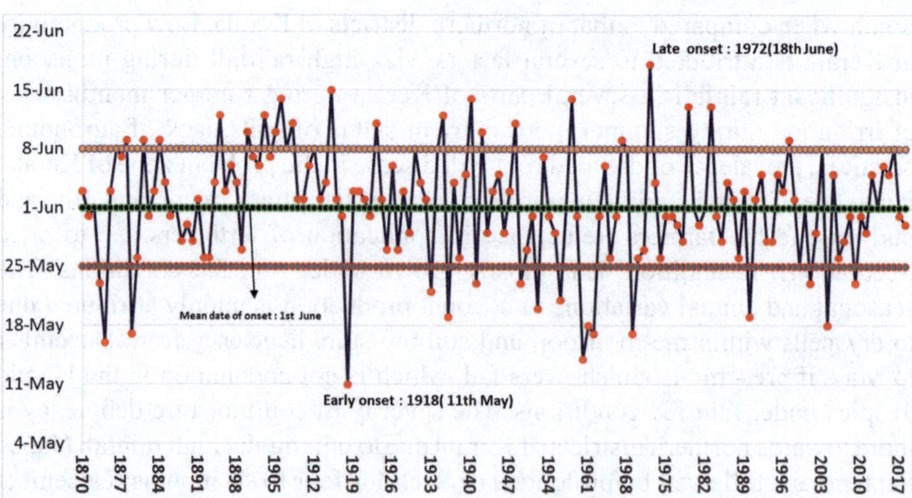

Fig. 3.1: Onset of monsoon over Kerala from 1870 to 2017 (148 years)

3.3.1 Rainfall over Kerala

The annual rainfall of the State of Kerala as a whole is close to 3000 mm. It is very high (5883 mm) at Nariamangalam (Ernakulam District) and low

(651 mm) at Chinnar (Idukki District). Rainfall increases from 1479 mm at Parasala in the South to 3562 mm in the North of Kerala. Similarly, it increases from the Coastal belt to the foot hills and then decreases on the hill tops. The monthly rainfall over Kerala appears to be bi-model (Figs. 3.2a and b) due to the influence of both Southwest and Northeast monsoons. The influence of Northeast monsoon is more significant towards South of Kerala, indicating that rainfall is uniformly distributed towards South when compared to that of North of Kerala. It is one of the influencing factors for better crop yields towards South. It is very true in the case of coconut and rubber. The northern Districts, Kasaragod and Kannur, experience uni-model rainfall, indicating that a prolonged dry spell is noticed from November to May if pre-monsoon fails. A strong dry wind also blows across the Palghat Gap, commencing from mid- November to mid-January, popularly known as 'Vrichika Kattu' in local language. It has tremendous negative impact on plantation crops. The other regions within Kerala are free from strong winds during the above said period. The long period average annual rainfall over Kerala is around 2828 mm with a coefficient of variation of 14.3 per cent. It indicates that the annual rainfall is stable and at the same time, rainfall during winter and summer months is not dependable. Hence the plantation crops' production is also variable depending upon the receipt of rains in both the monsoon seasons. Rainfall during the monsoon season (June-September) contributes 68% of the annual rainfall. It is 15.7% during post monsoon (October to November) while 13.9% in summer (March to May). It is insignificant (2.3%) in winter (Table 3.1). Floods are common in low-lying areas during the monsoon and post monsoon periods. It may not be a threat to plantation crops since they grow mostly on elevated lands/slopes. However, waterlogging is noticed in coconut gardens/homesteads in heavy rainfall zones across the State during the monsoon season. In such condition, deep drainage facility under the field condition is a solution to avoid lack of aeration in the root zone of plantations to avoid hypodermal thickening of root tips, especially in the case of coconut for a better crop yield.

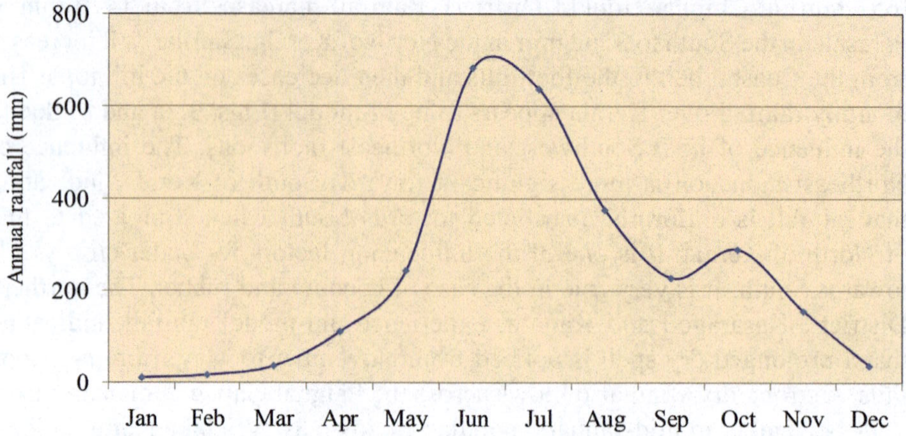

Fig. 3.2a: Monthly mean annual rainfall (mm) over Kerala

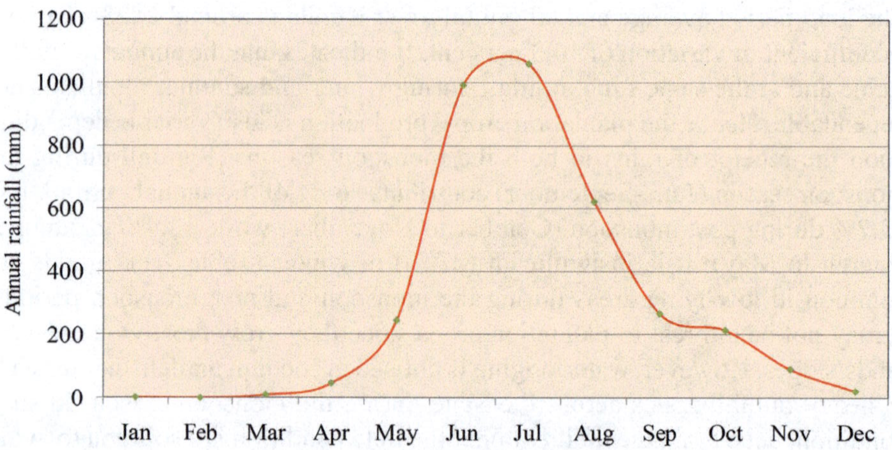

Fig. 3.2b: Monthly mean annual rainfall (mm) at Pilicode (Kasaragod District)

Table 3.1: Season-wise rainfall (mm) over Kerala

Season	Normal Rainfall (mm)	% contribution to annual rainfall	Standard Deviation	C.V (%)
Southwest monsoon (June-September)	1924.9	68.1	371	19.3
Post monsoon (October-November)	445.8	15.7	138	31.1
Winter				
(December-February)	64.3	2.3	46	72.3
Summer				
(March-May)	392.6	13.9	160	41.0
Annual/total (mm)	2828.2		405	14.3

3.3.2 Rainfall Trend

A decline in monsoon rainfall while increase in post monsoon is the trend across the State though cyclic trends of 40-60 years noticed in annual rainfall, which is evident since last 50-60 years (see Fig. 1.3). It also reveals that the percentage contribution of monsoon rainfall to the annual rainfall was declining while increasing during the post monsoon season. However, increase in post monsoon rainfall may not compensate rainfall decrease in monsoon season. Interestingly, there was a shift in climate as a result of changes in thermal and moisture regimes over Kerala. The State as a whole was moving from wetness to dryness within the Humid Climate from B_4 to B_3 since last 100 to 150 years. One of the major factors could be due to alarming deforestation that took place during the above said period across the Western Ghats.

3.4 Temperature Over Kerala

The average annual maximum temperature over Kerala is 30.5°C, recording high (32.9°C) at Punalur while low (24.9°C) at Pampadumpara. The seasonal maximum temperature varied between 35.9°C-35.2°C (Palakkad, Punalur and Pattambi) and 28.3-30.1°C (Pampadumpara and Ambalavayal) during summer, experiencing on an average of 32.8°C. Due to heavy rainfall from June to September, the average maximum surface air temperature is only 28.8°C, varying between 23.3°C (Pampadumpara) and 30.8°C-30.0°C (Punalur,Kottayam, Palakkad and Alappuzha) during the southwest monsoon season. It is obvious that the low land region experiences high temperature when compared to that of high ranges across the State of Kerala (Table. 3.2). The average annual minimum temperature over Kerala is 22.1°C, recording high (24.1°C) at Kochi, followed by Alappuzha (24.0°C) while low (17.3 and 17.4°C) at Ambalavayal and Pampadumpara. The average annual temperature is 26.3°C, varying between 21.2°C at Pampadumpara and 27.8°C at Palakkad across the State. On an average, the difference between maximum and minimum temperature is 8.4°C, varying between 6.7°C at Thiruvananthapuram, followed by 6.9°C at Kochi, Kozhikode (7.0°C) and 10.6°C at Punalur, followed by 10°C at Ambalavayal, Pattambi (9.9°C) and Palakkad (9.2°C). Rise in temperature is the trend noticed across the State of Kerala.

Table 3.2: Mean Annual Surface Air Temperature (°C) across Kerala at different locations

Name of the Station	Maximum Temperature	Minimum Temperature	Mean Temperature	Temperature Range	Seasonality (Summer-Southwest Monsoon)
Thiruvananthapuram	30.4	23.7	27.1	6.7	31.9-29.6
Punalur	32.9	22.3	27.6	10.6	35.2-30.8
Alappuzha	31.2	24.0	27.6	7.2	32.9-30.0
Kottayam	31.9	23.1	27.5	8.8	33.8-30.8
Cochin(Kochi)	31.0	24.1	27.6	6.9	32.4-29.5
Palakkad	32.4	23.2	27.8	9.2	35.9-30.2
Kozhikode	30.8	23.8	27.3	7.0	32.7-28.9
Pattambi	32.4	22.5	27.5	9.9	35.2-29.7
Ambalavayal	27.3	17.3	22.3	10.0	30.1-25.1
Pampadumpara	24.9	17.4	21.2	7.5	28.3-23.3
Average	30.5	22.1	26.3	8.4	32.8-28.8

3.5 Nut Yield Pattern in Coconut

The monthly coconut production at RARS, Pilicode, CPCRI, Kasaragod, Aralam farm, Kannur, CDB Farm, Vellanikkara, RARS, Kumarakom and CRS, Balaramapuram were collected for analyzing monthly yield pattern. The mean monthly maximum coconut production at CPCRI, Farm was observed during May, followed by June, March, April and July while the monthly coconut production was low in September. The seasonal coconut production was found maximum during summer followed by southwest and post monsoon seasons while the lowest seasonal coconut production during winter (Fig. 3.3). At RARS, Pilicode, the maximum monthly coconut production was noticed during April, followed by May, March, February and June while low in December. The mean monthly maximum coconut production was noticed during May at Aralam Farm, followed by June, July and August while low during February, followed by January and December. The seasonal coconut production at Aralam, Farm was high during the southwest monsoon, followed by summer and post monsoon and the minimum during winter season. The mean monthly coconut production at CBD, Farm, Vellanikkara was maximum in April, followed by March, May and June while minimum in December, followed by November and October. At RARS, Kumarakom, the monthly coconut production was low in September, followed by December, November, January and February while maximum in May, followed by April, June, March and July. The seasonal coconut production was maximum during summer (March–May), followed by southwest monsoon while low during winter, followed by post monsoon. At CRS, Balaramapuram, nut harvest is usually carried out once in two months. Therefore, there were six harvests in a year. The coconut yield was low during September–October, followed by November–December, and July–August while maximum during March–April, followed by May–June and January– February. The study reveals that the coconut yield during summer is high, followed by southwest and post monsoon seasons. The minimum coconut production is seen during the winter season. The monthly coconut yield pattern is well explained in the section 3.6.1 under coconut phenology.

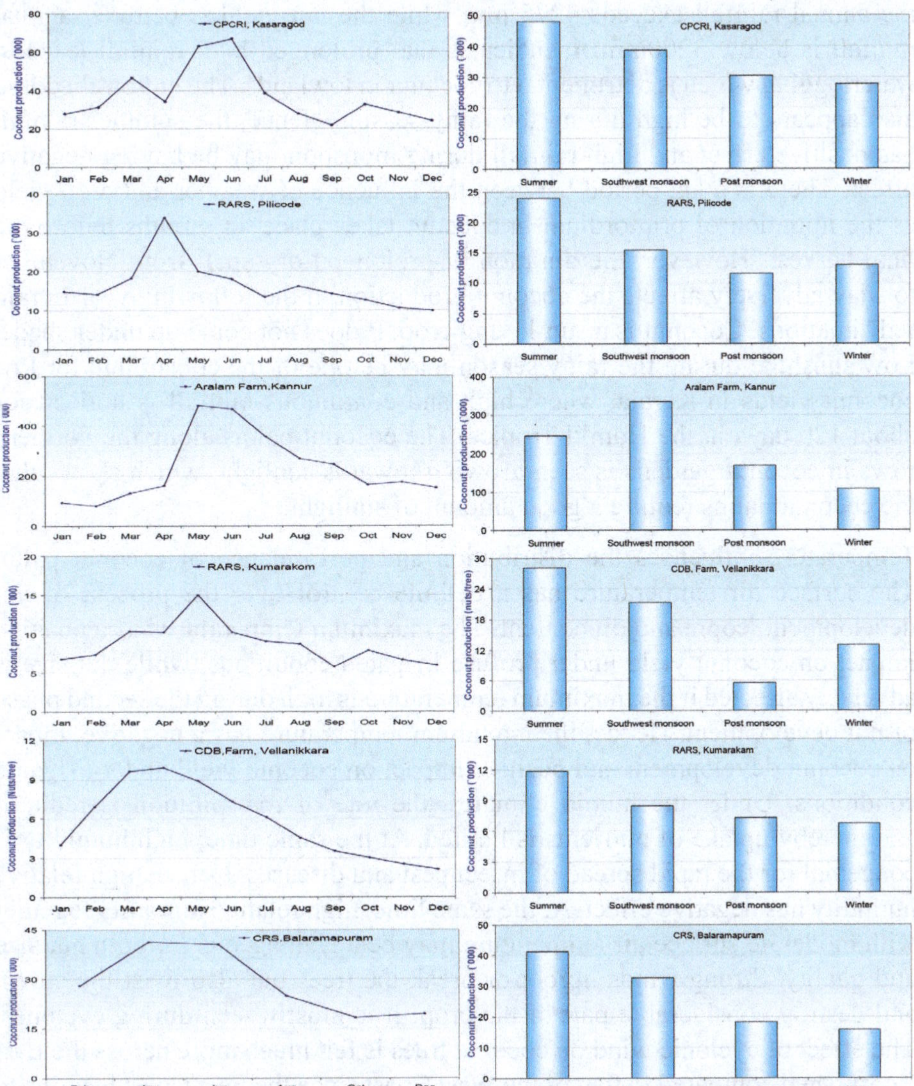

Fig. 3.3: Mean monthly and seasonal coconut production at different locations in Kerala

3.6 Weather and Coconut

Rainfall appears to be the most important weather element which determines coconut growth and development as it is grown under the rain fed conditions. Rainfall during summer months (March, April and May) augments coconut yield in the following year, while heavy rains lead to waterlogging in coconut gardens. High rainfall during the monsoon may adversely affect the coconut yield in the following year, as the nut yield is less than average whenever the annual rainfall exceeds 4275 mm while the nut yield is better if annual rainfall is below 3200mm. It indicates that prolonged high rainfall leads to waterlogging which is detrimental to coconut in lowlands. The button shedding also appears to be high during the rainy season. Hence, the summer rainfall has positive effect and high rainfall during monsoon may be having negative effect. There is a lag period between the influence of weather and crop yield as the initiation of primordium in coconut takes place 44 months before the final harvest. However, the duration of prolonged dry spell from November to May adversely affects the coconut production in the following year in rain fed situations. Coconut is a sun-loving crop. It does not come up under shade. Low sunshine during the rainy season may be one of the constraints for low coconut yields in Kerala, where high and continuous rainfall is noticed for about 120 days in the Humid Tropics. The coconut palms along the northern rows in coconut gardens is seen growing towards sunlight, which shows that the coconut palms require a good amount of sunlight.

Temperature influences the distribution and performance of coconut palm. The surface air temperature has a definite control over the process of nut development, copra and oil content. The maximum temperature has a positive impact on coconut yield under profuse irrigated conditions, while nut size is adversely affected if the maximum temperature is high during the second phase of nut development. Hence, the maximum temperature has a negative impact on coconut development and positive impact on coconut yield under irrigated conditions. Under the humid climates, the rate of transpiration is reduced and thereby uptake of nutrients is limited. At the same time, high humidity is congenial for the rapid spread of insect pest and diseases. Hence, high relative humidity has negative effect. At the same time, high relative humidity together with moderate surface air temperature may be advantageous for both nut size and quality. Strong winds uproot or break the trees but also twist the crown and destroy considerable part of the crop. It is mostly seen during cyclones. The effect of cyclonic wind on coconut trees is felt much more across the East Coast when compared to that of the West Coast since the East Coast is prone to cyclones. Hence, high wind has negative effect. High winds also lead to more evapotranspiration, which is detrimental to crops under rain fed conditions.

During Mundakan season (November to February), strong dry winds blow along the Palghat Gap in Thrissur and Palghat Districts. During the above season, the onset and cessation of strong winds (duration) are detrimental to crop production in seasonal, biennial, perennial and plantation crops to a considerable extent. This peculiar wind phenomenon during the winter season is not seen outside the Palghat Gap within Kerala.

3.6.1 Coconut Phenology and Nut Yield

Coconut takes 44 months from the primordium initiation to nut harvest. From the opening of spathe/spadix emergence, which is seen outside in the crown at petiole, to harvest it takes 10 to 12 months depending upon the prevailing temperature. With low temperature prevailing at higher altitudes the nut maturity takes about 14-16 months. The quality of copra is very poor in higher altitudes and coconut cultivation is not economically viable. Weather effects are evident at all the stages of nut development but the influence of weather depends on the stages of development. Rainfall, temperature of both soil and air, vapour pressure and wind speed are the major weather variables that influence the yield when other external factors such as female flower fertilization and crop management practices are not limiting. In view of the long duration (44 months) between the inflorescence initiation and nut maturation, the occurrence of dryspell in any year would affect the yield for the subsequent three to four years. Rajagopal *et al.* (1996) observed that the nut production under rainfed condition is influenced significantly by the length of dryspells at critical stages and the dryspell during the primordium, ovary development and button size is crucial for the production of nut yield. Since coconut has long reproductive phase, it is difficult to assess the critical stages which are influenced due to weather variables and soil moisture stress under field conditions. The nut yield in coconut during summer months always remained high irrespective of the good and poor coconut yield years. Poor nut yield from September to January, in spite of relatively very low dryspells, is indicative of the bunches with few nuts due to prolonged dryspell noticed during various critical stages of primordium initiation to nut development. Thus, there is lack of literature on interaction between the occurrence of weather abnormalities and their effects on various development stages in coconut. In order to find out the critical stages that are influenced due to prolonged dryspell during the coconut development, the primordium initiation period has been grouped into three groups viz., the primordium initiation from January to May (group A), June to September (group B) and October to December (group C). The monthly rainfall pattern, interspersed with dryspell between January 2002 and December 2006, is illustrated in the lower part of (Fig. 3.4) to understand the effect of dryspell on number of growth stages of coconut and its overlapping.

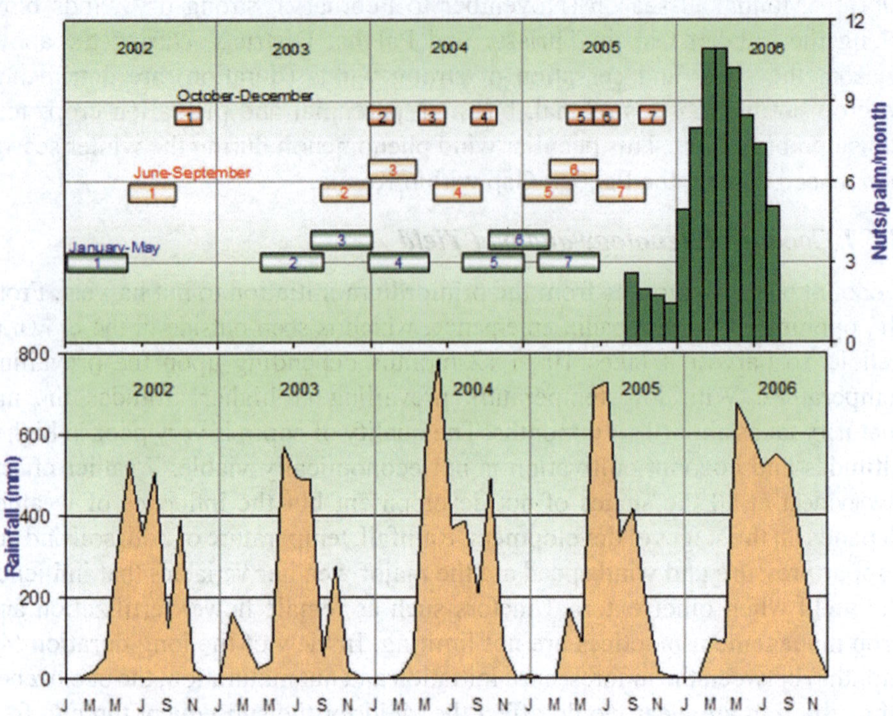

Fig. 3.4: Monthly rainfall versus coconut growth stages

The number of growth stages is divided in to seven critical stages viz., primordium initiation stage (1), branch development of inflorescence (2), male and female flower development stage (3), ovary development stage (4) spathe opening stage (5), button size nut (6) and mature nuts (7) stage. These stages are closely related to the initiation of inflorescence to coconut harvest. It indicated that distances or overlaps between the stages among the three groups were noticed. It was more so in the first two groups. It is noted that the length of dryspell at each stage was high in group A, followed by group B and group C. The nut yield per palm from September 2005 to August 2006 with three distinct trends is a direct reflection of the initiation of primordium and its development as coconut palm is subjected to the influence of rainfall or a dryspell. Total dryspell during the various development stages of nut development in the selected groups (A,B and C) varied between 33 to 71 weeks. The group A is characterized by maximum dryspells of 71 weeks from January 2002 to September 2005, followed by the group B (41 to 50 weeks) and the group C (33 to 39 weeks). As a whole, the primordium initiation from January to May accounts 34.4% of total reproductive stage prone to dryspell in group A while

24.7 % in group B and 19.4% in group C. The group B has a total dryspell of only 1 to 4 weeks during primordium initiation and ovary development stages and majority of growth stages do not experience dryspells. The mean monthly nut yield per palm of group A ranged from 2.6 to 3.8 and on an average of 8.8 nuts in group B while intermediary (6.2 nuts per pam per month) in group C. Though the number of dryspells is relatively less in group C, the nut yield is better in group B due to less dryspells of one, four and six during the critical stages viz., the primordium initiation, ovary development and button nut stage respectively and it is relatively free from drought. It revealed that the dryspell during the primordium stage is very critical. From the above, it is clear that the total rainfall, number of rainy days and the total dryspell are not as crucial as the length of dryspell at the critical stages such as primordium initiation, ovary development and button size nuts which ultimately determine the yield potential of rain fed coconut palms. The lag period between the occurrence of dryspells during the primordium initiation, ovary development and button size nut and monthly yield is well explained where one leaf axil in coconut crown produces one bunch of coconut every month under ideal conditions. This may not be true where coconut is grown in non-traditional areas under irrigated conditions. Therefore, the phenology of coconut is very complex one to understand under various coconut growing environments which needs more and more studies further in detail.

3.6.2 Button Shedding in Coconut

Button shedding in coconut decides the final harvest. It depends on the number of female flowers set in each bunch. It is obvious that the percentage of button shedding is more with the more number of nuts set in coconut. Normally, the button shedding is more during summer, followed by the rainy and post monsoon seasons while less during winter since the number of female flowers produced and set in the same order seasonally. Studies conducted at RARS, Pilicode indicates that the button shedding is more (83%) during southwest monsoon season, followed by summer (78.4%) and post monsoon (69.7%) and less (39.1%) during winter. It reveals that the button shedding is more during rainy season in high rainfall zones (Table 3.3a). It may be due to the fact that the button shedding may intensify during June and July depending upon the onset of monsoon after prolonged summer during which the buttons experience severe soil moisture stress under the rain fed conditions. The seasonal pattern of number of female flowers produced and set as well as the button shedding may vary under irrigated conditions. In another study at Vellanikkara, it is observed that the button shedding varies with the type of cultivar. The percentage button shedding is more (69%) in Komadan, followed by TipturTall (62%) and

Kasaragod (59%) while less in Kuttiyadi (55%) during summer 2002. Tiptur Tall also recorded relatively more (77%) button shedding during summer 2003 among the four selected cultivars. However, all the cultivars recorded more button shedding during summer 2003 when compared to that of summer 2002. The average percentage of button shedding during summer 2003 was 73% while 61% in summer 2002. It was due to relatively more soil moisture stress due to prolonged dry spell during summer 2003 when compared to that of summer 2002 (Table 3.3b). It reveals that prolonged soil moisture stress is one of the factors for button shedding in addition to the more number of female flowers produced and set in coconut.

Table 3.3a: Seasonal button shedding in coconut

Season	**% Button Shedding in Coconut**
Summer (Mar-May)	78.4
Southwest Monsoon (June-September)	83.0
Post Monsoon (October-November)	69.7
Winter (December-February)	39.1

Table 3.3b: Button shedding (%) in selected coconut cultivars during summer 2002 and 2003 at Vellanikkara

Cultivar	**2002**	**2003**
Tiptur Tall	62	77
Kuttiyadi	55	73
Kasaragod	59	71
Komadan	69	72
Mean	61	73

3.6.3 Drought Characteristics in Coconut

Coconut palms show the following characteristics under severe soil moisture stress depending upon the duration and intensity of summer drought as per the Rao et al., 2005.

Withering and mortality in the case of young seedlings under poor management.

Drooping, wilting and drying of lower whorl of leaves.

Breakages of leaves at petiole or just above it.

Spindle leaf breaking which lead to mortality in the case of senile palms under conditions of poor management.

Abortion of spadices, starts from October/ November onwards.

Button shedding and immature nut fall.

Nut size decline and

Finally decline in nut yield in the subsequent year up to fifty per cent depending upon the type of management.

3.6.4 Summer Droughts and Coconut Productivity

The summer monsoon failure or breaks lead to drought situation across the Country during kharif season may not be relevant in a State like Kerala which experiences copious monsoon rains. At the same time failure of northeast monsoon together with summer showers may lead to summer drought which is not uncommon in northern districts of Kerala. The summer 1982-83 experienced the unprecedented drought due to insignificant rains from 4th November 1982 to 13th June 1983. This led to a drastic decline in coconut crop production in the subsequent year 1984. The decline in coconut yield due to drought varied depending upon the yield groups. It is about thirty per cent in the subsequent year even under well managed coconut gardens in high yielding group, producing more than 80 nuts per palm per year. With the yield group of between 60 and 80 nuts per palm per year, the effect of drought in the following year is 24.5 per cent while 10.6 per cent in the poor yield group of 40 to 60 nuts per palm per annum (Table 3.4). On an average, the effect of drought on annual coconut yield is 23.7% in the following year irrespective of the yield groups. It is interesting to note that the number of actual coconuts produced is more in high yielding group when compared to poor yielders even under drought conditions though the effect of drought is more in high yielders.

Table 3.4: Percentage decline in coconut yield of different yield groups due to drought at RARS, Pilicode

Yield group	Decline in nut yield (%)
Group I (>100 nuts/palm)	29.3
Group II (80 to 100 nuts/palm)	30.4
Group III (60 to 80 nuts/palm)	24.5
Group IV (40 to 60 nuts/palm)	10.6
Mean	23.7

3.6.5 Effect of Drought on Monthly Nut Yield Across Kerala

The early withdrawal of northeast monsoon in 1982and failure of summer showers in 1983 led to disastrous summer drought and many plantation crops production was adversely affected in varied magnitudes depending upon the location and the type of crop. The drought period was over in the month of June after the commencement of southwest monsoon on 13th June in 1983. The decline in monthly coconut yield against normal was noticed from February 1984 onwards and it continued for 12 months till January 1985. The maximum decline in the monthly nut yield was noticed in the month of July

1984 (Fig.3.5a). The decline in monthly nut yield was noticed from the eighth month and reached maximum in the 13th month after the drought period was over and continued up to 19th month. Similar decline in monthly nut yield is noticed across Kerala (Figs. 3.5b and c). The percentage decline in monthly nut yield varied between 12.3% and 51.9% due to summer drought in the following year. A similar dryspell was noticed from December 1988 to May 1989. Decline in monthly nut yield was observed from February (8th month) onwards and it continued up to January 1991 (19th month). The maximum nut yield reduction was noticed in June (12th month) 1990. The percentage decline in monthly nut yield varied between 12.8% and 56.4% due to summer drought. The nut decline due to drought depends upon the different levels of management as well as the intensity of drought. The decline in the monthly nut yield varied between 11.2% and 82.6% due to summer drought during 1997-98 while it varied between 1.5% and 51.5% at the Aralam Farm Kannur. The decline in nut production was started in the 8th month after the drought period was over and it continued up to December 2003 due to summer drought in 2002 at Vellanikkara Coconut Farm. The maximum decline in the monthly nut yield was noticed in the month of June 2003. The percentage decline in the monthly nut yield varied between 13.8% and 31.5% . Siimilar was the case in summer drought during 2004 at Vellanikkara. Decline in yield reduction was noticed in the 8th month onwards and the maximum reduction was seen in the 13th month and the yield reduction continued up to 18th month after the drought period was over (Table 3.5). On examination of the effect of summer drought on monthly nut yield at various locations across Kerala, it is understood that the effect of drought on monthly nut yield commenced in the seventh, eighth or ninth month after the drought period was over in May or June, depending upon the receipt of pre-monsoon showers or onset of monsoon. The effect of summer drought on monthly coconut yield continued for a period of twelve months. It revealed that the effect of drought on monthly yield commenced between January and March after the drought period was over with the commencement of onset of monsoon and continued till January /February of the following year. It was the reason why the effect of summer drought on coconut yield is seen in the following year. The maximum decline in monthly nut yield is seen in twelfth or thirteen month after the drought period was over. It also revealed that the effect of drought on coconut yield under the field condition was noticed for a period of 12 to 14 months. The maximum decline on monthly nut yield varied between 30 and 80 per cent depending upon the management and cultivar across the State of Kerala. Even at Kumarakom, where the coconut palms are grown on bunds under the below mean sea level and the soil moisture in the root zone is not a constraint, the effect of severe atmospheric drought during summer on coconut yield varied between 13 and 26%. Though the summer droughts are not frequent, it is not uncommon in the northern districts of Kerala, where the area under coconut cultivation is more and grown in rain fed conditions. Therefore, the coconut palm does experience moderate to severe soil moisture stress during summer under rain fed conditions depending upon the receipt of pre monsoon showers or onset of monsoon. The studies also indicated that the effect of drought on coconut yield is likely to be less under better agronomic and crop protection practices. Hence, there is a need to follow pro-active measures for summer drought management in coconut gardens.

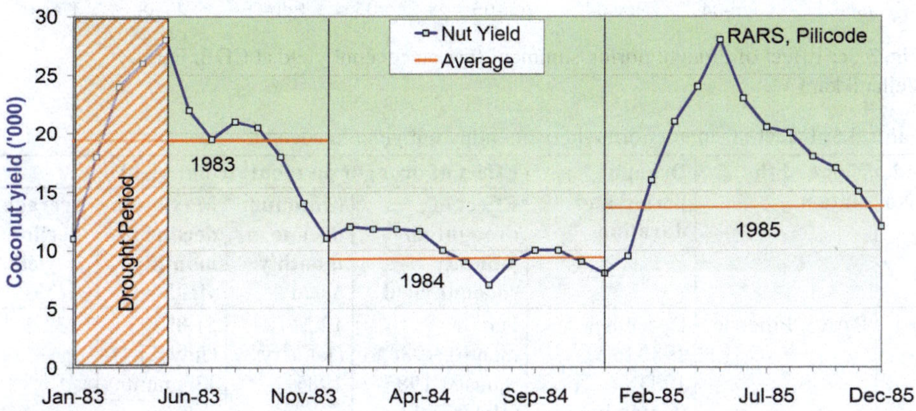

Fig. 3.5a: Effect of drought during summer 1983 on coconut yield at RARS, Pilicode

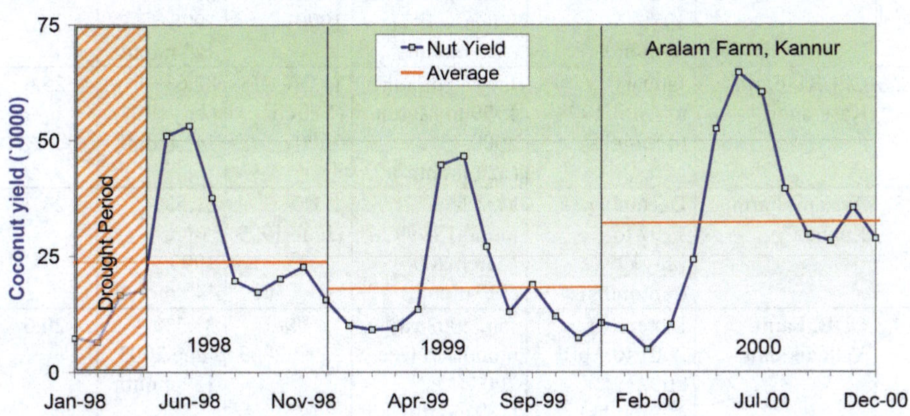

Fig. 3.5b: Effect of drought during summer 1998 on coconut yield at Aralam Farm, Kannur

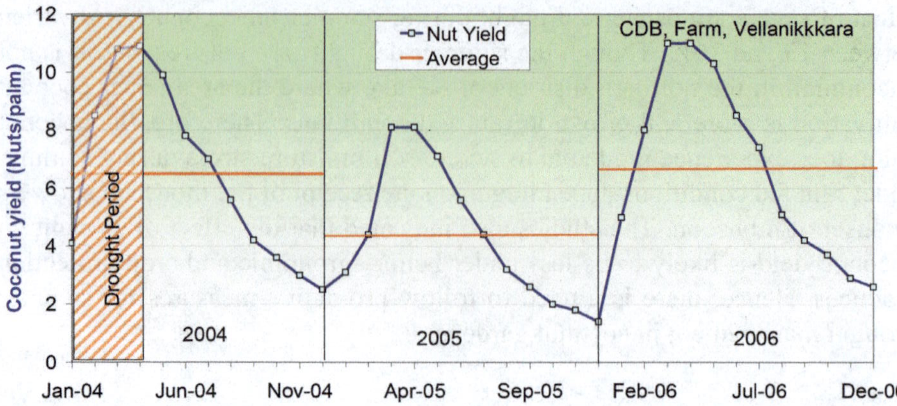

Fig. 3.5c: Effect of drought during summer 2004 on coconut yield at CDB, Farm, Vellanikkara

Table 3.5: Effect of summer drought on monthly nut yield in coconut

Sl. No	Name of the station	Drought period and duration	Effect of drought on monthly nut yield			
			Effect of drought on monthly coconut yield	Minimum decline in monthly Yield	Maximum decline in monthly yield	Average decline in yield (%)
1	RARS, Pilicode	December 1982 to May, 1983 (6 months)	Feb (8th month), 1984 to January, 1985 (19th month)	12.3% (January, 1985)	51.8% (July1984) 13th month	33.2
		December 1988 to May 1989 (6 months)	Feb (8th month),1990 to January 1991 (19th month)	12.8% (March, 1990)	56.4% (June 1990) 12th month	34.3
2	CPCRI, Farm, Kasaragod	January to April,1998 (4 months)	Jan (8th month) 1999 to March 2000 (22nd month)	11.2% (March 2000)	82.6% (July199) 14th month	25.6
3	Aralam, Farm, Kannur	December 1997 to May1998 (6 months)	March (9th month) 1999 to March 2000 (21st month)	1.5% (June 1999	51.5% August 1999 14th month	24.7
4	CDB, Farm, Vellanikkara	December 2001 to April 2002 (5 months)	Jan, 2003 (8th month) to Dec, 2003 (18th month)	13.2% April 2003	31.5% June 2003 13th month	20.6
		December 2003 to April 2004 (5 months)	Jan, 2005 (8th month) to Dec, 2005 (18th month)	16.4% March 2005	32.9% June 2005 13th month	24.4

Sl. No	Name of the station	Drought period and duration	Effect of drought on monthly nut yield			
			Effect of drought on monthly coconut yield	Minimum decline in monthly Yield	Maximum decline in monthly yield	Average decline in yield (%)
5	RARS, Kumarakom	December 1982 to May, 1983 (6 months)	Jan, 1984(7th month) to Nov, 1984 (19th month)	13.6% November 1984	16.2% July 1984 13th month	23.4
6	CRS, Balaramapuram	December 1982 to May, 1983 (6 months)	Jan - Feb,1984 (7/8th month) to Mar - Apar,1985 (22/23rd month)	4.2% Mar - Apr 1985	56.3% Nov-Dec 1984 18/19th month	31.8
		December 1988 to May, 1989 (6 months)	Jan-Feb, 1990 (7/8th month) to Nov-Dec, 1990 (18/19th month)	10.3% May-Jun, 1990	51.3% Nov-Dec, 1990 18/19th month	26.9

3.7 Microclimate of Coconut

Coconut is mostly grown in littoral sand, sandy loam and laterite soils. The coconut productivity is better in lateritic loamy soils under rainfed conditions. Of course, coconut yields are comparable even in sandy soils with frequent profuse irrigation once in three to four days during summer with better crop management. The coconut root zone temperature and soil moisture were monitored at RARS, Pilicode. The root zone temperature is intermediary in lateritic loam when compared to that of sandy and sandy loam soils in the same crop weather environment. It may be due to better moisture holding capacity of soils. The thermal regime within the root zone of coconut also varies with the root zone spread which depends on type of soils (Plates 3.2, 3.3 and 3.4). The surface roots of coconut in sandy soils may be relatively at lower depths when compared to other soils where no irrigation is practiced. The vertical radiation profiles also indicated that the net radiation available at the coconut canopy is less by around 30 % during the rainy season when compared to that of summer. It is similar within the coconut gardens between the two seasons. High rainfall and cloudy weather may be the reason for low incidence of solar radiation and thereby net radiation. Atmospheric temperature and vapour pressure profiles were also studied using Assmann Psychrometer. The second active surface is noticed at the coconut canopy level in terms of temperature and humidity. More and more information in detail on radiation and thermal regimes can be understood better from the published work by the author. Such basic studies need to be undertaken in coconut.

Plate 3.2: Micrometeorological tower of height of 56 feet used for measuring radiation profiiles at different heights inside the coconut

Plate 3.3: Soil-Earth thremometers at Agricultural Meteorological Station, RARS, Nileshwar (KAU)

Plate 3.4: Soil Thermometers Inside Coconut Garden

The coconut root zone temperature in sandy loam is high (31^{0}C) when compared to that of lateritic loam (29^{0}C) while intermediary (29.7^{0}C) in laterite soils. It is the same between the root zone of 60 cm and 90 cm depths in lateritic loam (29.2^{0}C) and laterites (29.7^{0}C) while varied in sandy loam (30.6-30.9^{0}C) depending upon the root zone density. It is obvious that the surface depths record high soil temperature irrespective of soil types and show declining with depth up to 10 -20 cm depth and thereafter increasing with depth (Table 3.6a). A sort of declining profile in surface soil depths and thereafter increasing profile is noticed depending on the root zone structure and texture. The seasonal soil temperature shows that there is steady increase from southwest monsoon (28^{0}C) to summer (32.3^{0}C) in coconut while it varied from 32^{0}C during southwest monsoon to summer (36.5^{0}C) in open. There is a seasonal difference of 4-5^{0}C between coconut and open (Fig. 3.6) while 3-4^{0}C difference depending upon the root zone depth in varied soils (Table 3.6b). Based on the radiation profiles in coconut, simple regression equations were worked out between sunshine hours and radiation of different types in and above the coconut canopy (Table 3.6c).

Table 3.6a: Mean annual soil temperature (°C) in coconut root zone and in open at different depths under laterite soils at RARS, Pllicode (12°12 N & 75° 10' E), Kerala

Deplh(ctn)	Open	Coconut	Difference
23	33.6	30.7	2.9
10	33.1	293	33
20	31.5	29.0	23
30	32.4	29.6	23
60	32.6	29.7	2.9
90	32.6	29.7	2.9
Mean	32.6	29.7	2.9

Table 3.6b: Mean annual coconut root zone temperature (°C) in different soils at RARS, Pilicode and Nlleshwar campuses (Kerala)

Depth(cm)	Lateritic loam	Laterite	Saidy loam
2.5	30.6	30.7	32.7
10	27.9	29-3	31.0
20	28.1	29.0	30.1
30	29.2	29.6	30.6
60	29.2	29.7	30.9
90	29.2	29.7	30.6
Mean	29.0	29.7	31.0

Table 3.6c: Relationship between bright sunshine hours and solar and net radiation in coconut garden

Location	Crop	Equation (W/m²)	r
RARS, Pilicode, Kasaragod district. Kerala (12° 12' N & A 75° 10' E)	Coconut	Rac 0.786 Rst-39.17	0.96
		Rnc 26-39n + 168.52	0.93
		Rsi=0.774 Rst-39.17	0.99
		Rsi=29.68n + 161.41	0.96
		Rt= 33.78n + 261.58	0.98

Rst-above coconut canopy (56' height)
Rsi-Direct solar radation inside coconut (10' height)
Rnc Net radiaton at the coconut canopy (38' height)
n- number of light sunshine hours per day

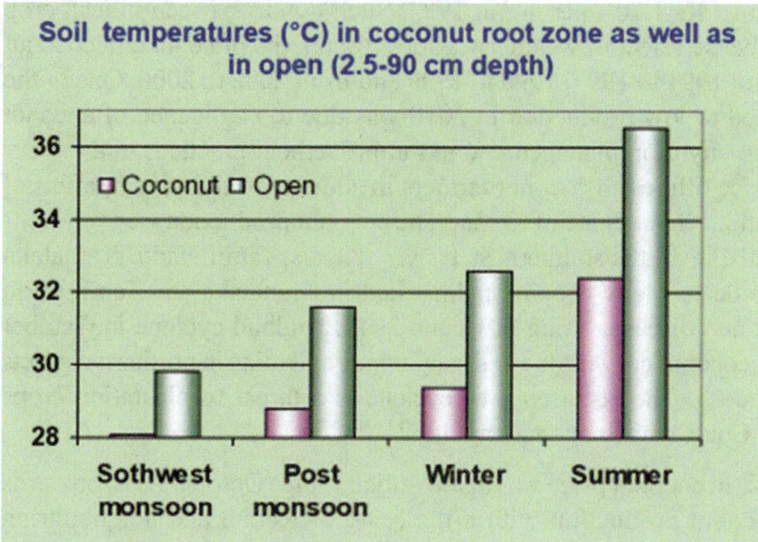

Fig. 3.6: Soil temperature in coconut and open

3.8 Weather and Insect Pest and Diseases

Yellowing, bud rot, leaf rot, stem bleeding and root wilt diseases are the common diseases in coconut. Cockchafer beetle (white grub), rhinoceros beetle, red palm weevil and leaf eating cater pillar are the common pests. Yellowing is mostly in high rainfall zones is noticed. Stem bleeding is seen in northern districts of Kerala while root wilt is prevalent in Kuttanad area and southern districts. The bud rot is seen in coconut crown and the moment spindle leaf falls it is due to the bud rot disease mostly. Continuous rainfall and high relative humidity may be the causal factors for incidence of bud rot in addition to the prevalence of microclimate within the crown. The white grub is common in sandy soils not in laterite soils. It indicates that the soil environment is conducive for white grub. Of course, rhinoceros beetle is very common in all the coconut growing areas. This section is highlighted to agromet group for association with the other related disciplines to work in detail on the influence weather factors on pest and diseases.

3.9 Weather Abnormalities and Coconut Prices

A steep increase in coconut price was seen in 1984 due to low coconut production in Kerala as a result of summer drought in 1983 and for the first time, the coconut growers got a high premium of more than Rs. 2200/- per quintal of copra. Under the open auction sale, on an average Rs. 3.07/- per coconut was obtained at the Regional Agricultural Research Station, Pilicode in 1984 as against Rs 1.70/- per nut in 1983. Such trend was seen in lean crop years due to adverse affect of weather vagaries. Again, the price hike in coconut was noticed from 1991 to 1993, 1996 to 1999 and from 2002 to 2006. One of the factors attributed to low production in 2010 was due to negligence of coconut gardens without any crop management and improvement practices due to very low coconut price offered to coconut farmers in addition to the adverse affect of weather aberrations in the State of Kerala. The coconut production was very poor in 2013 and 2014 in all the southern States viz., Kerala, Tamil Nadu, Karnataka and Andhra Pradesh. Many coconut palms died in Karnataka and Tamil Nadu due to drought and disease during 2013 and 2014. Hudhud cyclone in October 2014 damaged coconut crowns due to strong winds and rains in northern districts of Andhra Pradesh. The occurrence of cyclone is a threat to plantation crops along the East Coast while not so along the West Coast.

A steep increase in coconut price was again noticed since October 2010 onwards due to low coconut production within the State of Kerala and neighbouring coconut producing states and reached to a peak price in 2014 and continued in 2015 too. The coconut farmers get Rs. 6-10/- per nut depending upon the nut size. In addition, coconut price in Kerala may depends on several factors viz., import export policy, total coconut production outside Kerala, less edible oil consumption, coconut oil use in the industries as several substitutes are used in place of coconut oil. Import of palmolene oil may be one of the major contribution factors in declining oil price. The ever highest price of coconut was offered in December 2010 (Rs.6.50/nut), followed by 1999 July (Rs. 5.70/nut) and January 1997 (Rs. 5.67/nut) while the lowest price noticed in July 1979 (Rs 0.99/nut), followed by1982 January (Rs.1.00/nut) and 1981 October (Rs.1.06/nut). The price per coconut revolved only between Rs.3/- and 4/- from 2007 to the third quarter of 2010 (Table 3.7). The coconut price in Kerala was very high (more than Rs 8/- per nut) in the open market and coconut growers get reasonably good price for his farm produce from the fourth quarter of 2010. Another reason for price hike was that the oil-palm yields in Indonesia and Malaysia have shrunk this year because of dryness caused by El Nino and heavy rain brought on by La Nina. The import of coconut oil was also very low. The percentage decline in import of coconut oil in 2009-10 was 74.9% when compared to that of 2008–09. With all the above factors which influence the hike in coconut price, there was a strong linkage between the local coconut production and price as seen in the case of low coconut production during weather abnormalities like severe summer droughts in the State of Kerala. It indicates that pro - active measures in terms of EXIM policies and local price regulations at the State level are the need of the hour in the event of weather abnormalities which influence the coconut production adversely.

Table 3.7: Climate variability and coconut price at RARS, Pilicode

Year	Price of coconut in Indian Rupees			Year	Price of coconut in Indian Rupees		
	Highest	Lowest	Average		Highest	Lowest	Average
1979	1.10 (Match)	0 99 (July)	1 03	1995	3.45 (Dec)	2.75 (Jan)	3.00
1980	1.64 (Oct)	1.46 (Dec)	1.50	1996	5.82 (Dec)	3.00 (Aug)	3.86
1981	1.65 (May)	1.06 (Oct)	1.31	1997	5.67 (Jan)	4.21 (Sep)	4.82
1982	1.40 (Dec)	1.00 (Jan)	1.18	1998	4 45 (Apnl)	377(Sep)	4.00
1983	2.10 (Sep)	1.30 (May)	1.70	1999	5.70 (July)	3.75 (Jan)	4.91
1948	3.50 (June)	2.10(Jan)	3.07	2000	4.56 (Jan)	2.4 (Oct)	3.22
1985	2.27 (Feb)	1.30 (Nov)	1.69	2001	3 60 (Dec)	2 46 (Feb)	3.00
1986	2.25 (Nov)	1.37 (Feb)	1.89	2002	4.62 (Dec)	4.00 (April)	4.37
1987	2 80 (Doc)	2 30 (Jan)	256	2003	5.62 (Oct)	4.30 (May)	4.93
1988	3.20 (Dec)	2.35 (July)	2.77	2004	5.90 (Aug)	5.26 (May)	5.58
1989	2.15 (May)	1.72 (Oct)	1 93	2005	5.50 (April)	3.81 (Nov)	4.66
1990	3.00 (Dec)	1.78 (Jan)	2.26	2006	5.00 (Feb)	3.13 (Jul)	4.07
1991	3.80 (Feb)	1.10 (May)	3.16	2007	3.70 (Dec)	3.33 (Sep)	3.52
1992	4.20 (Mar)	3.85 (May)	3.98	2008	4.63 (Jul)	3.73 (Jan)	4.18
1993	4.51 (Jan)	3.20 (July)	3.90	2009	4.65 (Feb)	3.01(Oct)	3.83
1994	3 71 (Dec)	2.51 (Sep)	277	2010	6.50 (Dec)	3.30 (Jan)	4.90

Summary

Coconut is relatively tolerant to weather aberrations unlike other plantation/ seasonal crops. It tolerates high rainfall and rise in temperature to some extent. However, continuous and high rainfall during the monsoon season from June to September is not conducive and reflects on coconut yield in the following year. Coconut productivity under the rain fed conditions is influenced significantly by the length of dry spells at critical stages. The primordium initiation, ovary development and button-size nut development in coconut out of 44 months duration from primordium initiation to nut harvest are sensitive to dry spells, resulting a particular pattern in monthly coconut yield irrespective of good or poor yield years. The decline in annual coconut production due to severe summer drought could be seen in the following year under rain fed conditions. Similarly, good summer showers with less duration of dryspell are likely to influence the coconut yield favourably in the following year to a considerable extent. On examination of the effect of drought on monthly nut yield at various locations across Kerala, it is clear that the effect of drought on monthly nut yield commenced in the seventh, eighth or ninth month after the drought period was over in May or June, depending upon the receipt of pre-monsoon showers or onset of monsoon. The effect of summer drought on monthly coconut yield continued for twelve months. Decline in monthly nut yield was maximum in 12th or 13th or14th month after the summer drought period is over. The second stage of nut development between fourth and seventh month is sensitive to high temperature. High temperature that prevails during summer adversely affect nut size in coconut and thereby oil content per nut. High temperature is one of the reasons, why, the nut size and oil content per nut are relatively less in other coconut growing States viz., Tamil Nadu, Karnataka and Andhra Pradesh though they are the Coastal States along the East Coast. These States experience erratic low monthly distribution of rainfall round-the-year. The coconut crop performs better in these States under irrigated conditions only. Of course, coconut cannot come up and yields uneconomically with poor quality coconuts if irrigation is not provided throughout the year. At the same time, coconut comes up well in Kerala even under rain fed conditions with heavy monsoon rainfall and less dry spells except in summer. However, nut yield per palm is relatively low in Kerala since crop is mostly grown under rain fed conditions. However, the quality of copra and oil content are much better in Kerala when compared to that of other coconut growing States. The per palm coconut production is high in non-traditional areas due to profuse irrigation round-the-year.

Suggested Readings

Jan P.S. 2004. Leaf and spadix phenology of cocnut (Cocos nucifera L.). M.Sc. Thesis. Kerala Agricultural University.

Krishnakumar, K.N. 2011. Coconut Phenology and yield response to climate variability and Change. Ph.D. Thesis. Cochin University of Science and Technology. Kochi, Kerala. 217p

Martinez, R., Hemming, D., Malone, L., Bermudez, N., Cockfield, G., Mushtaq, S., Diongue, A. Hansen, J., Hildebrand, A., Ingram, K., Jakeman, G., Kadi, M., McGregor, G.R. Rao, P., Pulwarty, R., Ndiaye, O., Srinivasan, G., Eh.R., Seck, White, Nand Zougmore, R. 2012. Improving Climate Risk Management at Local Level – Techniques, Case Studies, Good Practices and Guidelines for World Meteorological Organization Members-Chapter 21 (477-532): In Risk Management-Current Issues and Challenges (Ed: Nerija Banaitiena). In Tech Publishers. Rijeka, Croatia,584p

Nybe, E.V. and Rao, G.S.L.H.V.P. 2005.Drought management in plantation crops. Kerala Agricultural University, Thrissur, Kerala, India, 176p

Rajagopal, V. Shivashankar, S and Jacob Mathew. 1996. Impact of dryspells on the ontogeny of coconut fruits and its relation to yield. *Agro Physiology*. 251–255.

Rao, G.S.L.H.V.P. 2021. Agricultural Meteorology (5th Reprint). PHI Learning Private Limited, New Delhi, India. 364p.

Rao, G.S.L.H.V. P. 2011. Climate change adaptation strategies in agriculture and allied sectors. Scientific Publishers (Jodhpur), India, 336p.

Rao, G.S.L.H.V. P. and Gopakumar, C.S. 2016. Climate change and plantations in the Humid Tropics.New India Publishing Agency. New Delhi. 461p

Rao, G.S.L.H.V.P. and Nair, R.R. 1988. Agrometeorology of plantation crops. Kerala Agricultural University.Thrissur, Kerala, India. 187p.

Rao, G.S.L.H.V.P. and Krishnakumar, K.N. 2011. Climate change and coconut productivity in Kerala. Kerala Agricultural University. Thrissur, Kerala, India. 219p.

Rao, B.V.R. and Rao, V.U.M. 2024. Agricultural Meteorology. Jain Brothers, New Delhi. 154p

Rao, G.S.L.H.V.P. and Siby Sebastian. 1994. Estimation of leaf area in tree crops. Journal of Plantation Crops. 22(1):44-46

Rao, G.S.L.H.V.P., Gopakumar, C.S. and Sebastian S. 1993. Effect of drought on nut yield in different yield groups of west coast tall. J. Plantation Crops. 21: 25-28

Rao, G.S.L.H.V.P. 1991. Agricultural drought with reference to coconut. Journal of Plantation crops. 18 (supplement): 166-170

Rao, G.S.L.H.V.P. 1991. Agrometeorological aspects in relation to coconut production. Journal of Plantation Crops, 19(2): 120-126

Rao, G.S.L.H.V.P. and Nair, R.R. 1988. Influence of weather on nut development in coconut. Journal of plantation Crops (Supplement) 16: 469-473

Rao, G.S.L.H.V.P., Sebastian, S. and Subash, N. 1995. Solar and Net Radiation Profiles in Coconut gardens in humid climates. Indian Coconut Journal. 25 (8):2-5

Rao, G.S.L.H.V.P.1994. Drought management in Coconut. Indian Coconut Journal. 24(6-8)

Rao, G.S.L.H.V.P. 1986. Effect of drought on coconut production. Indian Coconut Journal. 17(8):11-12

Rao, G.S.L.H.V.P. 1985. Drought and Coconut Palm. Indian Coconut Journal. 16(12): 3-6

Rao, G.S.L.H.V.P. 1988. Agrometerology of coconut (Chapter 5: 81-93). Six decades of Coconut Research (Eds.): M. Aravindkshan, R.R. Nair and P.A. Wahid. Kerala Agricultural University, Vellanikkara, Thrissur, Kerala, India. 151p

Questions

I. Elucidate on the following

1. Summer droughts and coconut productivity across Kerala
2. Micrometeorology of coconut
3. Dry spells and coconut phenology

II. Write short notes on

1. Onset of monsoon over Kerala
2. Rainfall distribution over Kerala
3. Thermal regime over Kerala
4. Weather and button shedding in coconut
5. Pest and diseases of coconut in relation to weather
6. Thermal regime of root zone in coconut
7. Radiation profiles in coconut
8. Drought characteristics of coconut
9. Nut yield pattern in coconut

III. Fill in the Blanks

1. The belated monsoon was on --------------------
2. The early monsoon was on ----------------
3. -------------- year the belated monsoon was noticed
4. -------------- year early monsoon was noticed
5. ------------- is the onset of monsoon with SD
6. ------------ is the rainfall model over Kerala
7. -------------- is the average annual temperature over Kerala
8. ----------------is the average annual maximum temperature over Kerala
9. --------------- is the average annual minimum temperature over Kerala
10. --------------- is the average annual temperature range over Kerala
11. -----------, --------------- and ------------ phases in coconut are sensitive to dry spells
12. Button shedding in coconut is more in -------------season
13. Button shedding in coconut is less in ---------season
14. ---------------- disease is specific in coconut in northern parts of Kerala
15. ------------------ disease is more prevalent in coconut in Kuttanad area
16. -------------coconut disease is common in rainy season across Kerala
17. ---------------- pest is common in sandy soils in coconut
18. Coconut root zone temperature is --------------- in laterite soils
19. ------------ is the trend in monsoon rainfall and - -----------is the trend in post monsoon rainfall
20. Coconut productivity is less in ------------season
21. ------------- is the first symptom in coconut due to summer drought
22. ----------- is the severe symptom in coconut due to summer drought in young coconut palms
23. -------------- is the severe symptom in yield bearing coconut palms
24. -------------- year was the disastrous summer drought year with reference to coconut
25. -------------- month shows maximum decline in coconut yield due to drought
26. -------------- which year yield decline is noticed in coconut due to summer drought
27. Coconut productivity is high in -------------while coconut production is high in --------States
28. ---------------range of rainfall is optimum for coconut trees

IV. State True or False

1. Coconut productivity is high in Kerala
2. Coconut productivity within Kerala is high in Kasaragod District
3. Tamil Nadu produces more coconuts in India
4. Female flower production in coconut leads to better harvest
5. True or False
6. Bud rot in coconut is seen in summer months
7. White grubs in coconut basin are noticed in sandy soils
8. Soil temperature in coconut root zone is moderate under lateritic loamy soils
9. Bi-model rainfall is conducive for coconut production in Kerala
10. Per palm coconut production is more in Andhra Pradesh
11. The monthly nut yield pattern in coconut is decided by dry spells in critical phases during the reproductive phase
12. The annual coconut yield is affected in the same year due to summer drought
13. The maximum nut yield reduction is noticed in 12th/13th month after summer drought period is over
14. 1981-82 summer was the summer disastrous drought year with reference to coconut
15. 2001-2010 was the warmest decade during the period from 1870 to 2010
16. True or False
17. Coconut price is to some extent controlled by weather extremes
18. A sudden increase in coconut price was noticed in 1984 for the first time in Kerala
19. Button shedding in coconut is more in southwest monsoon season
20. Heavy rainfall leads to low coconut productivity in the subsequent year
21. Net Radiation at the coconut canopy is less by 10 per cent when compared with Solar Radiation
22. The height of micromet tower in coconut garden was 40' height
23. Round the year warm and humid climate is not favourable for coconut growth and development

V. Distinguish Between

1. Uni-and Bi- model rainfall in Kerala
2. Coconut root zone temperature in sandy and laterite soils
3. Drought characters in young and yielding coconut palms

4. Monthly nut yield pattern in drought and no drought years
5. Maximum and minimum temperature across the coconut tract

VI. Match the Following

Coconut yield group	Drought effect on yield
Greater than 100 nuts/palm/year	24.5%
80-100 nuts/palm/year	10.6%
60-80 nuts/palm/year	24.5%
40-60 nuts/palm/year	30.4%
Less than 40 nuts/palm/year	9.0%
	34.0%

VII. Match the Following

Seasonal rainfall	C.V.%
SWM rainfall	72.3%
PM rainfall	14.3%
Winter rainfall	19.3%
Summer rainfall	31.1%
Annual rainfall	41.0%
	33.0%

4

Cocoa

Cocoa is popularly known as cacao in many cocoa growing nations. Cocoa beans are the source of commercial cocoa. The cocoa products viz., cocoa liquor, cocoa butter, cocoa cake, cocoa powder and cocoa chocolate are widely consumed across the World. Cocoa powder is essentially used as flavour in biscuits, ice cream, dairy drinks and cakes. Apart its use as flavour it is also used in the manufacture of coatings for confectioners or frozen desserts. Cocoa powder is also consumed by the beverage industry for example for the preparation of chocolate milk. Besides the traditional uses in chocolate manufacture and confectionery, cocoa butter is also used in the manufacture of tobacco products, soap and cosmetics. It is also a folk remedy for burns, cough, dry lips, fever, malaria, rheumatism, snakebite and wounds. It is reported to be antiseptic and diuretic. Looking at the industry needs for various purposes, there is a heavy demand for cocoa. At present, the gap between supply and demand in cocoa appears to be more at the national level. To fill the gap, various agro-technologies and crop improvement strategies are to be chalked out, for which the response of the crop to various environmental factors are to be understood. Weather aberrations during the flowering and fruiting result in the final cocoa production. Therefore, it is important to understand the effect of weather on flowering and fruiting pattern of cocoa to minimize the crop losses against the weather abnormalities. Though it is known that cocoa yield is highly variable and sensitive to various weather factors, the studies on crop weather relationships in cocoa and its response to climate change/variability are relatively scant.

The global production of cocoa is more than 40 lakh tonnes and the Ivory Coast leads in cocoa production in the world with a contribution of 38 per cent, followed by Ghana and Indonesia (4.1 Plate). Unlike in other crops, it appears that the production and demand of cocoa at the global level relatively match each other to a large extent. West Africa produces more cocoa, followed by South America and Asia. There was a sudden drop of more than ten per cent in 2006–07 (3.4 million tonnes) when compared to that of the previous year 2005-06(3.8 million tonnes) in 2005–06. It was attributed to unfavourable conditions in many cocoa producing areas. West Africa, the main cocoa producing region, was hit by a severe harmattan (dusty dry wind from November to March) and its inherent dry weather, which lasted from the end of 2006 to February 2007, had a strong negative impact on global cocoa production. In Asia and South America, El Nino-related weather conditions developed in September 2006 and continued until the beginning of 2007. It led to relatively low cocoa production in 2007. However, it was rebounced in 2008 to some extent though it was low (36.49 lakh tonnes) when compared to that of 2005–06. The favourable weather which prevailed over most of the cocoa growing regions during 2005-06 helped to achieve better production while dip in the global production in 2006–07 due to unfavourable weather. It reveals that the weather abnormalities like floods, droughts, cold and heat waves across the cocoa growing regions of the world adversely affect the Cocoa production and the cocoa industry is likely to suffer. Therefore, there is need to understand the effect of weather on reproductive phase of cocoa so as to manipulate the crop for sustenance of cocoa yield through better management practices.

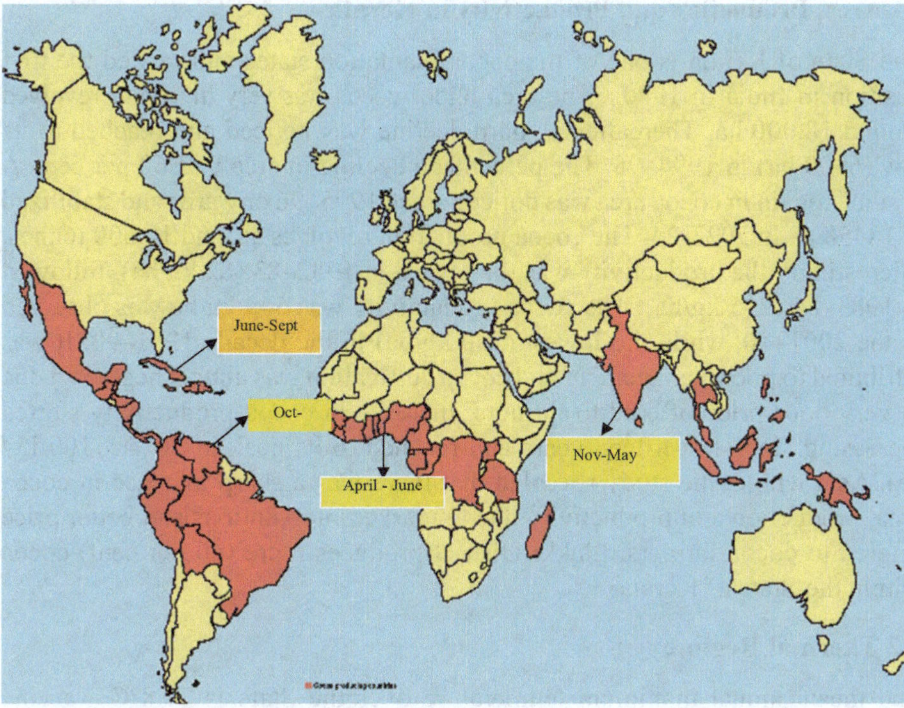

Plate 4.1: Cocoa growing regions over the world and peak period of flowering

Cocoa (*Theobroma cacao* L.), a beverage crop native to Amazon region of South America, has been introduced in India as profitable mixed crop in coconut and arecanut plantations in 1960s. Cocoa is mostly grown as an intercrop in coconut gardens. Andhra Pradesh stands first in cocoa production, followed by Kerala and both the States together contributes 80 per cent of the total cocoa production in India. There is a sharp increase in cocoa area in Andhra Pradesh while stagnation is noticed in Kerala during the last 15 years. Like other crops, there was stagnation in cocoa productivity due to weather aberrations like floods and droughts as cocoa is sensitive to waterlogging as well as soil moisture stress. High temperature during summer and heat wave are the weather constraints to low cocoa productivity. It is presumed that the low productivity in 2003–04 could be attributed to high summer temperature in 2004 over Kerala due to prolonged drought and heat wave that occurred in Andhra Pradesh and other southern states during 2003. The cocoa plantations needs equitable climate with well distributed rainfall. It requires an annual rainfall of 1500–2000 mm with a minimum of 90–100 rainfall per month. In majority of the regions, where cocoa is cultivated, a high but often unevenly distributed monthly rainfall occurs resulting in fairly well defined dry and wet seasons each year. The prolonged wet season is also not conducive for better cocoa production.

4.1 Area, Production and Productivity in Kerala

The State of Kerala is one of the cocoa plantation states and ranked the first position in India in 1980s. The area under cocoa was very high and revolved around 18,000 ha. Thereafter, a sharp decline was noticed and reached to its low (6900 ha) in 1994–95. The percentage decline in area was 62 per cent. A gradual regain in cocoa area was noticed since 1995–96 onwards and stabilized at 18458 ha in 2023-24. The cocoa production revolves around 10,500 tonnes. Interestingly the productivity was the lowest in 1982–83 (82 kg/ha), followed by1983–84 (215 kg/ha). The cocoa productivity was 549 kg/ha was observed in the 2001–10, while the lowest (330 kg/ha) in the decade 1981–90. It was attributed to poor management and cocoa cultivation was almost neglected due to very poor price offered to farmers. Increase in cocoa productivity started increasing since last fifteen years and reached to its peak in 2024 (1100 kg/ha). As a whole, the study revealed that there was a sharp increase in cocoa area, production and productivity due to market interventions and better price offered to cocoa farmers. Idukki District produces more (40 per cent) cocoa within the State of Kerala.

4.2 Thermal Regime

The mean annual maximum temperature over the State is 31.8°C, varying between the lowest in July (29.2°C) and the highest (34.5°C) in March. The maximum temperature starts increasing after the cessation of post monsoon rainfall in October and reaches its peak in March - April and decline by the receipt of summer showers. The dip in maximum temperature during the monsoon season over Kerala is typical due to heavy monsoon rains. The mean annual minimum temperature over the State is 23.4°C, varying between 21.7°C in January to 25.1°C in April (Fig.4.1). Of course, the crop is seen only in midlands/highlands and highranges as a homestead crop. Across the midlands or cocoa growing tract, the maximum temperature and minimum temperature vary between 31.8 °C and 23.4°C, respectively. However, under homestead conditions the crop is exposed to lower temperature when compared to that of the open crops. The long period trend in temperature of Kerala reveals that the mean annual maximum and minimum temperatures were in increasing trend.

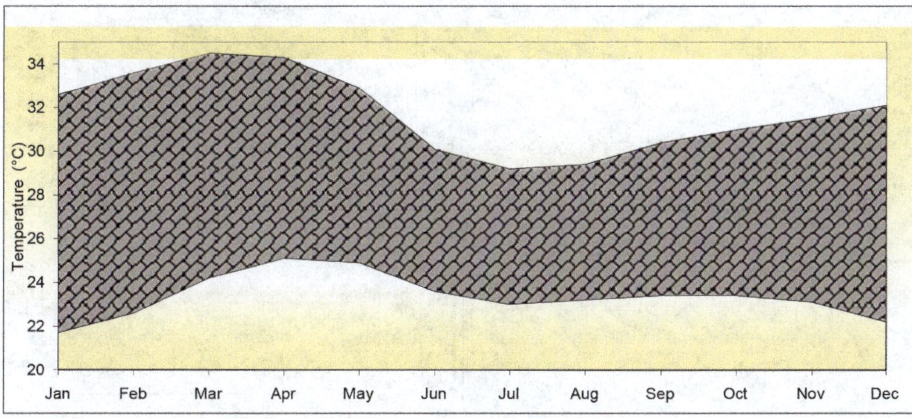

Fig. 4.1: Thermal Regime over cocoa growing tract of Kerala

4.2.1 Temperature and Cocoa Yield

The relationship with annual maximum temperature and yield (productivity) revealed that whenever the annual maximum temperature increases, cocoa yield decline and vice versa. For example, when the mean annual maximum temperature was high (32.8°C) in 1988, cocoa yield was low (421 Kg/ha).

Similar was the case in 1982 (32.2°C) and 1983 (32.4°C) during which the cocoa yield was low (82 kg/ha and 215 kg/ha), respectively (Fig. 4.2a). In contrast, when the mean annual maximum temperature was low (31.5°C) in 1999, the yield obtained was high (449 kg/ha). Similar relationships were found in 1985 (31.7°C), 1990 (31.9°C), 1992 (31.8°C), 1993 (31.8°C), 1994 (31.9°C), 1995 (32.0°C), 1996 (31.9°C) and 2003 (31.9°C) during which the yields were 31.6 kg/ha, 412 kg/ha, 570 kg/ha, 646 kg/ha, 623 kg/ha, 544 kg/ha, 569 kg/ha and 574 kg/ha, respectively. Cocoa yield during 1993 was high (646 kg/ha) when the annual maximum temperature was low (31.8°C).

Similar situations prevailed during 1995 (623 kg/ha), 2003 (574 kg/ha), 2004 (577 kg/ha) and 2007 (570 kg/ha) during which the maximum temperatures were 31.9°C, 31.9°C, 32.1°C and 32.3°C, respectively. Such situations were seen in majority of the years (89 per cent). The study indicated that the mean annual maximum temperature has a negative impact on the yield of cocoa. It also revealed that when the mean annual maximum temperature is above 32.2°C, cocoa yield was adversely affected.

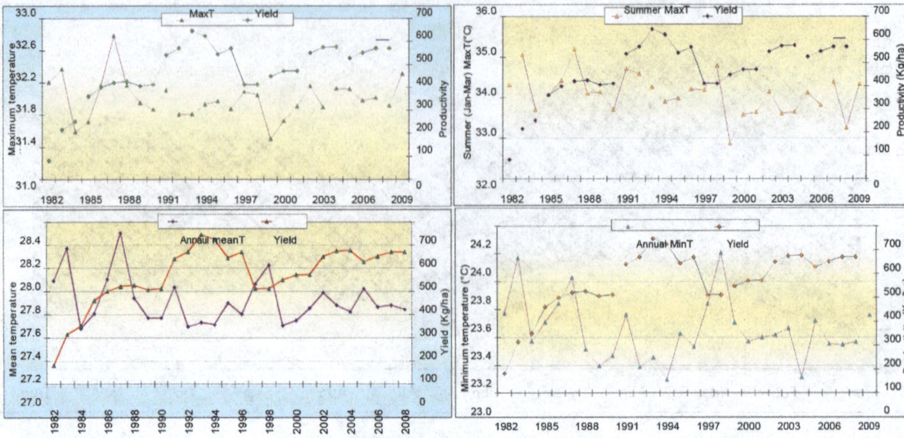

Fig. 4.2a: Temperature and cocoa yield in Kerala

4.2.2 Maximum Temperature and Yield

Similar trend was observed when the relationship between the summer (January-March) maximum temperature and cocoa yield was examined. When the summer maximum temperature was high (35.1°C) in 1983 cocoa yield was low (215 kg/ha). Similar events were noticed in 1987 and 2008 during which the summer maximum temperature was high (35.2°C and 34.8°C, respectively), cocoa yields were 421 Kg/ha and 412 kg/ha, respectively (Fig. 4.2b). In this case, when the summer maximum temperature was above 34.7°C, yield decline was seen. In contrast, when the summer maximum temperature was low (32.9°C), cocoa yield was high (449 kg/ha). Similar relationships were observed in 1990, 1993, 1994, 2004 and 2008 during which the maximum temperatures were 33.7°C, 34.3°C, 33.9°C, 33.6°C and 33.3°C, respectively and the yields were high (412 kg/ha, 646 kg/ha, 623 kg/ha, 577 kg/ha and 570 kg/ha, respectively). As observed in the case of annual maximum and summer maximum temperature, annual mean temperature also showed a similar trend with yield. It revealed that when the annual mean temperature was high (28.3°C), cocoa yield was low (421 kg/ha). Similarly, in 1982 (27.9°C), 1983 (28.2°C), 1986 (27.9°C) and 1998 (28.0°C) during which the annual mean temperature was high, cocoa yields were low (82 Kg/ ha, 215 kg/ha, 400 kg/ ha, 421 kg/ha and 412 kg/ha, respectively). In contrast, when the annual mean temperature was low (27.5°C), in 1992, cocoa yield was high (570 kg/ha). Such observations could be seen during 1993, 1994, 1996 and 2000 during which annual mean temperature was low (27.5°C, 27.5°C, 27.6°C and 27.5°C, respectively) while the cocoa yield was high (646 Kg/ ha, 623 kg/ha, 569 kg/ha and 449 kg/ha, respectively). It was also observed that when the mean annual mean temperature was above 27.8°C, cocoa yield was likely to decline and vice versa. As observed in the case of maximum and mean temperature, the relationship between minimum temperature revealed that whenever the mean annual minimum temperature was high, cocoa yield was low and vice versa. For example, annual minimum temperature was high (24.0°C each) during 1983 and 1998, during those years, cocoa yield was high (215 kg/ha and 412 kg/ha, respectively). In contrast, when the mean annual minimum temperature was low, cocoa yield was high. For example, during 1989 (23.2°C), 1990 (23.3°C), 1992 (23.2°C), 1995 (23.4°C) and 2004 (23.1°C) when the minimum temperatures were low, cocoa yields were high (406 kg/ha, 412 kg/ha, 570 kg/ha, 544 kg/ha, 577 kg/ha, respectively). Such situations were predominant in majority of the years (89 per cent).

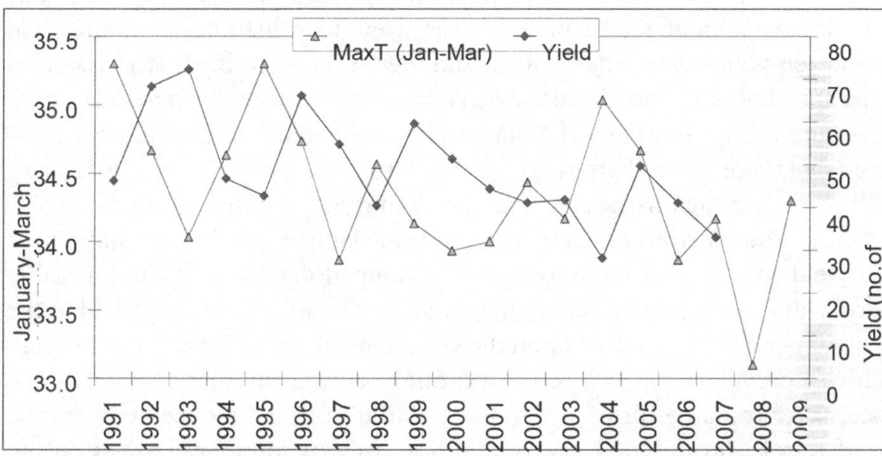

Fig. 4.2b: Summer maximum temperature and cocoa yield

4.2.3 Temperature and Cocoa Yield at Vellanikkara

In the case of temperature, the maximum temperature from January to March had a profound influence on annual cocoa yield. Whenever there was an increase in maximum temperature, the annual cocoa yield recorded was low. For example, the maximum temperature recorded between January and March was high (35.3°C) in 1995 against the normal (34.6°C) and the annual cocoa yield was low (42.7 pods/tree). It was also true during 2004 (35.0°C) in which low yield (28.2 pods/tree) was obtained. Whereas, the maximum temperature recorded in 1993 was low (34.0°C) and the annual cocoa yield was high (72.2 pods/tree). Similar was the case in 1997 and 2000 as the annual cocoa yield during the above years were 54.7 pods/tree and 51.4 pods/tree, respectively.

The maximum temperature during summer was high (34.6°C) in poor yield years when compared to that of good yield years (34.2°C). A difference of 0.7°C was noticed in April between good and poor yield years (Table 4.1). It revealed that high maximum temperature during summer is likely to affect the annual cocoa yield adversely. High maximum temperature in general and in particular during summer is detrimental to most of the plantation crops. When crops are exposed to increased temperature under rainfed situations, they are subjected to severe soil moisture stress, if pre-monsoon showers fail. In the event of erratic monsoon and scarce rainfall from post monsoon, situations will become more vulnerable. Such situations will adversely affect the plantation crops production. Yield loss due to unfavourable weather events is a concern in recent years due to global warming and climate change. Hence, an attempt was made to understand the effect of increased temperature during summer and its adverse impact on cocoa yield. Good and bad years were identified based on the yield data and temperature profiles during summer were examined. It is reported that the apical dominance was lost when cocoa was consistently exposed to high temperature during summer. The maximum temperature during summer 2004 showed that there was a sharp rise (1–3°C) during the period from 14^{th}January to 18th March as against the normal maximum temperature of 33°C-36.5°C (Fig.4.2c). The maximum temperature from January to March influences cocoa yield to a considerable extent. A prolonged dry spell from November to May with high maximum temperature adversely affects the pod yield to the tune of 40 per cent depending upon the crop management. Severe water deficit which was prevalent till May led to a drought during summer. Almost all the surface water resources dried up to due to failure of post monsoon across the State of Kerala during 2003–04. In addition, lack of summer showers during February and March worsened the situation and led to early severe drought, which was identical to late summer drought of 1982–83. The reduction in pod yield of cocoa was to the tune of 43 per cent in 2004. A decline of 39 per cent in annual cocoa yield was noticed when large field sample size was considered within the same farm (Amma *et al*., 2006). Similar results were found in low yields years viz., 1995 (42.7 pods /tree), 1998 (40.4 pods/tree), 2002 (41.1 pods/tree) and 2003 (41.7 pods /tree) when compared to that of annual yield of 49.6 pods/tree. The yield reduction was 14, 19, 18 and 16 per cent, respectively when compared to the mean pod yield.

Table 4.1 Maximum temperature (^{O}C) during summer in good and bad yield years

Months	Max. temp. in good yield years					Max. temp. in bad yield years					
	1992	1993	1996	1999	Mean	2004	1998	1995	2003	Mean	Normal
Jan	32.6	32.7	33.1	32.4	32.7	33.4	33.1	32.9	33.2	33.2	32.8
Feb	34.4	34.1	34.7	34.5	34.4	35.2	34.4	35.4	34.7	34.9	34.8
Mar	36.9	35.4	36.4	35.5	36.1	36.5	36.2	37.6	34.6	36.2	36.1
Apr	36.3	35.6	34.6	33.4	34.9	34.8	36.5	36.6	34.6	35.6	35.4
May	33.8	34.4	32.8	30.7	33.0	30.4	34.2	33.5	34.0	33.0	33.8
Mean	34.8	34.4	34.3	33.3	34.2	34.1	34.9	35.2	34.2	34.6	34.5

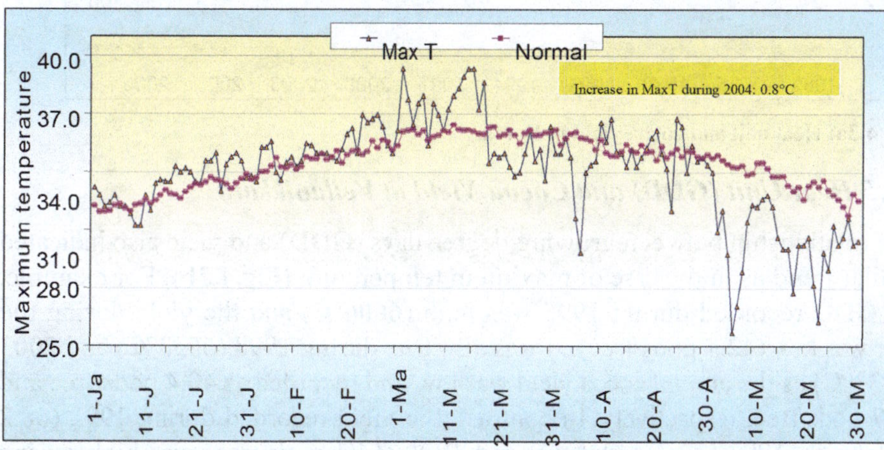

Fig. 4. 2c: Effect of summer maximum temperature on cocoa yield at Vellanikkara

4.2.4 Heat Unit (GDD) and Cocoa Yield

The relationship between the heat units (Growing Degree Days) and yield is similar to that of temperature as heat unit is a function of mean temperature. It indicated that when the GDD was high, yield was low and vice versa (Fig. 4.3a). For example, when the heat unit was high in 1982 (6530°C), 1983 (6632°C), 1987 (6682°C) and 1998 (6580°C), the yield was low (82, 215, 421 and 421 Kg/ha, respectively). In contrast, when the heat units were low in 1993 (6399°C), 1994 (6312°C, 1995 (6461°C) and 2003 (6453°C), cocoa yields were high (646, 623, 544 and 574 kg/ha, respectively).

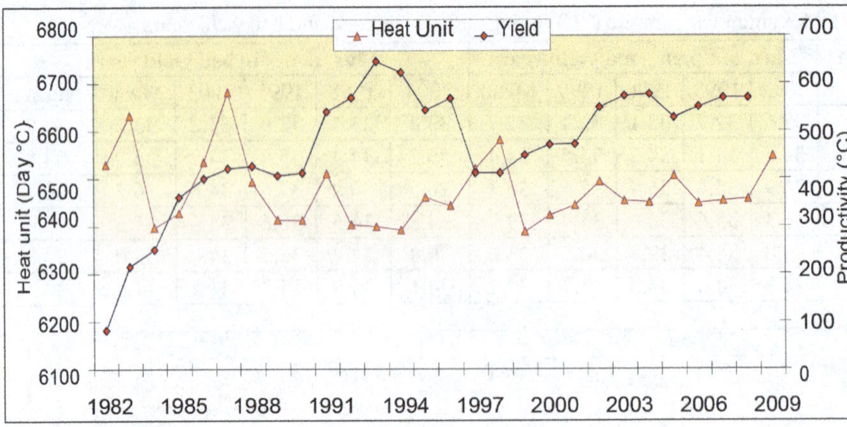

Fig. 4.3a: Heat unit and cocoa yield in Kerala

4.2.5 Heat Unit (GDD) and Cocoa Yield at Vellanikkara

The relationship between growing degree days (GDD) and yield also indicated similar trend as in the case of maximum temperature **(Fig.4.3b)**. For example, the GDD recorded during 1995 was high (6606°C) and the yield during the year was low (42.7 pods/tree). It was also true during 1998 (6533°C) and 2003 (6533°C) as the annual cocoa yield was low and recorded as 40.4 pods/tree and 41.7 pods/tree, respectively. In contrast, the yield recorded during 1992 (68.2 pods/tree), 1993 (72.2 pods/tree) and 1996 (66.1 pods/tree) was high during which low GDD was recorded viz, 6405°C, 6460°C and 6460°C, respectively.

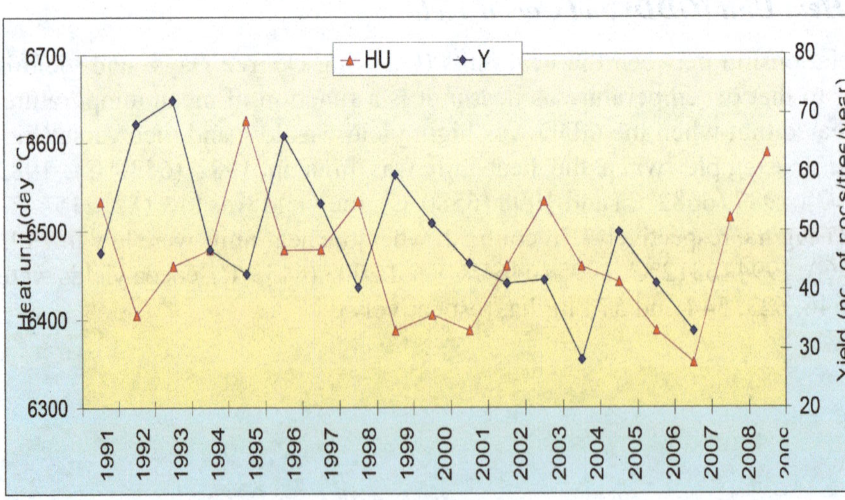

Fig. 4.3b: Heat unit and cocoa yield at Vellanikkara

4.3 Bi-Model Rainfall

Kerala has the fortune of having the two monsoons viz., southwest and northeast and hence the rainfall pattern of the State is bi-model having two peaks one during southwest monsoon (June to September) and another during Northeast monsoon period from October to November (Figs. 3.2a and 3.2b). However, the bi-model distribution of rainfall is not seen in the northern districts of Kerala as northeast monsoon is not prominent there. The major contribution (68%) of rainfall is from the southwest monsoon and northeast monsoon constitutes only 17 per cent of the total rainfall. In addition to the above, pre-monsoon showers are also a source of rainfall during the summer months. June is the rainiest month in the central and southern districts of the State, while the rainiest month is July in the northern Kerala. In general, rainfall increases from south to north of the State. While southern most part receives 1100 mm of rainfall, amount of rainfall received in northern region of the Kerala is 3600 mm. There exists wide variability in the receipt of rainfall across the State, varied from less than 1000 mm to more than 5000 mm. The bi-model rainfall may be favourable for better cocoa productivity. It is one of the reasons for low soil moisture stress during summer and the cocoa yields better in Idukki District and other cocoa growing southern districts

4.3.1 Rainfall and Cocoa Yield

It is a complex phenomenon to workout relationship between rainfall and the cocoa production at the State level as the crop had undergone a market crisis in 1980s and 1990s and cocoa plantations were cut and replaced with other profitable crops like rubber. It was a neglected crop for long because of the low price in the market. Recently only, farmers started to take care of cocoa plantations on commercial angle since last two decades. However, high rainfall appeared to be adversely affected the annual cocoa productivity of the State. For example, the annual rainfall recorded during 1994–95 was high (3159 mm) in which annual cocoa production was low (4300 tonnes). It was also true in the case of 1997–98 (3179 mm), 1998–99 (2966 mm) and 1999–2000 (2838 mm) as the cocoa production during the above years was recorded as 3500, 3500 and 4000 tonnes, respectively. In contrast, the annual cocoa production was more (6100 tonnes) when the rainfall recorded was low (2532 mm) during 1985–86 (Fig.4.4a). Similar trend was also noticed during 1986–87 (2147 mm), 2003–04 (2480 mm) and 2004–05 (2685 mm) and cocoa production recorded was 6000, 5870 and 5900 tonnes, respectively. It reveals that high rainfall, heavy cloudiness, low bright sunshine and high relative humidity appeared to be detrimental to cocoa. In the case of productivity also, similar was the trend noticed. Rainfall was high in 1997–98 (3179 mm) and the yield was low (412 kg/ha). Similar was the case in 1998–99 (2966 mm) during which the cocoa yield was low (412 kg/ha). In contrast, during low rainfall years, such as 1986–87 (2147 mm), 1993–94 (2740 mm), 1995–96 (2485 mm) and 2000–01 (2270 mm), the yield recorded was high (400 kg/ha, 646 kg/ha, 544 kg/ha and 471 kg/ha, respectively). Similar relationships were observed in the case of summer rainfall and yield also. At the Instructional farm,Vellanikkara also, whenever the annual rainfall was very high, the annual cocoa yield was low. For example, the annual rainfall recorded was high (3579 mm) in 1994 against the normal (2803 mm) and the yield during the year was low (46.8 pods/year) as evident from (Fig 4.4b). Similar trend was also noticed during 1998 (40.4 pods/tree), 2004 (28.2 pods/tree) and 2006 (41.1 pods/tree) during which the annual rainfall was 3435 mm, 2895 mm and 3460 mm, respectively. Whereas in 1993, 1996, 1999, and 2000, the annual rainfall recorded was relatively less (2439 mm, 2241 mm, 2619 mm and 2173 mm, respectively) as against the normal and the annual cocoa yield during the above years was high (72.2 pods/tree, 66.1 pods/tree, 59.6 pods/tree, and 51.4 pods/ tree, respectively). It revealed that the annual cocoa yield was relatively low when the annual rainfall was high against the normal rainfall and *vice-versa*.

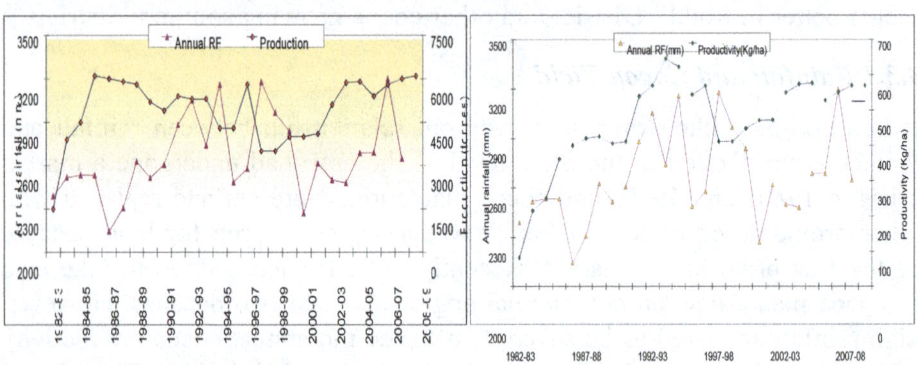

Fig. 4.4a: Rainfall and cocoa yield over Kerala

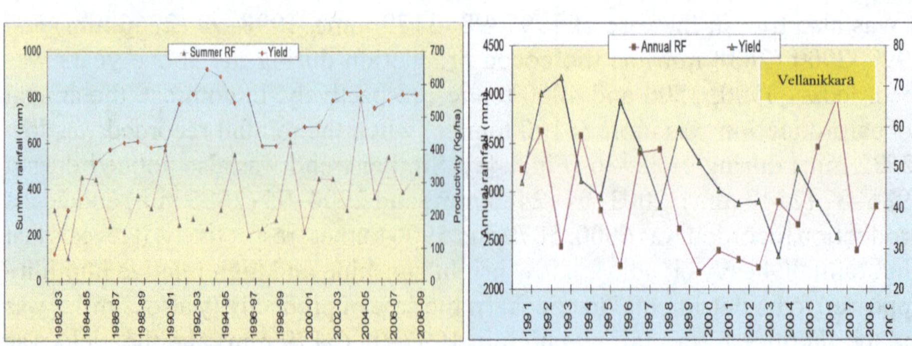

Fig. 4.4b: Rainfall and cocoa yield at Vellanikkara

4.3.2 Rainfall and Flowering

The biotic events such as flowering, cherelle production and pod set in cocoa are seen throughout the year with significant variation seasonally. However, the flowering appears to be very low during the rainy season (June to September). It normally initiates in October/November depending upon rainfall distribution from June to September and reaches to its peak by May. Similar trend is seen in the case of cherelle production and pod set. The above biotic phenomena is true only under rainfed conditions in the Humid Tropics, where heavy rainfall is noticed from June to September, followed by prolonged dry spell. Profuse flowering is also noticed during summer months if one or two good showers are received 15 days prior to commencement of flowering in cocoa. Interestingly, the number of flowers appears to be more under open conditions when compared to shade due to better availability of sunlight under good crop management practices. The number of cocoa (cacao) pods is high during summer, but pod weight is low. A harvest of five pods during October/November may be equal to eight/nine pods harvested in summer, showing superiority in cocoa pods harvested in October/November. High rainfall versus cocoa yield showed inverse relationship, indicating that high rainfall during rainy season may not be conducive in the case of cocoa due to waterlogging and lack of soil aeration. Overall, it reveals that high maximum temperature during summer, followed by heavy rains during the Southwest monsoon may not be conducive for obtaining good yield in cocoa under the Humid Tropics. High rainfall has a malevolent effect in places where the crop is grown under waterlogged conditions, while summer rainfall has benevolent effect on pod yield. On augmenting the dry season's rainfall by the use of irrigation, the results have not always been improved because the water lost from the leaves is so high with low humidity that the roots cannot match the loss. Crop yields are very low with heavy shade and increases with increasing light up to the 50% level. Cocoa yield is affected by the presence or absence of fertilizer if the level of light is above 50%. With added fertilizer, yields increase almost up to full light, whereas in the absence of fertilizer the yields fall off. The theory has therefore been advanced that the light regime for optimum yield of cocoa is a function of its mineral nutrition. The loss of apical dominance in cocoa is seen constantly at a temperature of above 32°C. The loss of apical dominance implies the development of side shoots of the plant. High maximum temperatures (> 36°C) during February and March, 2004 resulted in low pod yield in cocoa over the central region of Kerala. Cocoa is grown as an intercrop in coconut or part of homestead garden. Both the crops cocoa and coconut respond identically with reference to weather and climate under the Humid Tropics. Of course, the phenology of both the crops is totally different though the monthly crop yields are noticed round-the-year.

4.3.3 Soil Moisture and Cocoa Yield

The relationship between mean annual soil moisture and yield also indicated similar trend as in the case of rainfall. For example,4 the mean annual soil moisture recorded during 2003 was high (16.5%) and the yield during the year was low (41.7 pods/tree). It was also true during 2001, 2004 and 2006 (Fig.4.5a) as the annual cocoa yield was low and recorded as 44.4 pods/tree, 28.2 pods/tree and 41.1 pods/tree, respectively. In contrast, the yield recorded during 1999 (66.1 pods/tree) and 2005 (49.8 pods/tree) was high during which annual soil moisture was low and recoded as 11.4 % and 11.1 %, respectively. The same trend between the soil moisture and yield was reflected as in the case of rainfall.

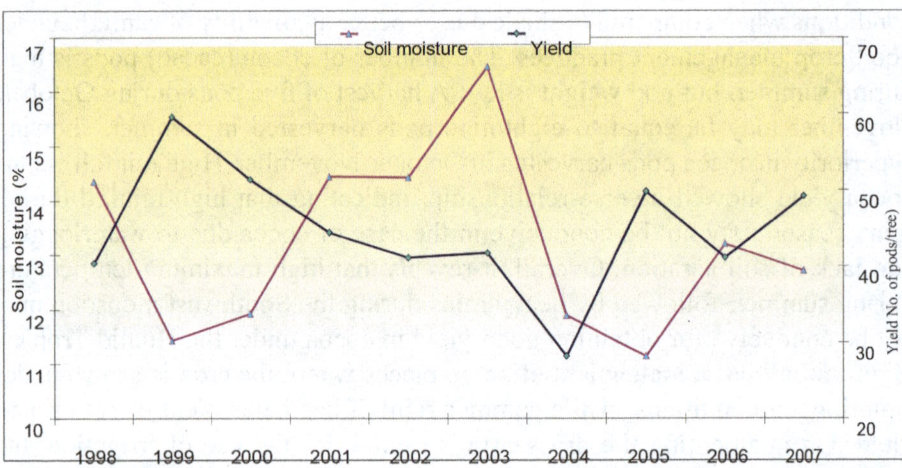

Fig. 4.5a: Soil moisture at 30 cm depth and cocoa yield at Vellanikkara from 1998 to 2007

It revealed that high rainfall may lead to waterlogging under the field conditions and remain seen four to six months. This type of situation may lead to lack of aeration in the root zone and restrict the uptake of fertilizers, finally leading to low yield and vice-versa. As the soil moisture is dependant on rainfall such relationship is expected. High rainfall lead to heavy soil moisture which may result in waterlogging or sometimes flood the fields where cocoa is grown. It is detrimental to cocoa crop. In contrast, as the yield pattern shows peak during summer, it had a negative relationship with soil moisture since it is minimal in the absence of rain during summer. In fact, any improvement in soil moisture during summer may result in good health of plant, in turn lead to better yield.

4.3.4 Water Deficit and Cocoa Yield

It revealed that there exists an inverse relationship between water deficit and yield. The yield was low, when the water deficit was high and vice versa. For example, when the annual water deficit was high in 1983 (451 mm), 1987 (378 mm) and 1990 (365 mm), the yields were low (215 kg/ha, 421 kg/ha and 412 kgha), respectively (Fig.4.5b). In contrast, when the annual water deficit was low in 1984 (69 mm), 1988 (225 mm), 1994 (220 mm), 1995 (208 mm), 1999(187 mm), 2001 (180 mm) and 2008 (187 mm), cocoa yields in the State was high (251 kg/ha, 426 kg/ha, 623 kg/ha, 544 kg/ha, 449 kg/ha, 451 kg/ha and 570 kg/ha, respectively).

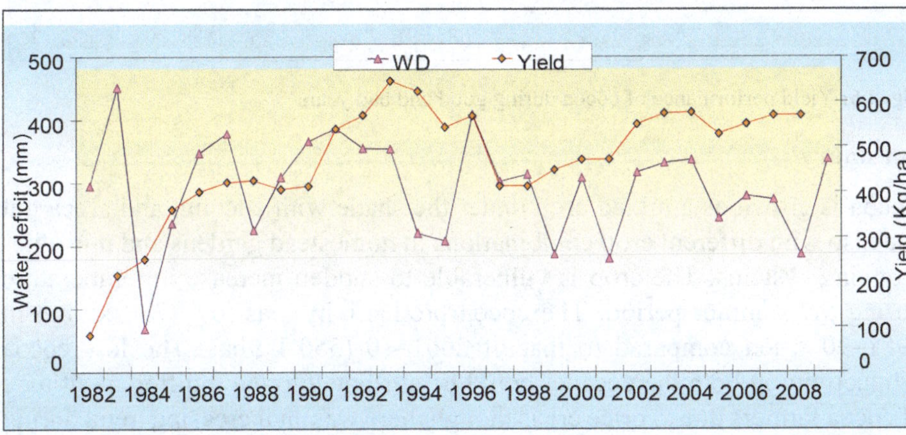

Fig. 4.5b: Water deficit and cocoa yield in Kerala

4.3.5 Cocoa Yield in Good and Bad Years

In order to understand the monthly and seasonal yield variability in cocoa, the monthly yield data of one good yielding year (1992) and one poor yielding year (2004) was examined. It was found that the difference in cocoa yield of good year during rainy months (June to September) was very significant (344.4 per cent over the poor year) followed by summer (222 per cent) and post monsoon season (28 per cent). There was a decline of 14 per cent in winter over the poor yielding year (Fig. 4.6). It is evident that in good and bad yield years, the difference in cocoa yield is more during the rainy season, followed by summer, post monsoon season and winter.

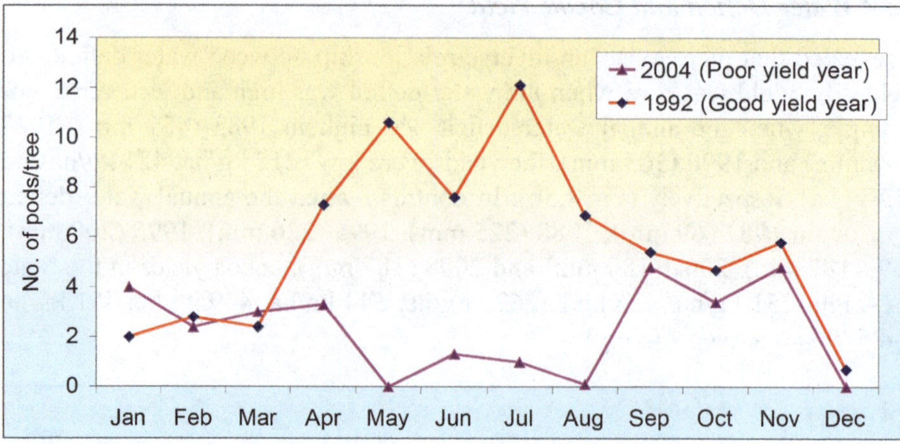

Fig. 4.6: Yield performance of cocoa during good and bad years

Summary

Cocoa is grown as a mixed crop under the shade with coconut and areca nut and also with different crop combinations in homestead gardens and multi-tier cropping systems. The crop is vulnerable to sudden increase in temperature during the summer period. The cocoa productivity was low (330 kg/ha) in 1981–90 when compared to that of 2001–10 (550 kg/ha). The low cocoa productivity in both the decades could be attributed to the indifferent attitude of cocoa farmers due to price crisis though there was an increasing trend during 2001-10. Thereafter, increase in cocoa area, production and productivity was noticed since last two-and-a-half decades. The increase in area, production and productivity in recent years in the case of cocoa could be attributed to adoption of superior clones evolved, technologies and farmers' interest due to stability in price. The bi-model rainfall appears to be conducive for cocoa under rain fed conditions. However, continuous high rainfall during the rainy season (June to September) is detrimental to the crop in the tropical humid climates. Warm summer temperatures and heat wave might be the constraints in addition to high rainfall during the rainy season for low cocoa production and productivity. Increase in maximum temperature varying between 34.3°C to 35.1°C, the difference of which was 0.8°C in 2004, adversely affected the cocoa yield and the overall percentage decline varied up to 32 per cent. The crop like cocoa, which is thermo-sensitive, may respond negatively to a steady increase in temperature during the summer period in the absence of soil moisture. It is quite interesting to note that the crop likes relatively open conditions to some extent when compared to that of shade. It indicates that cocoa is a sun loving crop rather than shade loving. With better management crop practices the crop performance is good in the open when compared to that of shade though the crop is seen cultivated as an intercrop in coconut/areca nut.

Suggested Readings

Manikandan, N. 2007. Climate Variability and Cocoa Production in the humid tropics M.Sc (Ag.) Thesis. Kerala Agricultural University, Vellanikkara, Thrissur, Kerala

Manikandan, N., Rao, G.S.L.H.V.P. and Amma, P. 2014. Influence of climate on yield of cocoa over Vellanikkara. International Journal of Innovative Agriculture. 1(1): 31-36

Nair, R.V., Mallika, V.K. and Swapna, M.1996. Response of cocoa to shade. J. of Plantation Crops. 24(Supplement): 99–101.

Nybe, E.V. and Rao, G.S.L.H.V.P. (Eds.). 2005. Drought Management in Plantation Crops. (Eds.). Proceedings of the National Workshop. Kerala Agricultural University, Thrissur, Kerala. 232p

Prameela, K.P. 1997. Effect of weather on cocoa and improvement of bean size through seasonal crop orientation. M.Sc.(Ag.) Thesis, Kerala Agricultural University, Thrissur. 119p.

Rao, G.S.L.H.V.P.2021. Agricultural Meteorology (5th reprint). PHI Learning (Pvt.) Ltd. New Delhi, India.364p

Rao, G.S.L.H.V.P. and Manikandan, N. 2009. Crop Weather Relationships of Cocoa in the Humid Tropics. Kerala Agricultural University. Thrissur, Kerala, 117p

Rao, B.V.R. and Rao, V.U.M. 2024. Agricultural Meteorology. Jain Brothers. New Delhi.154p

Rao, G.S.L.H.V.P. and Nair, R.R. 1988. Agrometeorology of Plantation Crops. Proceedings of the National Workshop. Kerala Agricultural University, Thrissur, Kerala. 187p

Rao, G.S.L.H.V.P. and Krishnakumar, K.N. 2024. Impact of weather and climate on plantation crops in the Humid Tropics. Chapter 21 (417-432pp). Agrometeorological Applications for Climate Resilient Agriculture (Eds.): Chattopadhyay, N., Stefanski, R., Attiri, S.D. and Rathore, L.S. Springer International Publishing AG. Singapore.445p

Questions

I. Elucidate on the Following

1. Crop weather relationships of coconut
2. Temperature versus cocoa yield
3. Rainfall versus cocoa yield

II. Write short notes on

1. Thermal regime of cocoa
2. Rainfall regime of cocoa
3. Heat Units versus cocoa yield
4. Summer temperature versus cocoa beans in good and bad yield years
5. Cocoa growing areas and flowering period across the world

III. Fill in the Blanks

1. ---------, ----------- and --------- are critical weather elements in cocoa production
2. ---------- is the peak cocoa flowering period in India
3. -------------, ----------, ---------, and ------ are the good yield years in cocoa
4. --------,----------,-----------, and ---------- are the bad yield years in cocoa
5. ------------ is the average summer maximum temperature in good yield years
6. -------------is the average summer maximum temperature in bad yield years
7. ----------- in cocoa area while ------- in cocoa productivity in Kerala
8. ------------ is the rainfall model across the cocoa tract
9. Cocoa is known as ------------ in other cocoa growing countries
10. The annual cocoa yield is ----------- when the annual rainfall is high
11. The annual cocoa yield is ------------ when the annual soil moisture is high
12. The difference in number of pods per tree is more in -----------season between good and poor yield years
13. Cocoa yield is ----------- when annual temperature is -----------
14. The reduction in coca yield was ------% in 2004 due to high summer temperature
15. When the heat units were high the pod yield was -----------
16. The increase in temperature was ----------^{0}C in 2004 during summer

IV. State True or False

1. Idukki District produces the highest cocoa pods in Kerala
2. Cocoa needs sunshine
3. Cocoa performs better under open when compared to that of shade
4. Cocoa exhibits flowering round-the-year
5. Summer rains between February and May influence favourably cocoa flowering
6. Rainfall during rainy season may influence cocoa flowering negatively
7. Rainfall in the Humid Tropics may be the important one in deciding cocoa flowering
8. The pattern of pod harvest is different between the shade and open
9. The number of pods harvested is more in shade when compared to that of open conditions
10. High maximum temperature during summer is detrimental to cocoa production
11. The coefficient of variation is very high among monthly cocoa pods harvested
12. Rodent is the major pest in cocoa plantations
13. Summer temperature followed by heavy rains are detrimental to cocoa production
14. Cocoa responds not only to temperature and rainfall but also to sunshine
15. V. Match the following

V. Match the Following

Monthly	C.V% of rainfall
January	34.9
February	33.0
March	32.7
April	36.1
May	34.4
	34.2

5

Cardamom

Small cardamom (*Elettaria cardamomum* Maton), popularly known as the "Queen of spices", occupies an important position among the foreign exchange earning commodities. It is cultivated on a commercial scale in Sri Lanka, Guatemala, Tanzania and Papua New Guinea. Guatemala is the largest producer of cardamom in the World. In India, it is confined to the States of Kerala, Karnataka and Tamil Nadu. Among these States, Kerala accounts for the major portion of production (58.4 %), followed by Karnataka (35.2 %) and Tamil Nadu (6.4 %). In Kerala, Cardamom is mainly grown in the erstwhile Travancore area in Idukki district and it is known as the Cardamom Hill Reserve (CHR). Cardamom is a native of the moist evergreen rain forests of the Western Ghats and is grown extensively in the hilly regions at elevations of 700 to 1300 metres as an under crop in forest lands. Cardamom thrives well within certain maximum and minimum thermal regimes. It is believed that diurnal symmetries in temperature trends, if any, have close links with the changes in cloudiness, humidity, atmospheric circulation pattern, winds and soil moisture. The land use pattern of cardamom tracts has changed to a great extent in recent decades. Crops like coffee, black pepper and areca nut are being cultivated at many locations in place of cardamom. With denudation of forests in many cardamom areas, the ideal cool humid microclimate has been destabilized and adversely affected the productivity of the crop. The recurring summer drought combined with extensive deforestation and resultant changes in cardamom ecosystem, outbreak of insect pests and diseases and lack of intensive management are the main reasons for the decline in production of cardamom across the Western Ghats. Cardamom, being a sensitive plant, any disturbance in the environment, especially the meteorological parametres, will adversely affect the growth, development and production. The projected global warming and rainfall changes may adversely affect forest ecosystems where thermo-sensitive crops like cardamom, coffee, tea and black pepper are grown. It is estimated that about 80 to 90 per cent of rainfall is received during the period from May to November. In 1983, the cardamom tract throughout the Western Ghats witnessed very low production due to failure of rainfall from November to May. Both the amount and distribution of rainfall throughout the year are also important to reap a better cardamom production. The aberrant weather conditions that prevailed during recent years coupled with poor management had resulted in low yields in India.

5.1 Rainfall Distribution

The mean annual rainfall across the high ranges of Kerala is 1901 mm, varying between 1865 mm at Pampadumpara and 1939 mm at Ambalavayal. The annual number of rainy days is 117. It is interesting to note that the number of rainy days did not show any significant difference between Pampadumpara and Ambalavayal. The monthly rainfall pattern at Pampadumpara (Idukki) and Ambalavayal (Wayanad) follows the same pattern. The distribution pattern of rainfall at Pampadumpara showed that the monthly rainfall is high during July (365.3 mm), followed by June (295.6 mm), August (271.1 mm) and October (262.2 mm). The least is in February (14.8 mm), followed by January (18.7 mm). In the case of Ambalavayal, similar rainfall pattern is evident. The lowest rainfall at Ambalavayal is noticed during January (12.1 mm), followed by February (16.6mm) and December (24.3 mm). Both locations receive high rainfall in June, July, August and October due to the influence of both southwest and northeast monsoons (Fig.5.1). Pamapdumpara receives relatively better rainfall from October to December when compared to that of Ambalavayal. It is attributed to the influence of northeast monsoon towards south of Kerala. It reveals that rainfall distribution across the cardamom tract of Kerala is bi-model, having a peak in July and another peak in October. The annual rainfall increases from South (Pampadumpara) to North (Karnataka) across the cardamom tract while the western parts of the Western Ghats receive high annual rainfall, ranging from 1873 mm to 2608 mm when compared to that of eastern parts (1452 mm). The coefficient of variation of annual rainfall indicated that it is dependable from year to year across the cardamom tract though winter rainfall is variable. The annual number of rainy days is high over western parts of the cardamom tract (113-127 days) and low over eastern part (81 days).

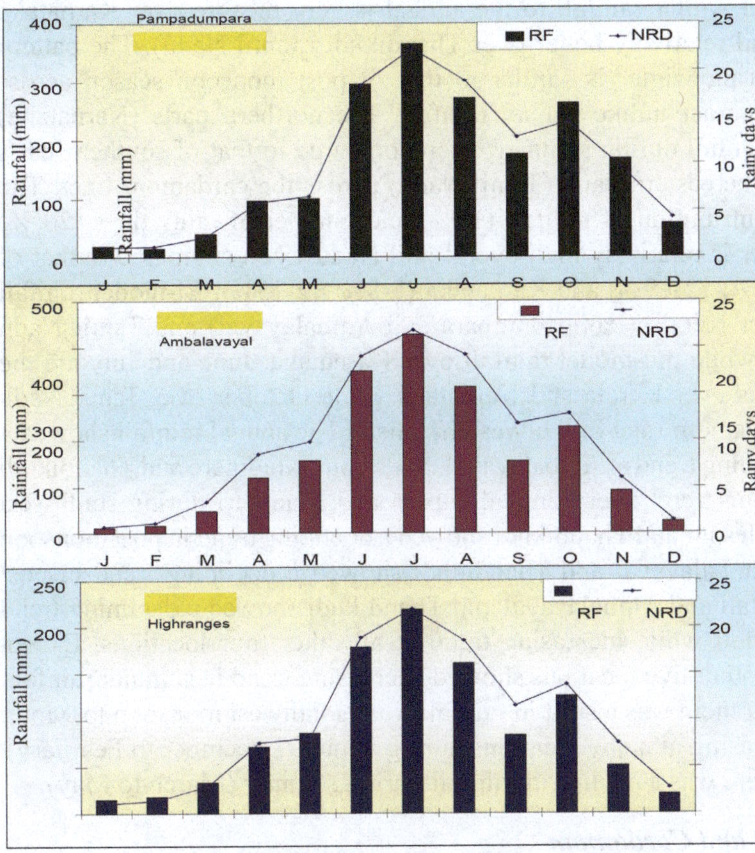

Fig. 5.1: Monthly rainfall across the highranges of the cardamom tract

The southwest monsoon rainfall is relatively dependable over the western region than that of the eastern region (Tamil Nadu) of the Western Ghats. The percentage contribution of southwest monsoon to annual is high over Karnataka (80-83%) and less over Tamil Nadu (33 %). It is intermediary (60-65%) over Kerala. It is true in the case of rainy days also. The post monsoon rainfall decreases from south (Pampadumpara) to north (Mudigere) unlike annual rainfall and rainfall during southwest monsoon. The eastern parts (Tamil Nadu) of Western Ghats receive high rainfall during the post monsoon (Oct to Nov), followed by southern parts (Pampadumpara) when compared to northern parts (Karnataka) across the cardamom tract. The percentage contribution of post monsoon rainfall to annual is high over Tamil Nadu (35.4%), followed by Kerala (16.1 – 23.1%) and less over Karnataka (9.7 – 11.1%). The number of rainy days is high over Tamil Nadu and Pampadumpara (Kerala) than the other locations during the post monsoon where cardamom is extensively grown. The contribution of winter rainfall to the annual is very meagre over Karnataka and Kerala and relatively better over Thandikudi (Tamil Nadu). The pattern of rainfall during winter is similar to that of post monsoon season across the cardamom tract unlike annual rainfall. The northern parts (Karnataka) receive less rainfall during summer when compared to that of southern parts (Kerala) and the eastern parts (Tamil Nadu) across the cardamom tract. The percentage contribution of rainfall (19%) and number of rainy days (20 %) during summer is relatively high over Tamil Nadu when compared to that of Kerala (12-16%; 12-19%) and Karnataka (7-8%; 12-13%). Bi-model rainfall is noticed over Kerala (Pampadumpara and Ambalavayal) and Tamil Nadu (Thandikudi) while uni-model rainfall over Karnataka. June and July are the rainiest months over Kerala and Karnataka while October over Tamil Nadu across the cardamom tract of the Western Ghats. The annual rainfall showed a marginal declining trend at Ambalavayal, Saklespur, Mudigere and Thandikudi while increasing trend over Pampadumpara and Madikeri during southwest monsoon. Saklespur and Thandikudi showed a declining trend in post monsoon rainfall and Ambalavayal and Thandikudi showed a declining trend in post monsoon rainfall and Ambalavayal and Thandikudi showed a declining trend in winter rainfall while increasing trend at all other four locations. Except Saklespur, all other five locations showed decreasing trend in summer rainfall. It revealed that there was a shift in rainfall from southwest monsoon to winter as it was increasing at many locations during winter (December to February). In contrast, there was a decline in rainfall during summer (March to May).

5.1.1 Rainfall and Cardamom

The cardamom yield showed an increasing trend since 1950 onwards though there was a decline in 2006–07 and 2007–08. However, the inter annual variation in cardamom yield was significant due to weather aberrations. When the annual rainfall was high in 1983 (2501 mm), the cardamom yield was low (33.7 kg/ha). It was true in 1967–68 (3036 mm), 1970–71 (2119 mm), 1977–78 (2211 mm), 1988–89 (2108mm) and 1993–94 (2226 mm) during which the yields were 34.2 kg/ha, 22.5 kg/ha, 27.5 kg/ha, 31.8 kg/ha and 59.13 kg/ha, respectively. In contrast, the cardamom yield was high when the rainfall was low. The above relationship was noticed in 1964–65 (1330.1 mm), 1975–76 (1587.5 mm), 1981–82 (1416 mm), 1985–86 (1475.1 mm), 1994–95 (1824 mm) and 1998–99 (1776.3 mm) during which the yields were high (56.1 kg/ha, 43.99 kg/ha, 54.98 kg/ha, 47 kg/ha, 100.1 kg/ha and 128.8 kg/ ha, respectively). However, it may not be true in all the cases as the yield was a complex one. Cardamom respond positively to summer rainfall and yields well whenever summer rainfall is noticed. In the humid tropics like Kerala, which experiences high rainfall with prolonged wetspell may not be conducive as it results in low crop productivity. The low productivity in high rainfall years could be attributed to waterlogging, lack of aeration, low evapotranspiration, low sunshine hours and less intake of nutrients. In addition, high rainfall may lead to incidence of crop pests and diseases. At the same time, the summer rainfall has positive influence on cardamom yield.

5.1.2 Impact of Water Deficit During Summer

The prolonged dryspells during the summer adversely affect some of the plantation crops like cardamom. The cardamom production was very low during 1983 (172.2 kg) when the water deficit from December to May was high (270 mm). Such trend was seen in 1987, 1989, 1992, 1994, 1998, 2004 and 2007 during which the water deficit was 211 mm, 140 mm, 185 mm, 111 mm, 200 mm, 134 and 125 mm, respectively. It revealed that there exists a strong negative relationship between water deficit during summer and cardamom yield. Higher the water deficit during the summer months (December-May), lesser is the cardamom yield and vice versa (Fig. 5.2). The yield variability was explained to an extent of 77 per cent based on soil moisture deficit during summer.

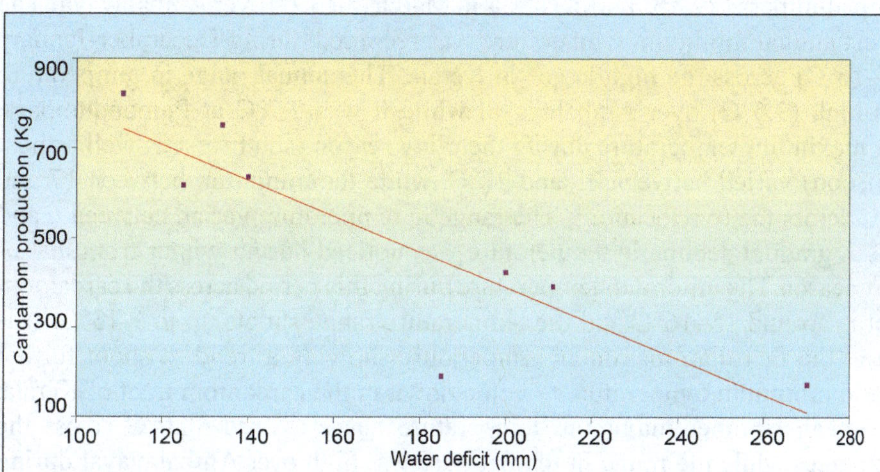

Fig. 5. 2: Water deficit versus yield relationship in cardamom

5.2 Thermal Regime

The mean annual maximum temperature across the cardamom tract is the lowest (24.1^{0}C) over Tamil Nadu and the highest (26.9-30.0^{0}C) over Karnataka while intermediary over Kerala (24.6-27.4^{0}C). The mean annual minimum temperature is high (16.9 -17.8^{0}C) over the western parts of the Western Ghats and low (15.4°C) over eastern parts (**Fig. 5.3**). The annual temperature range is high (10-12.4°C) over Karnataka and less over Kerala (7.1-9.5°C) and Tamil Nadu (8.7°C) across the Western Ghats. The maximum temperature varied between 23.7°C and 33.7°C over Karnataka while 21.3 – 26.2°C over Tamil Nadu. During southwest monsoon, Kerala recorded relatively low (22.6-25.1°C) due to heavy rains. In the case of minimum temperature, Tamil Nadu recorded the lowest (13.2-16.7°C) in all the seasons when compared to that of Kerala (15.8-19.0°C) and Karnataka (13.7-18.7°C). It was relatively high over Kerala during winter (15.9°C) and summer seasons (19.0°C) when compared to that of Karnataka and Tamil Nadu.

The mean annual maximum temperature varied between 25.2°C and 27.2°C, and the annual mean minimum between 17.3°C and 17.7°C, respectively at Pampadumpara and Ambalavayal. Irrespective of the seasons, variations of minimum temperature at both locations across the cardamom tract of Kerala are relatively less when compared to that of maximum temperature. However, the diurnal changes in minimum temperature during winter are predominant and in the case of maximum temperature, the diurnal changes are predominant during summer. April recorded the highest mean annual maximum temperature at Pampadumpara (29.3°C), while it was in March (30.3°C) over Ambalavayal.The lowest annual minimum temperature was recorded during December-January (15–16°C) across the highranges of Kerala. The annual range in temperature was high (9.5°C) over Ambalavayal while it was 7.7°C at Pamapdumpara. The maximum temperature during the rainy season (southwest as well as post monsoon) varied between 23 and 27°C, while the minimum between 17 and 18°C across the two locations. The range in temperature varied between 6 and 9°C. A gradual decline in temperature was noticed during winter from that of rainy season.The minimum temperature during this period across the cardamom tract is around 15–16°C and the temperature range shoots up to 9-13°C. It is due to the fact that, maximum temperature showing a trend to shoot further up and minimum temperature to come down in the cardamom tract of Kerala. The mean summer maximum temperature never crossed 30.0°C across the highranges while the range in temperature was high over Ambalavayal during summer (10.9°C) when compared to that of Pampadumpara (9.9°C). At the same time, the temperature range at Ambalavayal during summer was low (10.9°C) when compared to winter (12.1°C). Increase in maximum temperature and decline in minimum temperature resulted to increase in temperature range. It is a concern to the plantation crops grown across the highranges of Kerala in the context of projected climate change scenario.

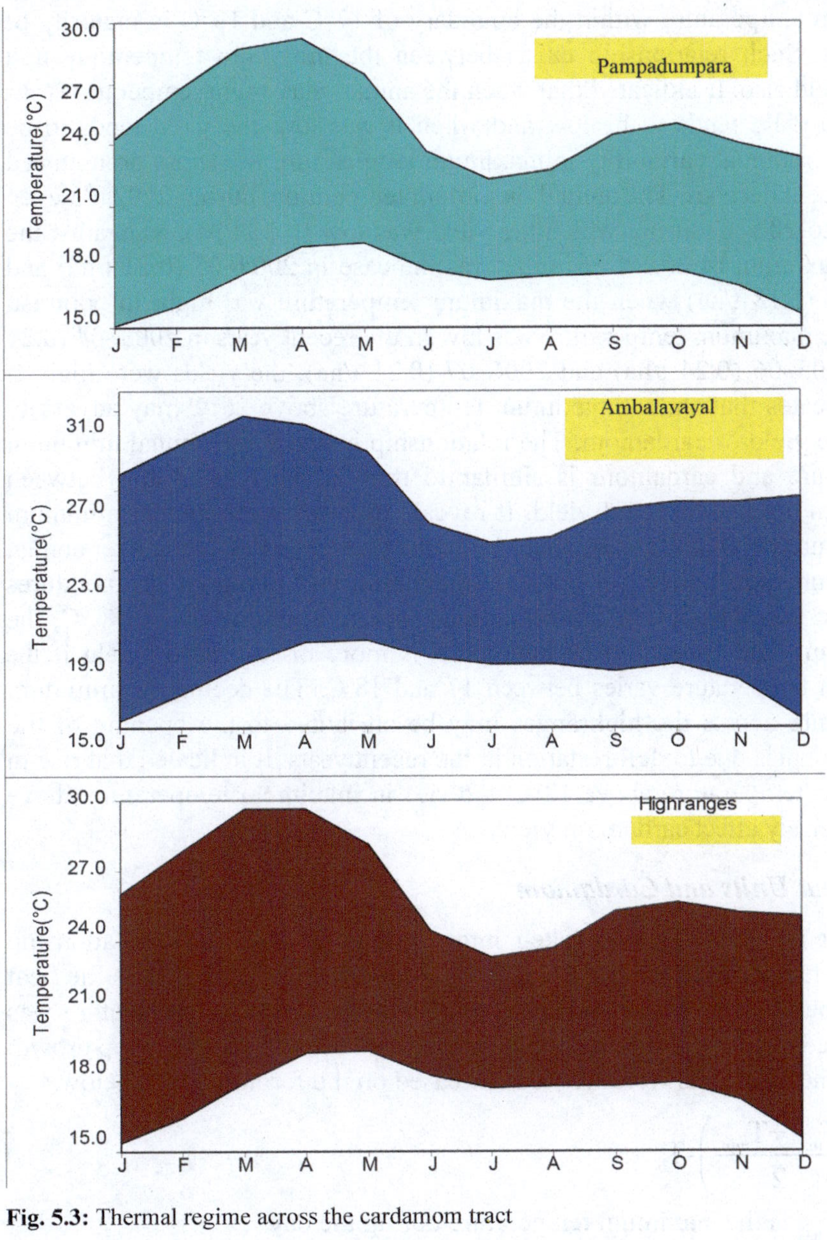

Fig. 5.3: Thermal regime across the cardamom tract

5.2.1 Temperature and Cardamom

In the case of cardamom, the upper limit of maximum temperature appears to be 26°C. No such trend was noticed in the case of minimum temperature. However, the increase in cardamom yield could be noticed with increase in minimum temperature within the boundary of 17°C and 18°C in majority of the years. Such relationship exists between the minimum temperature and coffee yield also. It indicated that when the annual maximum temperature was high, the yield tends to be low and when it was low, the yield tends to be high. Inter annual variability in maximum temperature was more pronounced in the recent decades. The annual maximum temperature during 2002–03 even shot up to 28.5°C during which the yield was low (0.21 t/ha), as against the mean maximum of 25.1°C. Similar was the case in 2004–05 (0.21 t/ha) and 2007–08 (0.18 t/ha) when the maximum temperature was high. In contrast, when the maximum temperature was low in the recent years in 2003–04 (0.21 t/ha), 2005–06 (0.24 t/ha) and 2006–07 (0.21 t/ha), the yields were high. It also indicates that annual maximum temperature above 26°C may adversely affect the yield of cardamom. The relationship between the annual minimum temperature and cardamom is similar to that of the relationship between maximum temperature and yield. It reveals that when the annual minimum temperature was high (greater than18°C), the yield was low. Wide inter annual temperatures are noticed in both the maximum and minimum temperatures since last two decades. If the minimum temperature goes above 18 °C, the cardamom yield appears to be low and it is more or less steady yield if the minimum temperature varies between 17 and 18°C. The decline in minimum temperature across the highranges may be attributed to the opening of the forest canopies due to deforestation in the recent years. It indicated that rise in minimum temperature above 18°C and rise in maximum temperature above 26°Cadversely affect cardamom yield.

5.2.2 Heat Units and Cardamom

The heat unit concept was taken into account to explain the relationship between temperature and crop yields of all the four crops tested as the heat units depend on the mean and base temperature. The base temperature was considered as 13°C in the case of plantation crops below which the crop growth ceases. The heat units were worked out based on the formula given below.

$$HU = \left(\frac{T_{max} + T_{min}}{2}\right) - T_b$$

Where T_{max} is the maximum temperature during the day

T_{min} is the minimum temperature of the day

T_b is the base temperature (13°C) below which no growth takes place.

Cardamom responded negatively with heat units. It indicated that whenever the heat units were more, the yields were low and vice versa as in the case of maximum and minimum temperatures. Of course, the relationship may not be true always as the yield is a complex one and depends on several other environmental factors. Interestingly, the yield in black pepper is low while coffee yield is high with the same amount of heat units. It indicated that the thermal environment is more suited to coffee when compared to that of black pepper and cardamom. It appears that the plantation crops viz., cardamom, tea, coffee and black pepper are thermo-sensitive and predominantly growing across the highranges to avoid high maxima and minima. The relationship between the Growing Degree Days (GDD) and yield also exhibited similar trend as in the case of annul minimum temperature. It was an inverse relationship. For example, GDD was high in 1985 (4442.3°C), 1998 (4511°C), 2003 (4562.7°C),2004 (4333.3°C), 2005 (4481°C), 2006 (4299°C), 2007 (4325°C) and 2008 (4402°C) and the yield recorded was poor. The Growing Degree Days (GDD) and cardamom yield followed the similar pattern to that of maximum and minimum temperature. When the GDD was high, the yield was low and vice versa. It was evident that when the GDD was high, the yield was always low. When the GDD was high (0.028 t/ha) in 1985, 1991 (0.064 t/ha),1998 (0.111 t/ha) and 2003 (0.21 t/ha), the yields were low. In contrast, during 2004 (0.215 t/ha), 2006 (0.240 t/ha) and 2007 (0.207 t/ha), the yields were high when the GDD was low.

5.3 Weather Abnormalities and Yield

The effect of climate change on plantation crops' production is not as evident as in the case of climate variability or rather, it may be having an indirect effect by way of climate variability, such as the occurrence of floods and droughts. The inter-annual variability in yield of plantation crops are affected by weather aberrations occurring from time to time such as abrupt changes in weather by way of increasing summer droughts due to no or scarce rainfall, high summer temperature, unusual summer showers and heavy and prolonged monsoon rains due to monsoon uncertainties. Since the economy of the State is to a greater extent determined by the plantation sector, decline in plantation crops' production is a concern. In the event of global warming and climate change, the occurrence of weather abnormalities and extreme weather events are likely to occur and reoccur in the event of normal water stress from December to May which is reflected on monthly moisture index which is negative during summer(**Fig.5.4**). It is obvious that the normal climatic conditions such as water deficit during summer and the present trend in global warming will deleteriously affect the plantation crops production in the State. Hence, it will be of immense use under the projected climate change scenario, if the impacts of abnormal weather events on plantation crops' productions are quantified based on the past data.

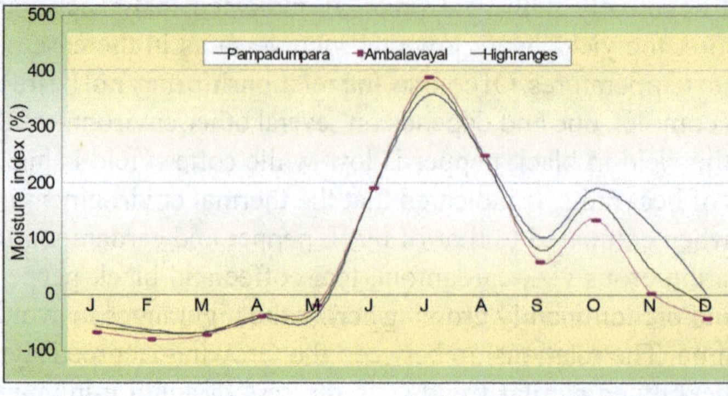

Fig. 5. 4: March of monthly moisture Index (%) across the highranges

5.4 Climate Variability and Climate Change

There was a decline in annual rainfall during the study period while increase in surface air temperature. The annual rainfall was significantly low in the decade 1981–90 and led to several summer droughts. As a result, increase in temperature was noticed in the above decade. The rate of increase in temperature was more evident since 1980s. Decline in rainfall and increase in temperature led to decline in moisture index and increase in aridity index, which is an indicator for dryspell. It led to more number of droughts (nine) in the present tri-decade of 1981–2009 when compared to that of 1951–80. The tri-decadal analysis also indicated that the decline in rainfall and moisture index, increase in temperature and aridity index were evident in the tri-decade of 1981–2009 (Figs. 5.5 and 5.6). All these factors reveal that the global warming and climate change are real in Kerala. Therefore, warming and dryness within the humid climate in the State of Kerala is a concern in the plantation sector of the State. The area under cardamom was in increasing trend from 1952 to 1990 and thereafter there was a decline in the recent decades. In the case of tea, the area decline was evident since 1950s and it was the lowest in the decade 1981–90. A marginal increase was noticed thereafter in the area under tea. However, the increase in area was evident in the case of coffee and black pepper during the study period while decline in area under tea and cardamom. Interestingly, the production of all the plantation crops was in tune with the area, it was so in the case of cardamom and tea. A steady increase was noticed in plantation crop production except in the case of black pepper during 1971–80. There was a sharp increase in capsule productivity in the case of cardamom and coffee beans in the recent decade of 2001–2008 while not so in the case of tea and black pepper. There was a sharp decline in productivity in the recent decades in the case of black pepper and tea. Overall, all the four plantation crops showed increase in productivity in the present tri-decade of 1981–09 when compared to that of 1951–80. Despite increase in temperature, decline in rainfall, increase aridity index and decline in moisture index, increase in plantation crop productivity was evident in the tri-decade of 1981–2009. It reveals that the plantation crops' productivity is vulnerable to climate variability rather than climate change. However, the crop simulation models in the case of tea and coffee indicated that the area is likely to decline if the temperature increase is by 2°C and some of the current areas under tea and coffee may disappear in the above situation of increase in temperature (IPCC, 2007). In fact, such trend was not noticed since last thirty years though the rate of increase in maximum temperature was 1.46°C during the present tri-decade. Instead of decrease in area, production and productivity due to increase in temperature, there was an increasing trend on long term basis.

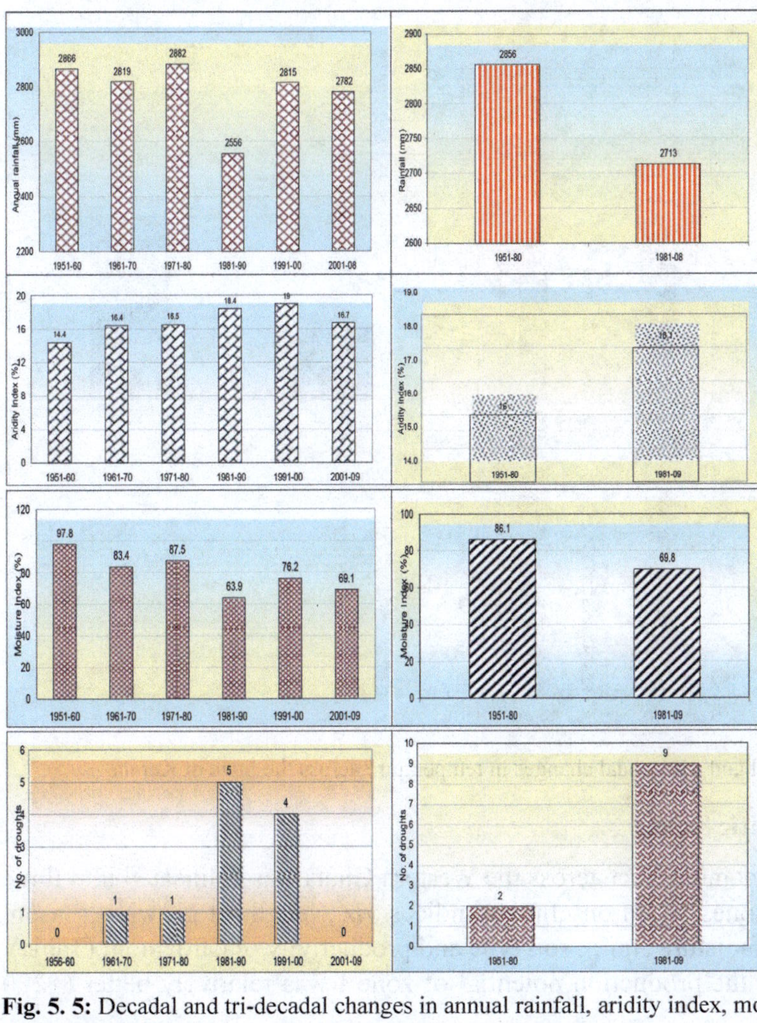

Fig. 5. 5: Decadal and tri-decadal changes in annual rainfall, aridity index, moisture index and droughts over Kerala

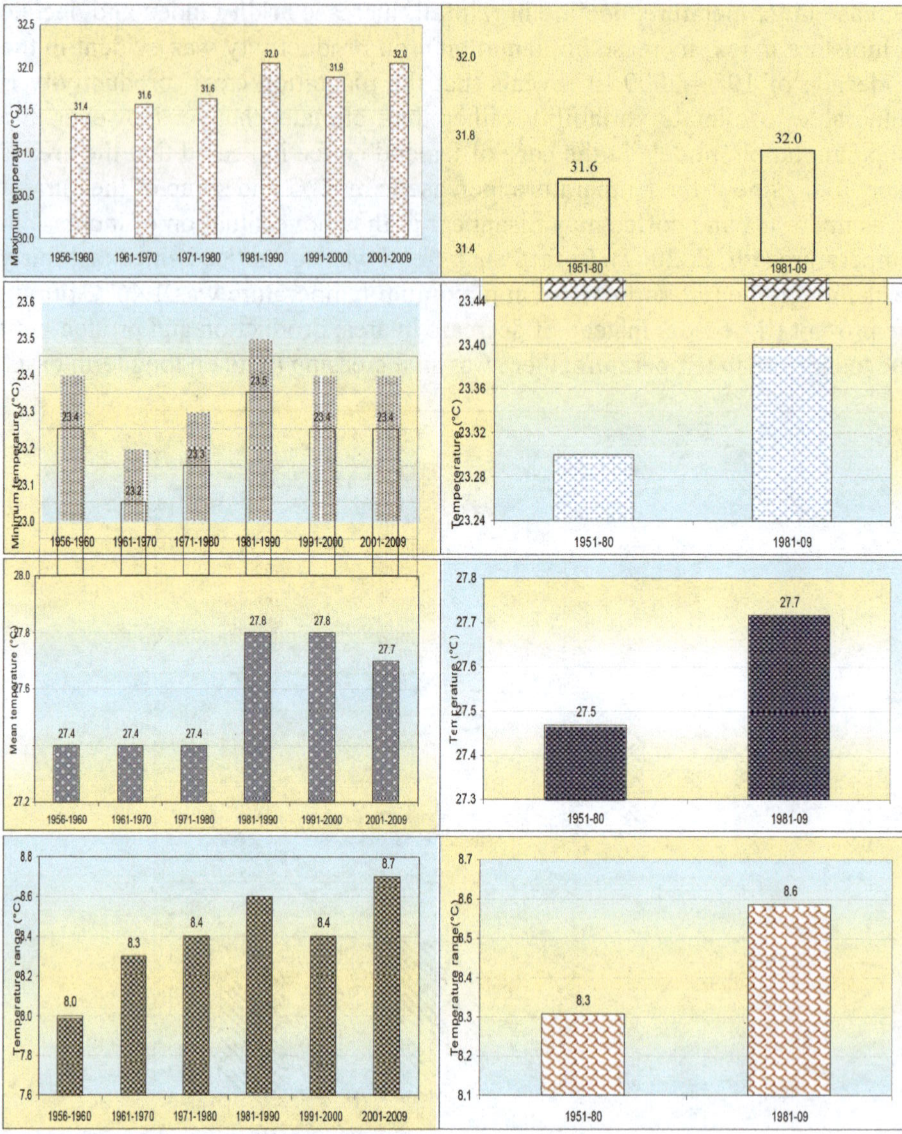

Fig. 5. 6: Decadal and tri-decadal changes in temperature across the State of Kerala

5.5 Agroclimatic Zones

The whole cardamom tract across the Western Ghats was delineated into three agroclimatic zones based on climate indices viz., length of growing season, available soil moisture (Ima), soil type and productivity of cardamom (Table). It is clear that the production potential of zone I was relatively better (>200 kg per ha) when compared to that Zone II and Zone III across the Western Ghats, where the length of growing season was more than 300 days with the annual Ima of more than 90 per cent. The annual temperature range was very low (14.1°C), and optimum across the Zone I. The production potential of small 5.5cardamom was low (100–150 Kg per ha) over Zone III (Karnataka) where the length of crop growing season was less than 250 days with annual Ima of 70–75 per cent. The annual temperature range was also high (19.0°C), which may be detrimental, in the case of cardamom. The Zone III falls under intermediary category (150–200 kg per ha), where the length of crop growing season was more than 250 days with annual Ima varied between 80 and 85 per cent. It is evident that the zone II and Zone III can be made better in terms of production of small cardamom by improving in situ soil moisture conservation practices as it experiences uni-model rainfall. It is understood that Tamil Nadu (Thandikudi) and southern parts of Kerala (Pampadumpara) are relatively better suited for cardamom in obtaining high yields as the areas are conducive in terms on soil and weather conditions under rain fed conditions, followed by northeast hill regions of Kerala **(Fig. 5.7)** whereas weather risk is involved in the case of Karnataka under the rain fed conditions.

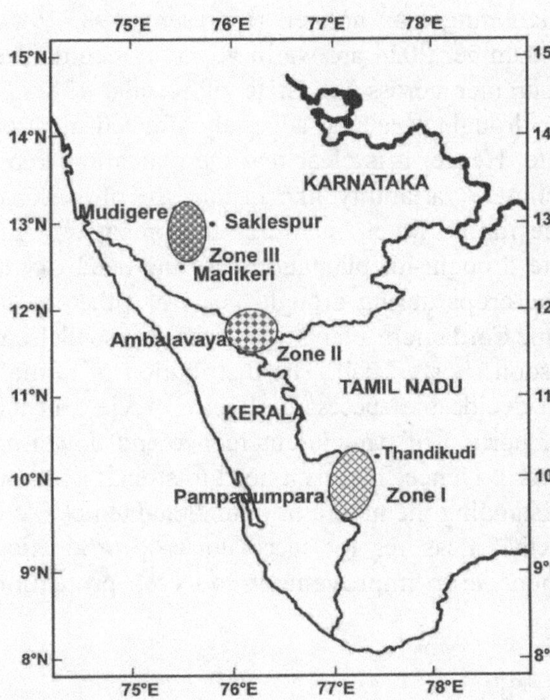

Fig. 5.7: Agroclimatic zones of small cardamom across the Westernghats

The study revealed that there was a shift in area among these four selected crops. At the cost of tea and cardamom area, the area under coffee and black pepper was increasing and it was phenomenal since last two decades. There may be several reasons for shift in area among the four plantation crops. From farmers' point of view, a crop mix with combination of coffee and black pepper may be beneficial under the current climate variability scenario as one crop like coffee suffers, another crop like black pepper gains in identical weather situation. It may not be so in the case of monocrop like tea and cardamom. It may be probably, one of the reasons why, shift was noticed in the cropped area of these four plantations. However, it needs further probe in depth from all the angles in view of the climate change scenario. The reason in high cardamom productivity could be due to the introduction of high yielding drought/ pest tolerant varieties by the faming community across the cardamom tract. Vagaries in monsoon, distribution of rainfall pattern, frequent occurrence of droughts, and increase in maximum and minimum temperatures, deforestation and shifting land use pattern have altered the natural habitat of cardamom to a greater extent. In the projected climate scenario, Cardamom Hill Reserves of Idukki district may pose a threat from decline in monsoon rainfall and rapid changes in thermal regimes. Such events are likely to accelerate the shift in cropping system across the highhranges as noticed at present. 1983,1987, 2014,2015,2016, 2023 and the summer 2024 are warm years on record. The year 1983 was the warmest summer across the State of Kerala. 1982-83 was an unprecedented summer drought year that adversely affected most of the plantation crops of the State. Hence, it is clear that the plantation crops' productivity is vulnerable to climate variability like in the case of seasonal crops. Methods to reutilize the runoff water, measures to combat drought and new schemes to rejuvenate drought-hit plantations are the need of the hour. This will help to save the crops during drought years and thereby to stabilize the yield of cardamom. Cardamom plantation faces dry spell from December to May if pre monsoon showers fail. The distribution of rainfall during December to April will decide the success or failure of yield in the case of cardamom since major portion of panicle emergence and flowering is noticed from December to April. Hence there is a need to strengthen crop weather relationships for understanding the nature of rainfall and temperature distributions to chalk out effective measures for successful crop production through better crop management, crop improvement and crop protection measures.

Summary

The two important agricultural ecocystems of Kerala viz., the Idukki and Wayanad are facing challenges due to warming and climate change. The weather extremes like floods, droughts, and heat waves are not uncommon in addition to the occurrence of landslides. The thermo-sensitive crop like cardamom is likely to be adversely affected in the ensuing decades due to increase in temperature and its range (difference between day and night temperatures). Wayanad is facing acute rainfall deficit in recent years. Groundwater exploitation is also a problem in Wayanad district. The studies conducted over the Wayanad District indicate that the occurrence of floods and droughts are not uncommon and it is likely that this District is prone to landslides as noticed the unprecedented landslide in July 2024 due to heavy rainfall within short duration in the early hours of 30 July, causing many people to die. Cardamom plantation faces dry spell from December to May if pre monsoon showers fail. The distribution of rainfall during December to April will decide the success or failure of yield in the case of cardamom since major portion of panicle emergence and flowering is noticed from December to April. Hence there is a need to strengthen crop weather relationships for understanding the nature of rainfall and temperature distributions to chalk out effective measures for successful crop production through better crop management, crop improvement and crop protection measures.

Suggested Readings

Cherian, A. 1977. Environmental ecology - an important factor in cardamom cultivation. Cardamom 9: 9–11.

Cocheme, J. and Franquin, P. 1967. An agrometeorological survey of a semiarid area in Africa south of Sahara. Technical Note No: 86. World Meteorological Organization, Geneva, p.136.

Critchfield, H.J. 1994. General Climatology. Fourth edition. Prentice-Hall of India Private Limited, New Delhi, p.453.

George, C.K. and Mathew, P.G. 1996. Drought management in small cardamom. Proceedings of the National Seminar on Drought Management in Plantation Crops, March 7–8, 1996. Rubber Research Institute of India, Kottayam, pp.15–18.

Jacob, J.A., Joseph, A. and Murugan, M. 1995. Cardamom and climate. Spice India 8: 12–14 &17.

Korikanthimath, V.S., Ankegowda, S.J., Yadukumar, N., Hegde, R. and Hosimani, M. M. 1998. Mixed cropping of arabica coffee and cardamom: Influence on microclimatic and physiological characteristics. J. Medicinal Aromatic Plant Sciences 20: 394–396.

Korikanthimath, V.S., Govardhan Rao., Hiremath, G.M 2000. Performance of Cardamom (Elittarai cardamomum Maton)under low elevation and assured irrigation conditions.

Korikanthimath, V.S., Hegde, R. and Kandiannan, K. 2002. Water management in spices-II. Spice India 15: 4–9

Korikanthimath, V.S. 1986. Influence of rainfall on the performance of cardamom under different level of management. In: Proc. Workshop on Impact of Drought on Plantation Crops. May 26–27, 1986. Central Plantation Crops Research Institute, Kasaragod, p.63.

Korikanthimath, V.S. 1987. Impact of drought on cardamom. Cardamom 19:5–12. Korikanthimath, V.S. 1991. Shade management for high productivity in cardamom (Elettaria cardamomum Maton). Spice India 4: 15–21.

Korikanthimath, V.S. and Padmini, K. 1999. Efficient management of evergreen forests for cultivation of cardamom (Elettaria cardamomum Maton). Spice India 12: 12–17.

Kurup, K.R. 1984. The forest - climate - cardamom relation. Cardamom 16: 9–11. Kuruvilla, K.M., Sudarshan, M.R., Mathusudanan, K.J., Priyadarshan, P.M.,

Mathew, P.G. 1989. Irrigation for cardamom: An insurance against drought. Spice India 2:7–9.

Murugan, M., Raj, N.M. and Joseph, C.R. 2000. Changes in climatic elements and their impact on production of cardamom (Elettaria cardamomum Maton) in the Cardamom Hills of Kerala, India. J. Spices and Aromatic Crops 9: 157–160.

Nair, C.K. 1988. The rainfall over the cardamom tract. Spice India 1:22–24.

Nanjan, K., Muthuswami, S., Thankaraj, T. and Sundararajan, V. 1981. Time of planting of cardamom at shevaroy hills. Cardamom 13: 13–15.

Natarajan, P. 1988. Irrigation studies on cardamom. Spice India 1: 25–27.

Parmeshwar, N.S. 1973. Floral biology of cardamom (Elettaria cardamomum Maton). Mysore J. Agric. Sci. 7: 205–213.

Radhakrishnan, V.V. and Naidu, R. 1992. Phenology of tiller and panicle in cardamom (Elettaria cardamomum Maton). J. Plantation Crops 20: 162–165. Manikandan, N 2004. Climate Variability and Small Cardamom production across the Western Ghats M.sc (Ag.) Thesis. Kerala Agricultural University, Thrissur, Kerala.

Ravindaran, K.C. and Kulandaivelu, G. 1998. Effect of light intensity on growth and yield on cardamom (Elettaria cardamomum Maton). J. Plantation Crops 26: 87–88.

Rao, B.V.R. and Rao, V.U.M. 2024. Agricultural Mete[or]ology. Jain Brothers, New Delhi. 154p

Rao, G.S. and Korikanthimath, V.S. 1983. The influence of rainfall on the yield of cardamom (Elettaria cardamomum Maton) in Coorg district. J. Plantation Crops 11: 68–69.

Rao, G.S.L.H.V.P. 2021. Agricultural Meteorology (5th Reprint). PHI Learning Private Limited, New Delhi, India. 352p.

Rao, G.S.L.H.V.P. and Krishnakumar, K.N. 2024. Impact of weather and climate on plantation crops in the Humid Tropics. Chapter 21 (417-432pp). Agrometeorological Applications for Climate Resilient Agriculture (Eds.): Chattopadhyay, N., Stefanski, R., Attiri, S.D. and Rathore, L.S. Springer International Publishing AG. Singapore.445p

Rao, G.S.L.H.V.P. and Manikandan, N. 2009. Climate Variability and small cardamom across the Western Ghats. Kerala Agricultural University. Thrissur, Kerala, India. 117p.

Rao, G.S.L.H.V.P. and Nair, R.R. 1988. Agrometeorology of plantation crops. Proceedings of National Seminar, held at RARS, Pilicode. Kerala Agricultural University. Thrissur, Kerala, India. 187p.

Sharangi, A.B., Rao, G.S.L.H.V.P., Das, S., Krishnamurthy, K.S.Upadhyay, T.K., Gopakumar, C.S. and Acharya. S.K. 2024. Brunt of Climate Change and Spice Crops: Scenario, Response, and Resilience. 755-812pp. Eds: Ravindran, P.V., Sivaraman, K.,Devasahayam,S., Nirmal Babu, K. Handbook of Spices in India: 75 Years of Research and Development. Springer Nature Singapore Pte. Ltd.535p

Sulikeri, G.S. 1987. Effect of light intensity and soil moisture levels on growth and yield of cardamom Elettaria cardamomum (L.). Ph.D thesis, University of Agricultural Sciences, Bangalore, p.360.

Questions

I. Elucidate the Following

1. Agroclimatological zones of cardamom across the Western Ghats
2. Thermal Regime of cardamom tract in Kerala
3. Rainfall Regime of cardamom tract in Kerala

II. Write short notes on

1. Soil moisture deficit versus cardamom productivity
2. Temperature and cardamom yields
3. Rainfall and cardamom yields
4. Heat Units and cardamom yields
5. Uni-and-bi-model rainfall in Kerala
6. Weather extremes and cardamom yields

III. Fill in the Blanks

1. ---------------District produces more cardamom in Kerala
2. ---------- rainfall is favourable to cardamom
3. ---------------- is known as the Cardamom Hill Reserve
4. ---------- and ---------is the range of rainfall conducive for cardamom
5. In place of cardamom, ---------,---------- and ----------- crops are cultivated at many locations
6. Bi-model rainfall having a peak in ----------- and in another peak in ------------
7. Rainfall trend across the highranges is ---------------
8. Variation in minimum temperature is ---------- when compared to maximum temperature
9. ----------- in temperature range is a concern in cardamom cultivation
10. Temperature range is --------------- at Ambalavayal when compared to Pampadumpara
11. Moisture index is negative in ---------, ----------, and ------------ months
12. The cardamom tract across the western ghats is delineated into ------- agroclimatic zones
13. Weather risk is more in ------------- zone for cardamom
14. Cardamom yield is ---------- when the annual rainfall is high

15. If the GDD is low cardamom yield is -------------
16. R^2 is ----------- between summer water deficit and cardamom yield
17. At the cost of tea and cardamom --------- and -------- cropped area is increasing
18. ---------- and ------------agroecosystems are vulnerable in Kerala to weather extremes
19. The area under cardamom over Kerala is ------------
20. Overall temperature range at Ambalavayal is -----------

IV. Expand

IPCC UNFCCC CC CHR HU

V. State True or False

Cardamom is a crop of Kerala across the Western Ghats

1. Cardamom productivity is not taken into account while delineating agroclimatic zones
2. The temperature range is optimum for cardamom in all the zones
3. Cardamom productivity is high in agroclimatic zone II
4. Cardamom is known as the king of spices
5. The mean annual number of rainy days over the cardamom tract 135
6. Rainfall pattern over Pambadumpara and Ambalavayal are similar
7. June, July, August and September are rainiest months over highranges
8. Minimum temperature is low over TN when compared to Kerala and Karnataka
9. 2004 was the worst summer drought over the cardamom tract
10. The base temperature in cardamom is considered as 10^0C while computing GDD
11. The annual maximum temperature above 30^0C may adversely affect cardamom yields
12. The length of crop growing season is more than 250 days in zone III
13. The monthly moisture index follows the bi-model rainfall across the highranges
14. Pampadumpara receives relatively more rainfall compared to Ambalavayal

VI. Match the Following

A	Dry Subhumid
B	Moist Subhumid
C_1	Semiarid
C_2	Arid
D	Perhumid
E	Humid
	Extreme Arid

6

Coffee

Arabica coffee is reported to have originated in the tropical rain forests in Ethiopian highlands while Robusta is a native of the humid tropical lowlands of Central Africa, probably, Uganda. The major coffee producing countries across the World are distributed between 23°N and 23°S of the equator. Beyond these latitudinal limits, temperature becomes a major limiting factor as coffee plants are sensitive to extreme temperature regimes. Thus within the latitudes, altitude becomes a major factor. In areas, nearer to the equator, coffee plants require higher altitudes when compared to farther north and south (6.1 plate). The world's largest coffee producer is Brazil with around 30% of total world output of coffee. Columbia comes in second with around two thirds of Brazil's production. India is the sixth largest (after Brazil, Columbia, Vietnam, Indonesia and Ethiopia) producer of coffee. Coffee is reportedly introduced into the Western Ghats regions of South India around 1600 AD. However, commercial cultivation of coffee was opened in South India with the British enterprise and investment during late 1820s. Coffee occupies an important position among the plantation crops in India. It is cultivated mainly in the traditional States of Karnataka, Kerala, Tamil Nadu and to some extent in non-traditional areas viz., Andhra Pradesh and Orissa as well as in North Eastern States viz., Assam, Manipur, Meghalaya, Mizoram, Tripura, Nagaland and Arunachal Pradesh. Coffee in India is unique. Almost all coffee plantation are grown under the shade of forest/planted trees, where as it is grown under open conditions in all the major producing countries viz., Brazil, Vietnam and Columbia. In Kenya, Northern Tanzania and Columbia the presence of two rainy and dry seasons result in two cropping cycles in a year. Still, in some Latin American Countries, there are no such period of moisture stress and the coffee flowers sporadically throughout the year resulting in more than two rounds of flowering. While cultivation under shade provides distinct advantageous, it also inhibits overall productivity compared to coffee grown in open conditions. Owing to the presence of shade trees in coffee plantations, mechanization of field operations becomes difficult and hence coffee cultivation in India still remains a traditional agriculture activity. Shade trees not only provide suitable microclimate to the coffee trees, but also prevent soil erosion, improve soil fertility and act as a powerful tool for managing the major pests and diseases. The coffee plantations under shade play a major role in preserving the original eco-system of Western and Eastern Ghats of the Country.

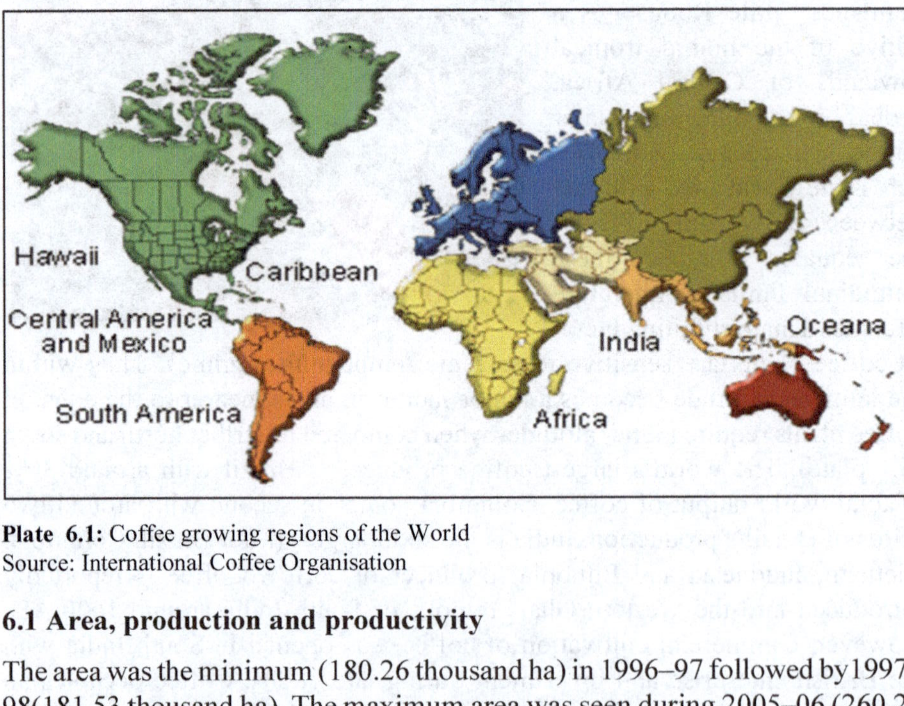

Plate 6.1: Coffee growing regions of the World
Source: International Coffee Organisation

6.1 Area, production and productivity

The area was the minimum (180.26 thousand ha) in 1996–97 followed by1997–98(181.53 thousand ha). The maximum area was seen during 2005–06 (260.23 thousand ha) followed by 2004–05(256.29 thousand ha). Area shows a gradual increase since 1996–97 but showed a decline during 2006–07 (245.97 thousand ha). At the same time, production data shows wide inter annual variability. The highest production was observed during 2005–06 (92.94 thousand tonnes) followed by 1998–99(79.84 thousand tonnes). It was the lowest in 1999–2000 (50.13 thousand tonnes). Though the maximum production was observed in 2005–06, it fell down to 69 thousand tonnes in2006–07. The highest productivity of 417 Kg/has was observed in 1998–99 followed by 357 Kg/ ha in 2005–06. Trend in area and production of coffee in Kerala are towards increasing. During 1950s, the area under coffee cultivation in the State was only 12600 ha. Since then, coffee area exhibited a gradual increase over the years and it was about 85000 ha in 2007–08. Area under coffee registered substantial increase during the last two decades with an annual growth rate of over 2 per cent. The increase in production recorded during the period was much higher. Coffee provides opportunities for livelihood to nearly one lakh families including agricultural labourers. In Kerala, coffee is also one of the small holder plantation crops with nearly 76,000 holdings coming under the category with an average size of 2 ha. It is noted that since1990s, area under coffee has stagnated and it was revolving around 85000 ha. Production also gradually increased from 5110 during 1952–53 tonnes to 5925 tonnes in 2009–10 (Fig. 6.1a). Since 1990s, the highest coffee production of7060 tonnes was recorded in 2000–01 followed by 2001–02 (6670 tonnes). Coffee production in the State in recent years was in declining trend. Decadal analysis of area and production of coffee over the State shows that the area under coffee has not increased significantly (82–83 thousand tonnes) during the decades 1991–2000 and 2001–2008. At the same time, production has shown an increasing trend during the above period (Fig.6.1b). In tune to the increase in area and production, productivity of coffee in the State was increasing (Fig. 6.1c). Kerala is the second largest producer of coffee in India. It produces 21 per cent of the total coffee out in the country. In Kerala, coffee is grown in an area of 84,976 ha of which 67,366 ha (79.4 per cent) and 84 per cent of production is from Wayanad district. It is interesting to note that Arabica coffee is very marginal in Wayanad district, the Coffee bowl of Kerala (Table 6. 1). Travancore and Nelliyampathy hills produce very small quantity of coffee. In the case of coffee, blossom and backing showers are very important. As seen, if any one them fails, the coffee yield is adversely affected. Of course, there was a negative relationship between high annual rainfall and coffee yields as seen in the case of tea. It was true in the case of black pepper also. Interestingly, both the crops coffee and black pepper respond differently to summer rainfall because it is good in the case of coffee as it requires blossom and backing showers between February/March and March/April while it is not so in the case of black pepper as it needs dry spell. In fact, it is not the dry spell needed to black pepper but continuous rains to complete its floral biology and berry formation and its development.

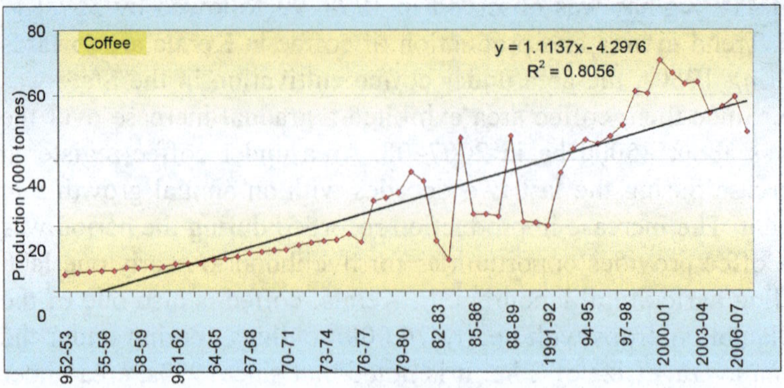

Fig. 6.1a: Trend in production of coffee over Kerala

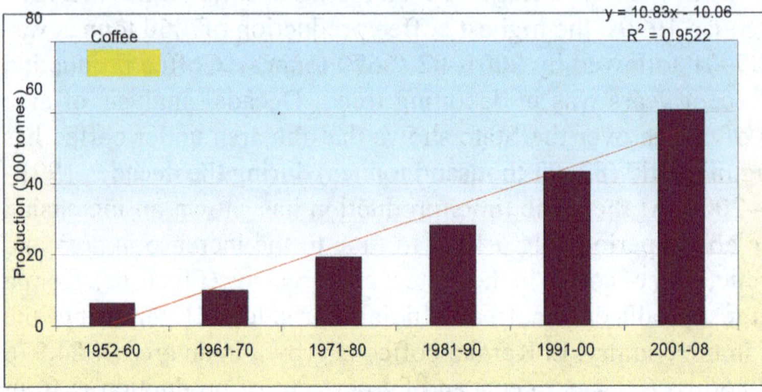

Fig. 6.1b: Decadal trend in coffee production over Kerala

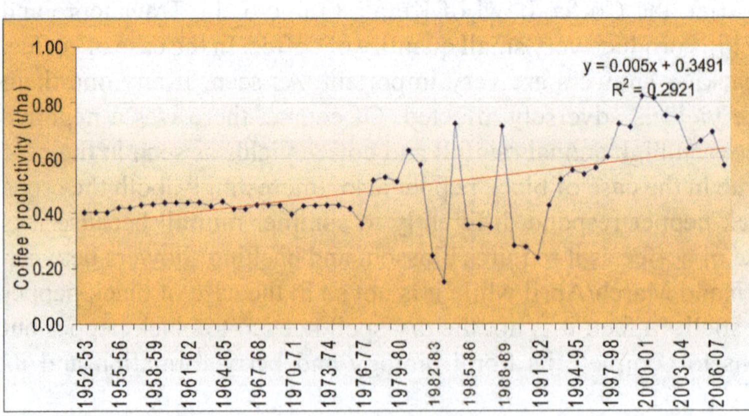

Fig. 6.1c: Trend in coffee productivity over Kerala

Such environment lacks in the case of pre- monsoon showers during summer because the dry spell which follows after pre- monsoon showers adversely affects the floral biology and yield of black pepper is likely to be low under such a situation. However, both the crops responded negatively to high annual rainfall. The low productivity in high rainfall years could be attributed to waterlogging, lack of aeration, low evapotranspiration, low sunshine hours and less intake of nutrients. In addition, high rainfall may lead to incidence of crop pests and diseases.

Table 6.1: Area and production coffee in Kerala during 2009-10

	Area(ha)			**Production(MT)**		
	Arabica	**Robusta**	**Total**	**Arabica**	**Robusta**	**Total**
Wayand	2 (0.0)	67364 (83.1)	67366 (79.4)	- (0)	49950 (86.3)	49950 (84.3)
Travancore	1909 (51.4)	10871 (13.4)	12780 (15.1)	775 (56.4)	6650 (11.5)	7425 (12.5)
Nelliyampathis	1800 (48.5)	2850 (3.5)	4650 (5.5)	600 (43.6)	1275 (2.2)	1875 (3.2)
Total	3711	81085	84796	1375	57875	59250

( )- figures in parenthesis indicate the percentage contribution to the total

6.2 Weather and Coffee

The traditionally grown two coffee varieties respond differently to environmental factors. Arabica is relatively more sensitive to weather fluctuations when compared to that of Robusta. Arabica is relatively tastier when compared to that of Robusta where black coffee is preferred. The blossom and backing showers during summer are critical for coffee flowering. Across the Western Ghats, blossom showers are required in case of Arabica during February/March while March/April in case of Robusta. Rains on the opening day of blossom lead to partial to complete failure of blossoms in Arabica while partial in case of Robusta. The crop set is poor in case of Arabica in the absence of backing showers during March/April while April/May in case of Robusta. In both the varieties, one or two wet spells during October and November may be good for next coffee crop. Partial to complete failure of the crop is seen in both the varieties in case of severe soil moisture during summer or excessive wetness and waterlogging during the rainy season. Both the varieties are sensitive to strong winds and hailstorms. To some extent, Robusta can withstand against moderate hailstorms (Table 6.2).

Table 6.2: Effect of weather factors on Arabica and Robusta

Weather factors	**Arabica**	**Robusta**
Lack of blossom showers	Absence of rains in February/ March	Absence of rains in March/April
Rain on opening day of blossom	Partial to complete failure	Partial failure
Hailstorm	Injury to the floral and vegetation phase	Partial failure
Backing showers	Absent in March/April - poor crop set	Absent in April/May – poor crop set
Showers in October/November	Good for next crop	Good for next crop
Severe western exposures (with lowering of sub-soil moisture)	Partial to complete failure	Partial to complete failure
Excessive wetness and waterlogging	Partial to complete failure	Partial to complete failure
Wind	Sensitive to wind disturbance	sensitive to wind disturbance

Coffee growing areas receive a rainfall of 1500 mm to 2500 mm per annum. Rainfall should be well distributed for a period of nine months and a dryspell of three months coinciding with the period of harvest. A period of moisture stress may be beneficial to coffee for bringing about a concentrated flowering leading to a clearly defined harvest season. Owing to wide peculiar agroclimatic conditions prevailing in India such as wide range of fluctuations in day temperatures and monsoon rainfall pattern with a drought period of three to five months, there is a great compulsion to grow coffee under shade so as to maintain an ideal microclimate for coffee bushes. Arabica requires cool, equable climate where as Robusta coffee prefers warm humid conditions. Arabica performs at altitudes above 1000m above msl and Robusta from 500-1000 m above msl. Rainfall distribution is very important for the healthy growth and development of coffee plant. A fairly well distributed blossom showers (the first summer showers received in the coffee areas are popularly known as 'blossom showers' as they trigger uniform flowering of coffee bushes throughout the area which receives these showers) and backing showers (the follow up showers that are received after completion of flowering and fruit set, are important for supporting the retention of the newly set coffee fruits and that encourage fresh growth of plants) is desirable for Arabica and Robusta coffee for obtaining a better harvest. If these factors are fulfilled, other factors like temperature, light intensity and relative humidity could be moderated by shade trees. Geographical parameters like elevation, aspect and slope as well as environmental factors like rainfall, temperature and relative humidity can influence the economic cultivation of coffee. High temperature combined with poor sub soil moisture can be a limiting factor in south Indian conditions while, cold winter temperature can be a limiting factor in non traditional area and north eastern region. Unusual climatic factors like hail storms, rain on the day of flowering would adversely affect the production of both Arabica and Robusta Coffee. Eastern winds during December - February months can cause injury to coffee plant. Shade trees protect the crop from the wind damage. Wind breaks can be grown to prevent wind damage to coffee plant. Climate variability has always been the main factor for the fluctuations of coffee and black pepper yields and climate change may result to shifts in coffee and black pepper growing areas in future. Temperature and rainfall are the major driving factors which determine the production. The major varieties cultivated around the world viz., Arabica and Robusta differ in their agro- meteorological and ecological requirements in the case of coffee. Arabica coffee since evolved in the cool, shady environment survive a temperature range between 15°C to 24°C. Much higher temperatures tend to impact negatively on yield and quality. Rainfall requirements vary between 1500-2000mm per annum, though now-a-days the Arabica variety is grown with the help of irrigation in areas with insufficient rainfall. Robusta coffee was originally evolved in the lowland Equatorial Africa. It grows in areas with abundant rainfall of around 2000mm per annum and altitudes up to 800m. Rainfall must be well distributed throughout the year because Robusta has a shallow root system. Optimum temperature ranges from 22–26°C and the species is less tolerant to very high or very low temperature than is Arabica. It is reported that as temperature increases, coffee ripens more quickly leading to a fall in inherent quality. Extreme high temperatures during the sensitive period of the life cycle of the crop coupled with low rainfall during flowering and fruiting period will adversely affects the yield. Increase in temperatures will not only favour the spread of certain pests and diseases but also results in spreading the disease to new areas/regions where the pest-disease was absent previously. Studies suggest that incidence of inset-pest and disease will increase depending up on the future temperature fluctuations. Increased evaporation as a result depleted soil moisture content in areas which do not require irrigation, may have to be irrigated in future. In India also, under the projected climate change scenario, rainfall is dramatically changing the ecosystem and crop growing environments. Such scenarios may aggravate the situation and cultivators may have to depend upon deep bore wells due to the depletion of water table in the event of prolonged rainfall deficiency. In the Wayanad District of Kerala, the major coffee producing centre in the State, rainfall decline is obvious.

6.2.1 Rainfall

Rainfall seems to the most important factors among various weather elements influencing flowering and yield of coffee. Kannan et al., (1987) observed that blossom showers of 20–40 mm during February-March and backing showers of 50–75 mm during March - April are considered to be critical for flowering and fruit set in Robusta coffee. Failure of any of these rains will lead to poor coffee yields. Ideal time for receipt of blossom showers for Robusta coffee is mid February to mid March. Delay in blossom showers beyond March would affect the fruit set. A blossom rainfall of about 20–40 mm received either in one day or in 2–3 consecutive days during ideal time (February-March) is adequate for inducing normal blossom in Robusta coffee. Inadequate blossom showers result in poor growth of flower buds leading to a condition called'pinking' denoted by the pinking colour of the failed flower buds. Backing showers should be received within one month after the receipt of blossom showers. 70–75 mm of backing rains is desirable for normal retention of newly set fruits. Any delay or absence of backing showers also would result in drying up of newly set fruits and thereby affects the final yield. It indicated that whenever the annual rainfall was high, coffee recorded poor yield as in the case of tea and vice versa. For example, annual rainfall was high in 1985-86, yield was low (0.36 t/ha), followed by 1987-88 (0.35 t/ha), 1989-90 (0.28 t/ha),1991-92 (0.24 t/ha), 1994-95 (0.56 t/ha) and 1996-97(0.57 t/ha). In contrast, when the annual rainfall was low, productivity was comparatively high. For example, when the rainfall was low during 2000-01, the yield was high (0.83 t/ ha). Similar was the case in 1997-98, 1998-99, 2005-06 and 2006-07 during which the yield was high. In majority of the years (86 per cent) during the study period, the inverse relationship with annual rainfall was evident. It is true that annual rainfall do not have significant role in deciding crop performance, though it responds negatively. It is the blossom and backing showers which are critical to coffee yield. As in the case of many plantation crops, high annual rainfall is not conducive to coffee.

6.2.2 Blossom and Backing Showers on Coffee Production

It was observed that during 1976 and 1983, there were no blossom showers either in February or March, which resulted to poor flowering in coffee growing tracts of Wayanad. The yield recorded during the ensuing crop season was 172kg/ha in 1976 and 315 kg/ha in 1983 respectively (**Table 6.3**). Similarly, comparatively poor yield (598 kg/ha) was noticed during 2006 when no significant rains were seen received in February. At the same time, an amount of 85.4 mm of rains were received in March. During 2000 also, no significant blossom rains were received during February and March which also led to comparatively poor yield (656 kg/ha). The yield was comparatively higher in1977, 1981 and 1984 during which the yield recorded was 684 kg/ha, 753 kg/ha and 811 kg/ha, respectively. The years 2001 (802 kg/ha), 2003 (739 kg/ha),2007 (705 kg/ha) and 2008 (703 kg/ha) also recorded comparatively higher yields when the blossom showers were received. During all the above years, blossom showers were seen received during the critical periods (Feb-Mar), which resulted to better coffee yield (Fig. 6.3).

Table 6.3: Effect of blossom and backing showers on coffee production

Year	Rainfall (mm) during March	Rainfall (mm) during April	Rainfall (mm) during March - April	Yield of dry pepper (Kg/ ha)	Yield of coffee beans(Kg/ha)
1979	4.4	57.7	62.1	1011	377
1980	65.2	208.5	273.7	1480	368
1981	63.1	78.6	141.7	956	753
1982	40.0	78.6	118.6	3599	585
1983	1.5	21.5	23.0	2375	315
1984	114.5	130.0	244.5	185	811
1985	75.8	93.0	168.8	2966	514
2000	1.4	110.8	112.2	387	656
2001	12.0	206.2	218.2	398	802
2002	29.0	93.4	122.4	326	639
2003	104.2	131.4	235.6	285	739
2004	35.8	132.4	168.2	324	690
2005	54.2	147	201.2	334	660
2006	85.4	97	182.4	318	598
2007	5.2	94.8	100.0	357	705
2008	178.8	75.2	254.0	198	703

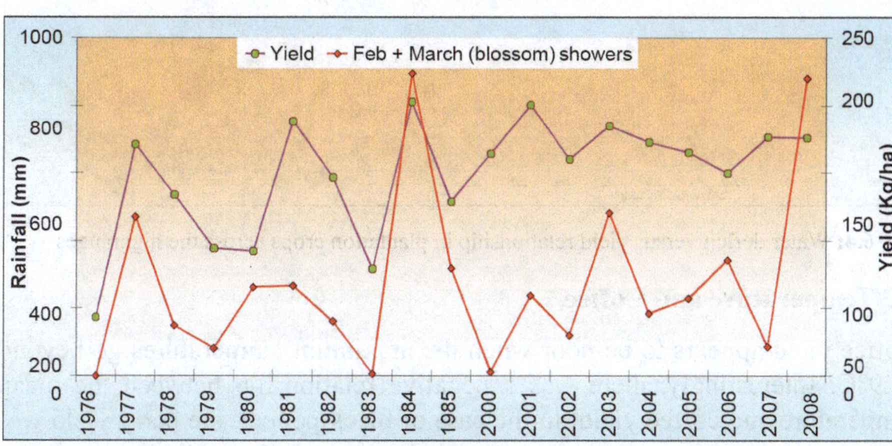

Fig. 6.3: Effect of blossom showers on coffee yield

On critical examination of the rainfall data as well as yield, it was evident that better yield could be obtained even if the rainfall during February was scanty but make up with rainfall in March, as observed in 1981 along with backing showers. It is also true that if blossom showers are enough and no backing showers during April/May, the yield is likely to be less and vice versa

6.2.3 Water Deficit During Summer

The dry spells during summer between blossom and backing showers may not adversely affect the crop like coffee. Of course, the effect of soil moisture status during summer could be felt to some extent in coffee. In the case of coffee, it only needs timely blossom and backing showers in February/March and March/ April as noticed. Coffee thrives well and produces better under optimum and timely blossom and backing showers. That is the reason, why, the effect of water stress on coffee is not felt much unlike other plantation crops **(Fig. 6. 4).**

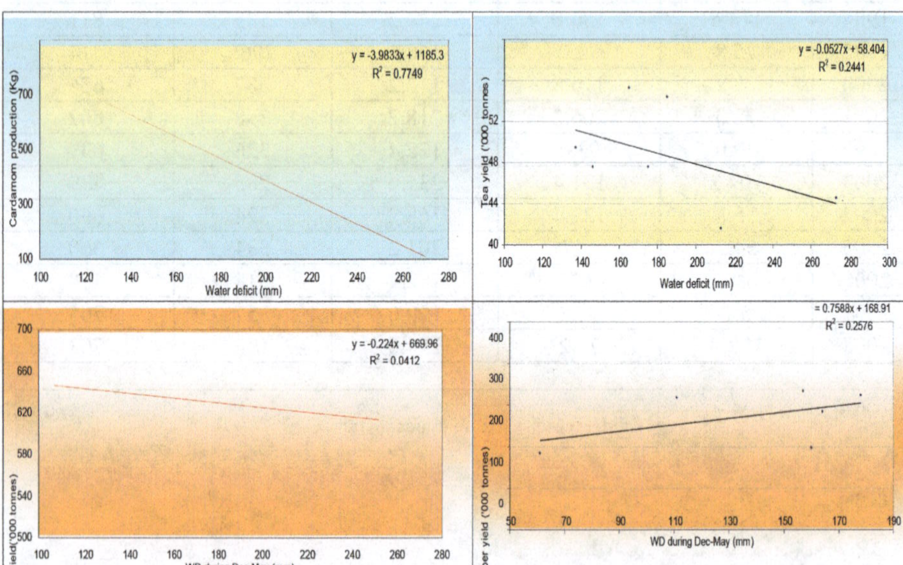

Fig. 6.4: Water deficit versus yield relationship in plantation crops across the highranges

6.3 Temperature and Coffee

Coffee yield appears to be poor when the maximum temperatures go beyond 26.9°C. Interestingly, there was a negative relationship between the mean temperature and coffee yield. In the case of black pepper, the berry yield was low when the maximum temperature goes beyond 27.4°C while it can tolerate an upper limit of minimum temperature as 18°C. The growing degree days (heat unit) versus crop yield appears to be in the same fashion as in the case of temperature since it is a function of temperature. Being thermo-sensitive crops, any increase in temperature beyond the upper limits will adversely affect the yield. The upper limit of maximum temperature for all the crops tested vary between 26 and 27°C except in the case of black pepper which can tolerate up to 27.4°C. The boundary of minimum temperature varied between 17°C and 18°C for all the four crops tested. It revealed that maximum temperature and coffee yield has an inverse relationship in majority of the years. When the maximum temperature was high, the coffee yield was low. It was evident in 1985–86 (0.36 t/ha), 1990–91 (0.28 t/ha), 1991–92 (0.24 t/ha) and 1992–93 (0.43 t/ha). Similar was the case in 1986–87, 1987–88,1989–90, 1994–95, 1995–96, 1996–97, 1997–98 and 1998–99 during which the coffee yield was low. In contrast, comparatively better yields were obtained during 2000–01(0.83 t/ha), 2001–02 (0.83 t/ha) and 2006–07 (0.71 t/ha), when the maximum temperature was comparatively low. It was also revealed that whenever maximum temperature above 26.9°C, coffee yields were low in majority of the years under the study period. On an average, majority of the years followed the inverse relationship with maximum temperature. It was also evident that whenever the maximum temperature was above 26.9 °C, the coffee was low. In contrast, high yields were obtained only in 2000–01 (0.83 t/ha) and2001–02 (0.83 t/ha), when the maximum temperature was 27.3°C and 27.2°C, respectively. A sharp decline in maximum temperature was evident from2003–04 onwards. It reveals that maximum temperature above 26.9°C may adversely affect the coffee yield. Wide inter-annual variability was observed in minimum temperature. It revealed that there exists an inverse relationship between minimum temperature and productivity. The productivity was low when the minimum temperature was high during 1989–90 (0.28 t/ha), 1990–91 (0.28 t/ha), 1991–92 (0.24 t/ha) 1992–93 (0.43 t/ha) and 1993–94 (0.53 t/ha). Similar was the case during 1997–98 (0.61 t/ha) and 1998–99 (0.73 t/ha). In contrast, when the minimum temperature was low, coffee yields were high. It was evident since 2001–02 onwards. The yield was high when the minimum temperature was high. It is interesting to note that the minimum temperature and yield followed similar trend from 2001–02 onwards. It was also observed that only four years recorded minimum temperature above 18°C, during which coffee yield was low. It indicates that minimum temperature above 18°C may adversely affect the coffee yield. Minimum temperature range between 17 and18°C may be conducive for coffee plant as is evident from the data.

6.3.1 Heat Units and Coffee

As in the case of maximum and minimum temperature, Growing Degree Days (GDD) or heat unit showed an inverse relationship with coffee yield. Whenever the GDD was high, coffee yield was low and vice versa. For example, when GDD was high, low yields were obtained. It was evident during 1985–86 (0.36 t/ha), followed by 1986–87 (0.36 t/ha), 1987–88 (0.35 t/ha), 1989–90 (0.28 t/ha), 1990–91 (0.28 t/ha), 1991–92 (0.24 t/ha), 1992–93 (0.43 t/ha),1997–98 (0.61 t/ha) and 1998–99 (0.73 t/ha) as against the normal of 0.57 t/ha. Comparatively high yields were obtained during 1984–85 (0.73 t/ha),1988–89 (0.73 t/ha), 1994–95 (0.56 t/ha) and 1995–96 (0.55 t/ha as against the normal of 0.57 t/ha when the GDD was low. It is interesting to note that from 1999 –2000 onwards, GDD and yield followed similar pattern as noticed in the case of minimum temperature.

6.4 Weather Abnormalities and Yields

The effect of climate change on plantation crops' production is not as evident as in the case of climate variability or rather, it may be having an indirect effect by way of climate variability, such as the occurrence of floods and droughts. The inter-annual variability in yield of plantation crops are affected by weather aberrations occurring from time to time such as abrupt changes in weather by way of increasing summer droughts due to no or scarce rainfall, high summer temperature, unusual summer showers and heavy and prolonged monsoon rains due to monsoon uncertainties. Since the economy of the State is to a greater extent determined by the plantation sector, decline in plantation crops' production is a concern. In the event of global warming and climate change, the occurrence of weather abnormalities and extreme weather events are likely to occur and reoccur. Such events will deleteriously affect the plantation crops production in the State. Hence, it will be of immense use under the projected climate change scenario, if the impacts of abnormal weather events on plantation crops' productions are quantified based on the past data.

6.5 Coffee-not An Alternate Bearer

The yield data of coffee was examined to understand whether coffee exhibits alternate bearing tendency. It revealed that coffee did not exhibit alternate bearing tendency and it is a regular yielder like tea (Table 6.4). The inter-annual variability in yield pattern could therefore be attributed to fluctuations in weather. For example, the yield obtained in years 1976, 1982 and 1983 were poor. The reason could be attributed to receipt of poor rainfall during the flowering and fruit setting phase which resulted to poor coffee yields from the coffee bowl of Kerala, the Wayand district. It may be noted that 1983–83 was a disastrous drought year over Kerala. Poor rainfall during the critical phases of the coffee crop resulted to poor yield. 2002, though was an all India drought year, during which coffee yield were not much affected since the distribution of rains were satisfactory during the critical phase though rains during February was absent. Yield during 2007 was also poor. It was a flood year as far as Kerala is concerned.

Table 6.4 Coffee yields in alternate years*

Even years	Yield('000 tonnes)	Odd years	Yield('000 tonnes)
1952	5.11	1953	5.67
1954	5.97	1955	6.25
1956	6.22	1957	7.21
1958	7.07	1959	7.41
1960	7.41	1961	8.28
1962	8.48	1963	8.57
1964	9.68	1965	9.88
1966	10.51	1967	11.46
1968	11.98	1969	12.47
1970	15.57	1971	14.11
1972	14.92	1973	15.46
1974	15.78	1975	17.53
1976	15.03	1977	27.65
1978	28.62	1979	30.18
1980	36.48	1981	33.66
1982	15.38	1983	9.56
1984	46.88	1985	23.55
1986	23.55	1987	22.93
1988	47.43	1989	21.25
1990	20.91	1991	20.04
1992	36.17	1993	43.47
1994	46.24	1995	45
1996	47.32	1997	50.66
1998	61.15	1999	60.5
2000	70.6	2001	66.7
2002	63.32	2003	63.9
2004	54.3	2005	56.5
2006	59.5	2007	48.65
Mean	28.271	Mean	26.73

*Not significant

6.6 Impacts of Climate Variability and Climate Change

There was a decline in annual rainfall during the study period while increase in surface air temperature. The annual rainfall was significantly low in the decade 1981–90 and led to several summer droughts. As a result, increase in temperature was noticed during the above decade.The rate of increase in temperature was more evident since 1980s. Decline in rainfall and increase in temperature led to decline in moisture index and increase in aridity index, which is an indicator for dryspell. It led to more number of droughts (nine) in the present tri-decade of 1981–2009 when compared to that of 1951–80. It may be continuing even in the current tri-decade since the decade 2011-2020 was the warmest decade (Figs. 6.5 and 6.6). All these factors reveal that the global warming and climate change are real in Kerala. Therefore, warming and dryness within the humid climate in the State of Kerala is a concern in the plantation sector of the State. However, the crop simulation models in the case of coffee indicated that the area is likely to decline if the temperature increase is by 2°C and some of the current areas under coffee may disappear in the above situation of increase in temperature (IPCC, 2007). In fact, such increasing trend in temperature was noticed since last thirty to 40 years its impact is not seen in terms area and productivity across the Western Ghats. However, the rate of increase in maximum temperature was 1.46°C during the present tri-decade. Instead of decrease in area, production and productivity due to increase in temperature, there was an increasing trend. The area under coffee was increasing and it was phenomenal since last two decades. From farmers' point of view, a crop mix with combination of coffee and black pepper may be beneficial under the current climate variability scenario as one crop like coffee suffers, another crop like black pepper gains in a same weather situation. Vagaries in monsoon, distribution of rainfall pattern, frequent occurrence of droughts, and increase in maximum and minimum temperatures, deforestation and shifting land use pattern have altered the natural habitat of coffee to a greater extent.

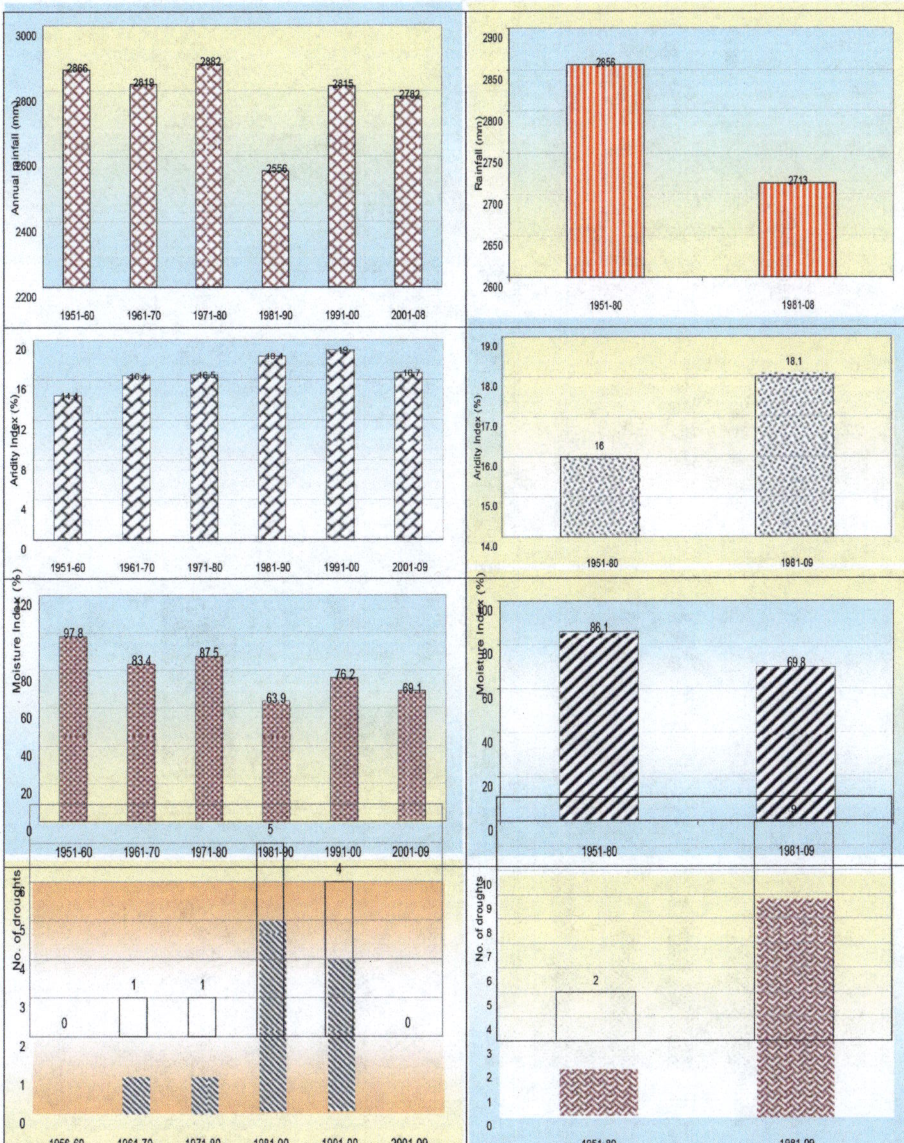

Fig. 6.5: Decadal and tri-decadal changes in annual rainfall, aridity index, moisture index and droughts over Kerala

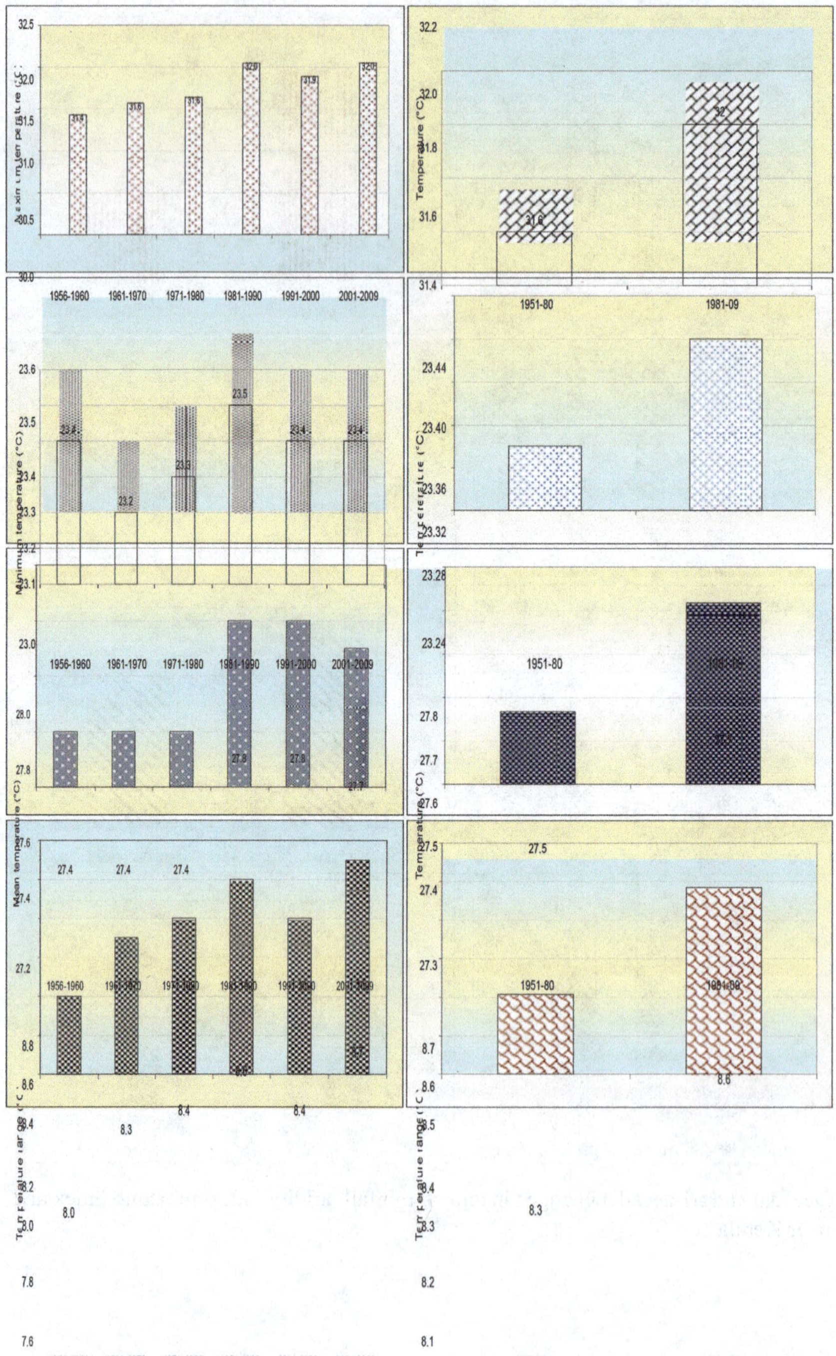

Fig. 6.6: Decadal and tri-decadal changes in temperature across the State of Kerala

Summary

Coffee requires summer showers during February-March (blossom showers) and March-April (backing showers) for better blooming and fruit set. It is advisable to cultivate coffee and black pepper as a mixed cropping system in coffee growing tracts in order to assure better returns. Coffee performs better if black pepper fails and vice versa in an identical weather situation in majority of the years. Of course, both the crops perform better in some years depending upon the rain fall distribution in summer from February to May. Temperature range (difference between maximum and minimum temperatures) is increasing during summer and post monsoon season over the coffee growing tract. In recent years, warm nights are noticed over the high ranges and also coffee matures early. The bean weight in coffee is also declining.

The quality of speciality coffee is likely to decline if the monsoon rains are deficit with erratic distribution from June to September during which coffee beans are processed along the Malabar Coast to obtain better flavour in coffee. During the years 2002, 2012 and 2015, the monsoon rains were deficit by more than 20% and rainfall distribution were uneven, which adversely affected the quality of beans in terms of increasing acidity and decrease in sugar content. Coffee farmers believe that rains in October and November are likely to be beneficial for the next coffee crop to follow. These observations need further study in detail to understand the effect of rainfall and temperature on quantity and quality of coffee beans.

Suggested Readings

Anonymous 2010. Database on coffee. Economic and Market intelligence Unit, Coffee Board of India. May/June 2010. www. Indiacoffee.org.

Carlos, H., Crisosto, H.C, , Grantz, A.D and Meinzer, F.C. 1991. Effects of water deficit on flower opening in coffee (Coffea arabica L.) Tree Physiology 10, 127–139.

Coffee Guide 2010. International Trade Centre (ITC). www.the coffeeguide.org 2010.

Climate Change and the Coffee Industry. Abstract for trade information services, Geneva: ITC, 2010. vi, p.28 (Technical paper).

Economic Review. 2008. Chapter 4. Agriculture and Allied Sectors pp.30–112. Kerala State Planning Board, Govt. of Kerala, Thiruvananthapuram.

Economic Review. 2010. Chapter 4. Agriculture and Allied Sectors pp.36–103. Kerala State Planning Board, Govt. of Kerala, Thiruvananthapuram.

International Trade Centre (ITC). www.the coffeeguide.org. 2010. Climate Change and the Coffee Industry. Abstract for trade information services. Geneva: ITC, 2010. vi, 28 pages (Technical paper).

Kannan, K., Devadas, V.S and Thomas, C.G. 1987 Effect of weather parameters on the productivity of coffee and pepper in Wayanad. In: Proc. of National Seminar on Agrometeorology of plantation crops held at RARS, Pilicode during 12–13, March 1987. P.187.

Korikanthimath, V.S., Ankegowda, S.J., Yadukumar, N., Hegde, R. and Hosimani, M. M. 1998. Mixed cropping of arabica coffee and cardamom: Influence on microclimatic and physiological characteristics. J. Medicinal Aromatic Plant Sciences 20: 394–396.

Krishnakumar, K.N., Rao, G.S.L.H.V.P. and Gopakumar, C.S. 2008. Climate change at selected locations in the humid tropics. J. Agrometeorol., 10(1): 59–64.

Lorena Soto-Pinto., Ivette Perfecto., Juan Castillo-Hernandez and Javier Caballero- Nieto. 2000. Shade effect on coffee production at the northern Tzeltal zone of the State of Chipas, Mexico. Agriculture, Ecosystems and Environment. Vol. 80. Issues 1–2. August 2000. pp-61–69.

Rao, G.S.L.H.V.P., Pushpalatha, P.B., Gopakumar, C.S and Mahesh Chandran. 2005. Influence of monsoon on quality of coffee beans in the humid tropics. In proceedings of National Workshop on Drought Management in Plantation Crops. pp.176.

Rao, G.S.L.H.V.P., Mohan, H.S.R., Gopakumar, C.S and Krishnakumar, K.N. 2008. Climate change and cropping systems over Kerala in the humid tropics. J. of Agrometeorol. (Special issue-Part 2): 286–291. pp. 286–291.

Rao, B.V.R. and Rao, V.U.M. 2024. Agricultural Meteorology. Jain Brothers, New Delhi. 154p

Rao, G.S.L.H.V.P. 2021. Agricultural Meteorology (5th Reprint). PHI Learning Private Limited, New Delhi, India. 364p.

Rao, G.S.L.H.V.P. and Krishnakumar, K.N. 2024. Impact of weather and climate on plantation crops in the Humid Tropics. Chapter 21 (417-432). Agrometeorological Applications for Climate Resilient Agriculture (Eds.): Chattopadhyay, N., Stefanski, R., Attiri, S.D. and Rathore, L.S. Springer International Publishing AG. Singapore.445p

Sharangi, A.B., Rao, G.S.L.H.V.P., Das, S., Krishnamurthy, K.S.Upadhyay, T.K., Gopakumar, C.S. and Acharya. S.K. 2024. Brunt of Climate Change and Spice Crops: Scenario, Response, and Resilience. 755-812pp. Eds: Ravindran, P.V., Sivaraman, K.,Devasahayam,S., Nirmal Babu, K. Handbook of Spices in India: 75 Years of Research and Development. Springer Nature Singapore Pte. Ltd.535p

Sunil, K.M. and Devadas, V.S, 2009. Effect of weather on the productivity of black pepper and coffee under Wayanad conditions. In the proceedings of the National Seminar on Climate change Adaptation strategies held at KAU Main Campus during 3–4, December 2004. pp.171–175..

Questions

I. Elucidate on the Following

1. Coffee growing environment
2. Weather and coffee
3. Thermal Regime of Coffee
4. Rainfall Regime of Coffee
5. Coffee not an alternate bearer

II. Write Short Notes on

1. Blossom and backing showers
2. Weather abnormalities and coffee
3. Coffee and Black pepper as mixed crops
4. Rainfall and coffee yields
5. Temperature and coffee yields
6. HU and coffee yields
7. Water deficit during summer and coffee yields

III. Fill in the Blanks

1. ---------------- is the favourable altitude for Arabica cultivation
2. ---------------- is the favourable range of altitude for Robusta cultivation
3. Coffee in India is grown under the ----------
4. India is in the ------------ largest position in coffee producing countries
5. --------- and ------------ at the time of coffee flowering are detrimental
6. ----------- and ----------- are detrimental to coffee production in traditional areas
7. ----------- is detrimental to coffee production in non-traditional areas
8. -------------- is the coffee variety mostly grown in Wayanad District
9. -------------- is known as the Coffee Bowl of Kerala
10. ---------- model rainfall is conducive for coffee growth and production
11. Coffee yield appears to be poor when the maximum temperature shoots beyond--------^{0}C
12. If the minimum temperature is above -----------^{0}C coffee yield appears to be poor

13. ------minimum temperature and ------maximum temperature are not conducive for coffee
14. Coffee yield was low in ------------ due to floods
15. ------------ is the relationship between summer soil moisture and coffee yields

IV. State True or False

1. Coffee is a crop of Wayanad
2. Coffee likes bimodal rainfall
3. Coffee and black pepper performs well during summer drought
4. Soil moisture during summer may not influence coffee yields
5. Temperature range is widening in both the seasons (SWM and PM)
6. Blossom showers are sufficient for better flowering in coffee
7. Rains during November is good to coffee plantations during the next crop
8. Coffee can be an intercrop in coconut plantations
9. Sprinkler irrigation is a substitute in the absence of blossom and backing showers
10. Soil moisture round the year in coffee plantations is favourable for better coffee production

7

Black Pepper

Black pepper (*Piper nigrum L.*), famous as "Black Gold" and also known the "King of Spices" is the most important agricultural commodities of commerce and trade in India. It is the most important, most popular and most widely used spice in the World. Of the total spices traded internationally, black pepper accounts for 34 per cent. South West India is the traditional home of this important spice, particularly the Malabar Coast. Black pepper is also grown in tropical zones such as the Asia Pacific Region, mainly India, Indonesia, Malaysia, Sri Lanka, Thailand, China, Vietnam and Cambodia. Outside the Asia Pacific region, the crop is distributed in Brazil, Mexico, Guatemala, etc., totaling about 26 countries. It is grown successfully between 20°N and 20°S latitude and from mean sea level to 1500 meters above sea level. World's black pepper production varies in the range of 2.7 to 3.35 tonnes annually. The major production share is contributed by Vietnam with an annual production of 80,000 tonnes. About 85 per cent of the World trade of pepper is determined by the largest pepper producing countries in the International Pepper Community (IPC). World export fluctuates around 1.5–2.5 lakhs tonnes per annum. Global black pepper supply has shown a drastic increase over last ten years except in 2008–09. Vietnam's sudden increase in production has resulted in the massive global out put of 3–3.5 lakhs metric tonnes from 1.9–2.0 lakhs metric tonnes in late nineties. At present Vietnam holds the top position of World's largest producer with a contribution of one-third of total out put and largest exporter too. It is followed by India, Brazil and China. Other major producing countries are Indonesia, Malaysia and Sri Lanka (Fig. 7.1). In black pepper growing regions of Indonesia, Thailand and Vietnam, the rainfall is well distributed having practically no drought period. Global pepper prices are influenced by output of black pepper in Vietnam, India and Brazil. World pepper production during 2007 was estimated at 2.71 lakh tonnes compared to 2.89 lakh tonnes reported during 2006. Production is slightly better in 2008 when compared to 2007. India is the second largest producer of black pepper in the World after Vietnam. In India, black pepper is cultivated mostly in homestead gardens as a mixed crop and is seldom grown as pure plantation. Our annual production is in the range of 50000–80000 metric tonnes. India has produced record quantity of 80,000 tonnes in 2002. The average annual production for the last ten year (1999–2008) has been slightly higher compared to the previous ten year (1989–1998). In the last three years, India's production is stable around 50,000 tonnes. India's pepper production constitutes 19 per cent to World's production. As major producer of this spice, India is a key exporter to World market. India's annual export is fluctuating between 14,000 to 30,000 metric tonnes in last few years. During early 1970s, India dominated in world export market. From 1978, Brazil overtook India as major exporting country. Once again India regained first position in 1986 and remained top exporter only in 1993 and 1997. India exports pepper to countries like USA (42%), Germany (5%), UK (5%), Italy (5%), Canada (4%) and others includes UAE (39%). USA is the single largest importer of Indian pepper with annual import quantity of 14,500 MT. In India, domestic consumption varies from 50,000 to 60,000 tonnes while global consumption varies 2–2.5 lakh tonnes. As far as productivity is concerned Thailand ranks first (4089 kg/ha) and India is in the last position (311 kg/ha). The low productivity is mainly due to poor generic potential of the vines, high population of senile and unproductive vines, losses caused due to abiotic and biotic stresses, non availability of quality planting material of improved varieties and non adoption of appropriate agronomic practices (Thangaselvabal *et al.*, 2008).

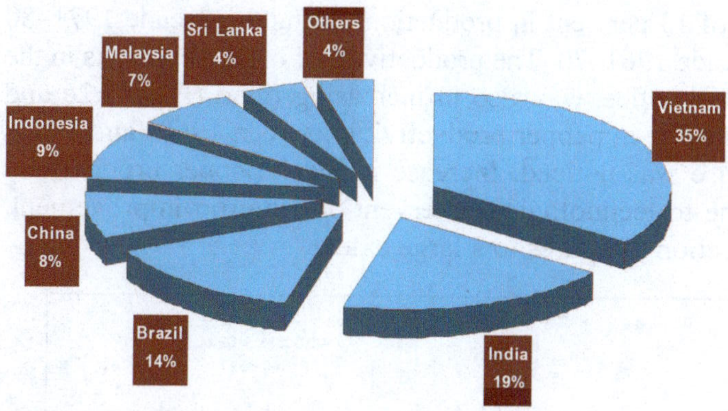

Fig. 7.1: Contribution (%) of leading pepper producing countries
Source: International Pepper Community

7.1 Area, Production and Productivity of Black Pepper

In India, Kerala is the leading producer of black pepper accounting for more than 90%. Long term trend analysis showed that area and production of black pepper in the State has been increasing. The data collected from the Directorate of Economics and Statistics, Govt. of Kerala revealed that the area under pepper cultivation in the State was just 78.8 thousand ha during 1952–53. It gradually showed an increasing trend until 1974–75 (118.4 thousand ha). Since then area declined all on a sudden and it was 108.3 thousand ha during 1975–76 and did not exhibit any increase for about one decade (Fig. 7.2a). From 1985–86 onwards, area under pepper cultivation increased slowly. It increased to 121.6 thousand ha in 1985–86 and reached to 253.04 thousand ha in 2007–08. The decadal area in black pepper showed a gradual increase from 1952–60 to 2007–08. It has increased from 87.4 thousand ha in 1952–53 to 230.1 thousand ha in 2007–08, registering an increase of 163.2 per cent over a period of 56 years (Fig. 7.2b). In the case of production, it was just 22.63 thousand tonnes during 1952–53. The production was just 17.35 thousand tonnes during 1984–85, which was the lowest production received in any year. It crossed 30 thousand tonnes in 1986–86 and then gradually reached to 63.9 thousand tonnes in 2007–08.The highest (69.02 thousand tonnes) was noticed during 2002–04, followed by 1998–99 (68.5 thousand tonnes). Wide inter-annual variability could be observed in production as in the case of area. The lowest production was obtained in 1984–85 while the highest in 2005–06 (75.7 thousand tonnes). Continuous summer rainfall received during summer 1984 might have led to the lowest production of pepper during that year. At the same time, production also showed an increase of 163 per cent during the same period. It increased from 25.7 thousand tonnes in 1952–60 to 67.6 thousand tonnes in 2001–08.

There was a decline of 13 per cent in production during the decade 1971–80 over the previous decade 1961–70. The productivity of black pepper, as in the case of cardamom, and coffee, was also in increasing trend (Figs. 7.2c and 7.2d). There was a decline in pepper productivity between 1960s and 1980s and thereafter increase was noticed. Increase in black pepper productivity in recent years is due to technological interventions in crop improvement, production and protection measures to a large extent.

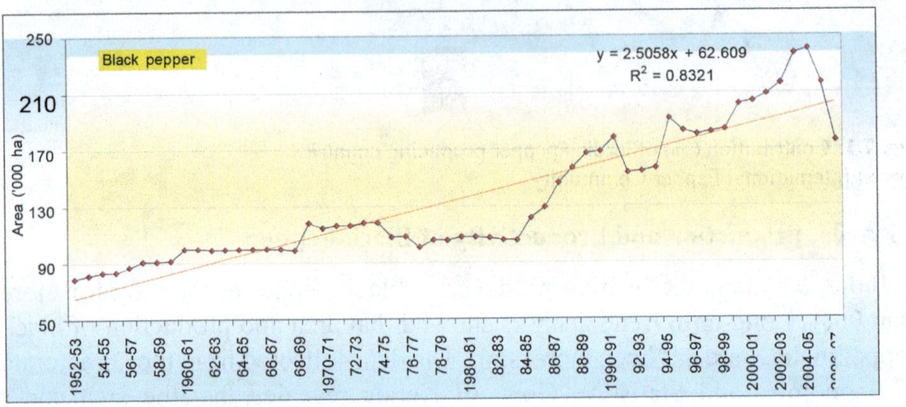

Fig. 7.2a: Trend in annual area of black pepper in Kerala

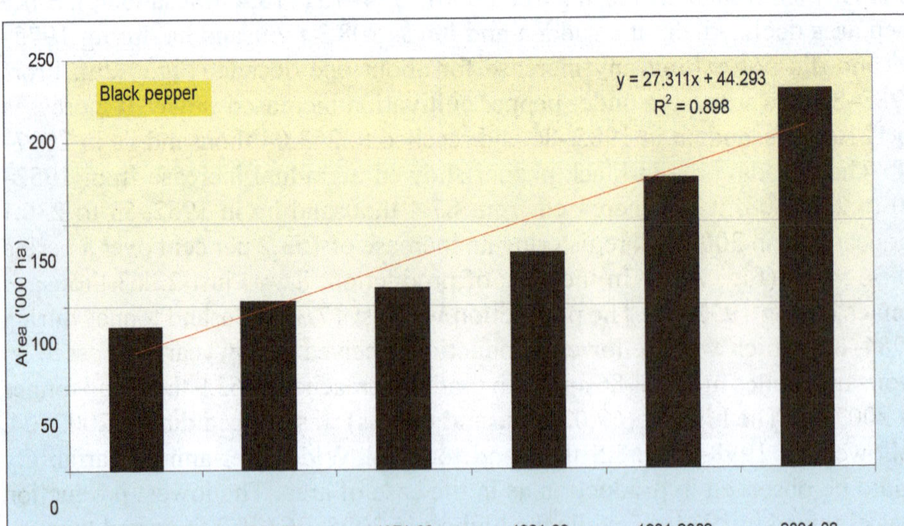

Fig. 7.2b: Trend in decadal area of black pepper over Kerala

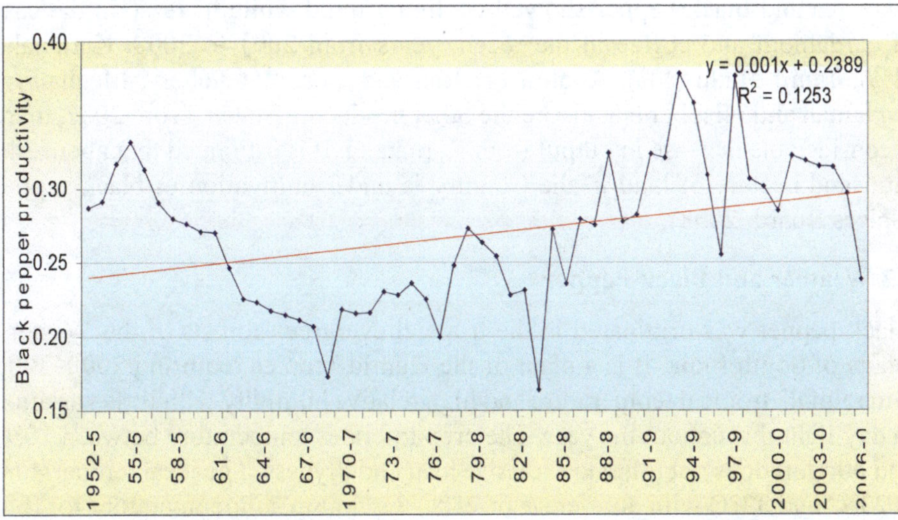

Fig. 7.2c: Trend in annual productivity of black pepper over Kerala

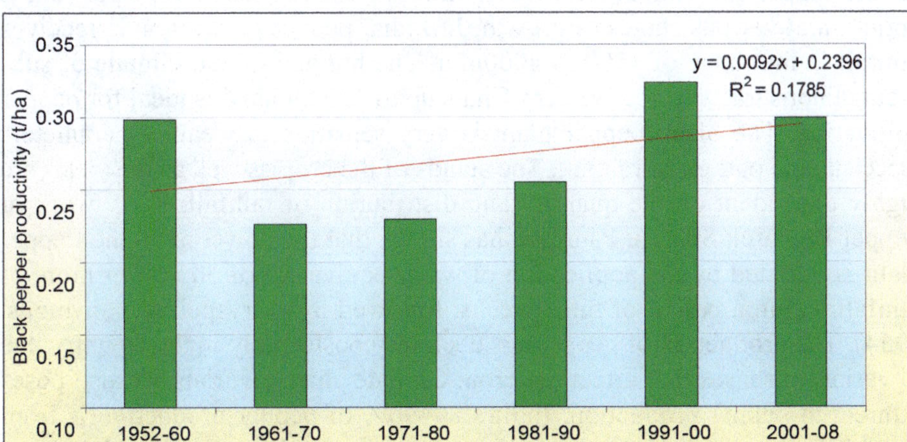

Fig. 7.2d: Trend in decadal productivity of black pepper in Kerala

As a whole, the area under cardamom, coffee and black pepper across the Highranges of Kerala was increasing while declining in the case of tea. Interestingly, all the four selected crops showed increase in production during the study period though inter-annual variations in plantation crops' production and productivity was insignificant. Though increasing trend was noticed in plantation crops' production and productivity during the study period, the decline in recent years is the concern. Similar was the decadal production trends in the case of cardamom, coffee and black pepper. In the case of productivity on decadal-wise, tea and black pepper showed declining trend while increase in the case of cardamom and coffee in the recent years from 2001 to 2008. Karnataka (4%), Tamil Nadu (1%), Andhra Pradesh and Goa, Pondichery, Meghalaya, Andaman and Nicobar Islands are the other producing States. From 2007, there is considerable increase in output from Karnataka. It is estimated that about 246 thousand hectare of land in the Country is under cultivation of black pepper (Spices Board, 2008).

7.2 Weather and Black Pepper

Black pepper was originated in the tropical evergreen forests of the Western Ghats of South India. It is a plant of the Humid Tropics requiring 2000–3000 mm rainfall, tropical temperature and high relative humidity with little variation in day light throughout the year. The crop tolerates temperature between 10°C and 40°C. It does not tolerate excessive heat and dryness. The ideal temperature is 23°C and 32°C with an average of 28°C. Optimum soil temperature for root growth is 26°C and 28°C. A well distributed rainfall of 1250–2000 mm is ideal for the pepper cultivation. Relative humidity in the range of 60–95 per cent is optimum at various stages of growth. In India, pepper growing area receives rainfall in the range of 1500 to 4000mm. The hot and humid climate of sub-mountainous tracts of the Western Ghats up to 1200m amsl is ideal for pepper cultivation. The black pepper plant is very sensitive to weather parameters especially the pattern of rainfall. The stands of the crop as well as the yields are highly dependent on the quantity and distribution of rainfall. Early work of Pepper Research Station, Panniyur has shown that the flowering in the Pepper plant is initiated by the application of water equivalent of 70 mm or more of rainfall within a period of three weeks, followed by a dryspell (Anonymous, 1954). The production of pepper has declined considerably in the Country due to various diseases that affect the crop. Climate change/variability also poses a threat to pepper production. India's 80–90% of pepper production is from Kerala. The crop is mainly grown in the highranges of Kerala viz., Idukki and Wayanad districts. Wayanad is considered as the land of spices. Black pepper produced from this hill station is famous for its aroma and quality. The pepper plant is very sensitive to weather parameters. The flowering and fertilization is synchronized with rainy season. Rainfall in May-June initiates the flushing and flowering process in pepper. Once the process starts, there should be good precipitation until fruit development is over. A break in rainfall for even a few days at a stretch occurring during any part of the critical period (reproductive phase of the crop) will affect the pepper yields considerably (Pillai, *et al.,* 1988). Profuse blooming and fruiting of pepper vines in Sumatra after a continuous drought of eight months has been reported. Nalini (1983) also noted a positive correlation of rainfall with flower bud differentiation process which started during April-May with the receipt of pre-monsoon showers. Variation in maturity pattern depends on factors such as variety, rainfall, altitude and ambient air temperature. Pepper berries mature in about 6–8 months after flowering. The period generally coincides with dry weather in India. The season of flowering and harvest varies from Country to Country. India, Sri Lanka and Thailand have long harvesting period when compared to other producing nations. In India, harvesting starts from November/December and continues till the end of March. In Indonesia, it starts from July and ends in October. Brazil, it starts from August and continues till October. Vietnam pepper harvested during May-July. However, there are two harvesting seasons in Sri Lanka because of the bimodel rainfall distribution with 60% is harvested in November to January and remaining produce is harvested from May to July. Harvesting period in Thailand is almost same as India, starts with December and extend till April (Table7.1). Long spells of dry weather are unfavourable for the crop growth. According to Menon (1981) and Nalini (1983), a dryspell before flowering is advantageous for better crop production. Pepper yield is significantly correlated with the rainfall received during the first half of May and the cumulative total rainfalls in the second half of the year. The distribution of rainfall, moisture holding capacity of the soil and drainage status of the soil is more important than the total rainfall (Sadanandan, 2000). Vagaries of monsoon are felt in the State quite often in the recent years. Rainfall decline and rise in night temperature are felt in Wayanad. At the same time, rise in temperature and its range poses a threat to thermo-sensitive crops grown in Idukki. Delay in onset of monsoon, distribution pattern, amount of rainfall etc will determine the pepper production during a particular year. Annual as well as monsoon rainfall are showing a declining trend across the State. Post monsoon rainfall is increasing. Rainfall decline is heavily felt in Wayanad district since last few years. Maximum temperature showed an increasing trend while minimum a declining trend at Idukki. The temperature range is thus increasing. This situation is detrimental to a thermo-sensitive crop like pepper. Variations in thermal as well as moisture regimes are felt across the State at several locations where pepper is grown. Climate change and global warming will adversely affect the pepper production across the State due to its effect on thermal and moisture regimes. As the crop is intimately related to the monsoon rainfall for its fertilization and further growth, vagaries in monsoon is a concern across the pepper growing areas of the State. Apart from rainfall, temperature also plays an important role in deciding the berry development and maturity. Studies on the effect of weather parameters and pepper production are scarce in and outside the Country.

Table 7.1: Global harvesting calendar of black pepper

Country	J	F	M	A	M	J	J	A	S	O	N	D
India												
Indonesia												
Brazil												
Malaysia												
Vietnam												
Sri Lanka												
Thailand												

7. 3 Rainfall Distribution

The mean annual rainfall across the high ranges of Kerala is 1901 mm, varying between 1865 mm at Pampadumpara and 1939 mm at Ambalavayal. The annual number of rainy days is 117. It is interesting to note that the number of rainy days did not show any significant difference between Pampadumpara and Ambalavayal. The monthly rainfall pattern at Pampadumpara (Idukki) and Ambalavayal (Wayanad) follows the same pattern. The distribution pattern of rainfall at Pampadumpara showed that the monthly rainfall is high during July (365.3 mm), followed by June (295.6 mm), August (271.1 mm) and October (262.2 mm). The least was in February (14.8 mm), followed by January (18.7 mm). In the case of Ambalavayal, similar pattern was evident **(Fig. 7.3)**. The lowest rainfall at Ambalavayal was noticed during January (12.1 mm), followed by February (16.6mm) and December (24.3 mm). Both the locations receive high rainfall in June, July, August and October due to the influence of both southwest and northeast monsoons. Pamapdumpara receives relatively better rainfall from October to December when compared to that of Ambalavayal. It was attributed to the influence of northeast monsoon towards south when compared to that of north of Kerala. It reveals that rainfall distribution across the hill crops of Kerala is bi-model, having a peak in July and another peak in October.

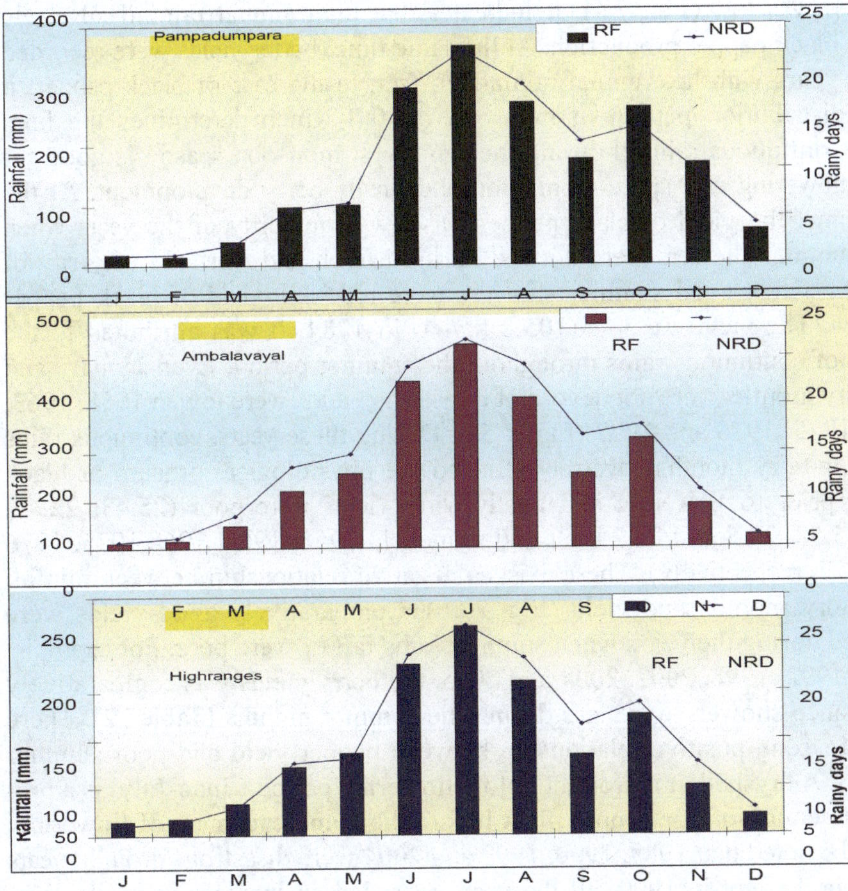

Fig. 7.3: Rainfall regime of black pepper across the highranges

7.3.1 Rainfall and Black Pepper

In majority of the years (58 per cent), an increase in production was noticed in accordance with increase in rainfall in the case of black pepper. For example in 1986, 1988, 1992 and 2003 during which a hike in yield was noticed when the annual rainfall was high. Likewise a decrease in production was also noticed when the annual rainfall was low as noticed in majority of the years (46 per cent) during the study period. It showed that in majority of the years, there was a direct relationship with annual rainfall. It also revealed that high annual rainfall results to low yield in black pepper and vice versa. For example, high rainfall was recorded in 1984, 1991, 1992 and 2007 during which the black pepper yield was poor (231 kg/ha, 278 kg/ha, 282 kg/ha and 257 kg/ha, respectively). Similarly, when the yield was high in 1990 (324 kg/ha), 1993 (324 kg/ha), 1995 (378 kg/ha) and 1999 (376 kg/ha) the annual rainfall received was low (Fig. 7.4). It indicates that high annual rainfall adversely affects black pepper production. At the same time, better yields were recorded during years with low annual rainfall. In fact, in the case of black pepper, it is the distribution pattern of monsoon rainfall which determines the final yield. Continuous rainfall during the southwest monsoon season is good for better flowering and spike elongation and finally berry development. At the same time, the yield of black pepper was low in majority of the years when good summer showers were noticed during March and April. In majority of the years when good summer showers were received, yield of black pepper was low. The yield was low (105.8 kg/ha) in 1984. It was attributed to the receipt of continuous rains throughout the summer period. Even January and February months were not devoid of rains. The yields were low in 1958, 1963, 1968, 1975, 1976 and 1977 (Fig. 7.5a). During these years, continuous rains during the dry months adversely affected the physiological process of black pepper prior to flowering and finally berry yields were poor (25.43, 22.42, 20.4, 25.58, 24.5 and 20.2 thousand tonnes, in 1958, 1963, 1968,1975, 1976 and 1977, respectively). There exists a negative relationship between rainfall during dry months and yield (Fig. 7.5b). Comparatively good yields were obtained during the years when summer rains fail or were poor. For example, during 1992, 1998, 2002, 2004 and 2005, the berry yields was comparatively better since showers were less during the summer months (Table7.2). There exists a strong positive relationship between pepper yield and poor summer showers. A dryspell or a drought before flowering period (June-July) is a pre-requisite condition for pepper plant before the commencement of flowering. It may be noted that 1983, 1990, 1992 and 2004 were disastrous drought years in Kerala. Except in 1990, all the years recorded an increase in production, revealing that black pepper production is not affected even during disastrous drought years, to a certain extent while all other plantation crops production are adversely affected. The yield reduction in 1990 could be attributed to incidence of insect, pest and diseases. This reveals that drought/continuous water stress/long dryspells are not an adverse condition as far as pepper plants are concerned. At the same time, severe summer conditions with acute water stress coupled with high maximum surface air temperatures may devastate the young and senile pepper vines as reported in 2004 in Wayanad district. However, during 2004, no yield reduction is recorded in the pepper yield data of the State. The years 1961, 1967, 1979, 1986 also recorded comparatively better yields. It reveals that ups and downs in the productivity curve are mostly attributed to the occurrence of summer showers during the critical periods of the crop.The receipt or non receipt of summer showers especially during March-April will determine the success or failure of black pepper.

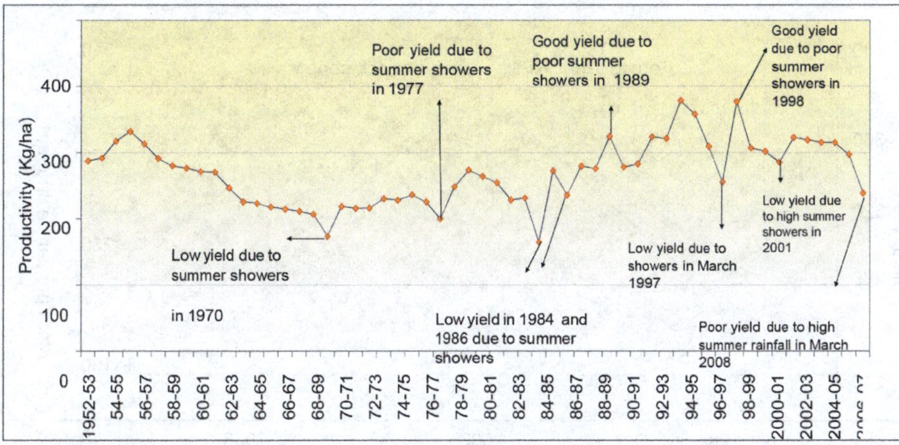

Fig. 7.4: Effect of summer rainfall on black pepper productivity

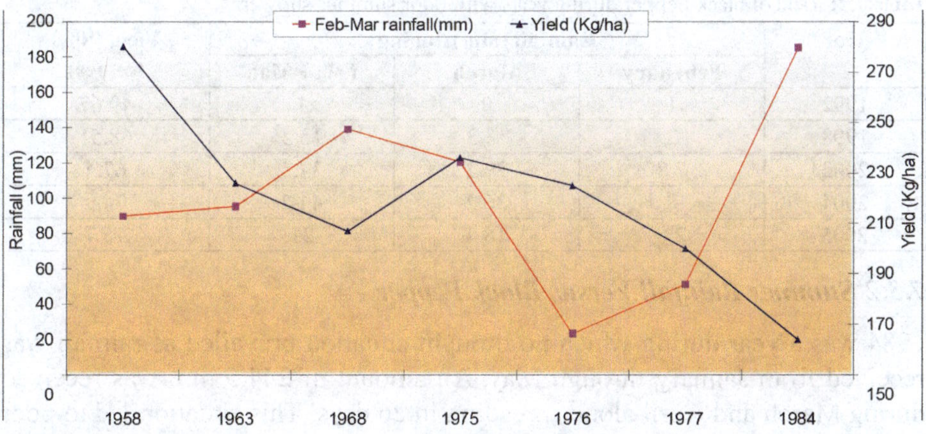

Fig. 7.5a: Pepper yield over Kerala during years with good summer showers

If the summer showers are good, black pepper yield will be less and vice versa. In sub mountainous regions of Wayanad district, coffee and black pepper is planted as a mixed farming system as black pepper and coffee behave antagonistically to summer showers. That is to say, when one crop (coffee) is benefited with good yield on receipt of summer showers, the other crop (black pepper) yield will be adversely affected.

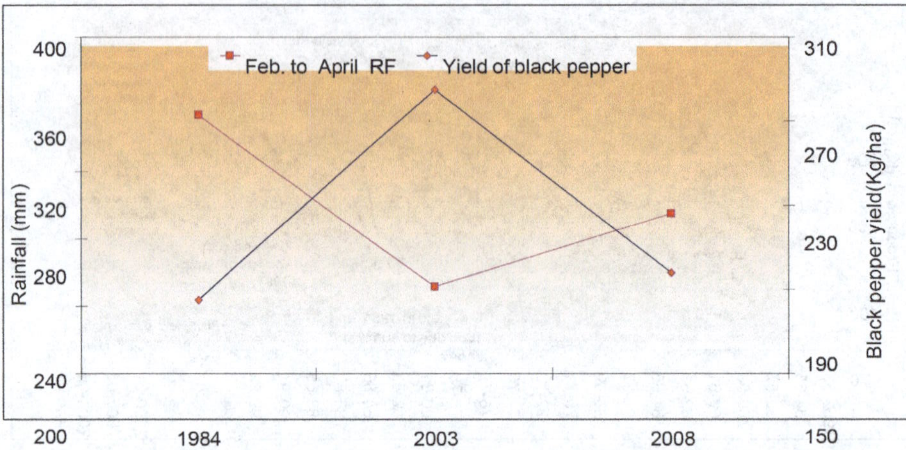

Fig. 7.5b: Black pepper yield at Ambalavayal during years with good summer rains

Table 7.2: Yield of black pepper during years with poor summer showers

Year	Rainfall (mm) during			Yield ('000 tonnes)
	February	March	Feb. + Mar.	
1992	1	0	1	49.67
1998	2	13.3	15.3	68.51
2002	7.8	29.7	37.5	67.4
2004	8	36.2	44.2	74.9
2005	2.6	18.4	21	75.7

7.3.2 Summer Rainfall Versus Black Pepper

1984 was a year during which no drought situation prevailed as rainfall was received from January through May. An amount of 244.5 mm was received during March and April alone spreading in 26 days. This situation led to poor harvest in black pepper. The yield of pepper was only 185 kg/ha while coffee yield was high during that year (811 kg/ha). Similarly, during 2003 (285 kg/ha) and 2008 (198 kg/ha) good summer showers were received during March and April which resulted to poor pepper harvest. The yield of coffee was comparatively better, 739 kg/ha and 703 kg/ha, respectively in 2003 and 2008. The pepper yield (productivity) was the highest during 2001 (398 kg/ ha), followed by 2000 (387 kg/ha) and 2007(357 kg/ha). In the above years, rainfall during March was meagre which resulted to dry weather during March. After this drought period, fairly good summer showers were received and this resulted to better yield (Table 7.3). Trend analysis revealed that the rainfall across the highranges of Kerala is declining. Since southwest monsoon is the main rainy season in Kerala, the decline in rainfall during monsoon is a concern to the highranges. It revealed that post monsoon rainfall is increasing across the highranges in tune to the increase in post monsoon rainfall across the State as a whole. While Pampadumpara showed an increase in rainfall except during southwest monsoon, rainfall across Ambalavayal is declining (Table 7.4). The decline is more so since the last one and-a -half- decades.

Table 7.3: Effect of rainfall on black pepper production in Wayanad District

Year	Rainfall (mm) during March	Rainfall (mm) during April	Rainfall (mm) during March - April	Yield of dry pepper (Kg/ha)	Yield of coffee beans(Kg/ha)
1979	4.4	57.7	62.1	1011	377
1980	65.2	208.5	273.7	1480	368
1981	63.1	78.6	141.7	956	753
1982	40.0	78.6	118.6	3599	585
1983	1.5	21.5	23.0	2375	315
1984	114.5	130.0	244.5	185	811
1985	75.8	93.0	168.8	2966	514
2000	1.4	110.8	112.2	387	656
2001	12.0	206.2	218.2	398	802
2002	29.0	93.4	122.4	326	639
2003	104.2	131.4	235.6	285	739
2004	35.8	132.4	168.2	324	690
2005	54.2	147	201.2	334	660
2006	85.4	97	182.4	318	598
2007	5.2	94.8	100.0	357	705
2008	178.8	75.2	254.0	198	703

Table 7.4: Trend in annual and seasonal rainfall (mm) across the highranges

Station	Rainfall trends				
	Annual	SWM	PM	Winter	Summer
Pampadumpara	+	-	+	+	+*
Ambalavayal	- **	- **	+ **	- **	- **
Highranges	- **	-**	+**	-**	-**

* - significant at 5% level, ** - significant at 1% level,

7.4 Thermal Regime

The mean annual maximum temperature varied between 25.2°C and 27.2°C, and the annual mean minimum between 17.3°C and 17.7°C, respectively at Pampadumpara and Ambalavayal. Irrespective of the seasons, variations of minimum temperature at both locations across the cardamom tract of Kerala are relatively less when compared to that of maximum temperature. However, the diurnal changes in minimum temperature during winter are predominant and in the case of maximum temperature, the diurnal changes are predominant during summer. April recorded the highest mean annual maximum temperature at Pampadumpara (29.3°C), while it was in March (30.3°C) over Ambalavayal. The lowest annual minimum temperature was recorded during December-January (15–16°C) across the highranges of Kerala. The annual range in temperature was high (9.5°C) over Ambalavayal while it was 7.7°C at Pamapdumpara. The maximum temperature during the rainy season (southwest as well as post monsoon) varied between 23 and 27°C, while the minimum between 17 and 18°C across the two locations. The range in temperature varied between 6 and 9°C. A gradual decline in temperature was noticed during winter from that of rainy season. The minimum temperature during this period across the cardamom tract is around 15–16°C and the temperature range shoots up to 9–13°C (Fig. 7.6). It is due to the fact that, maximum temperature showing a trend to shoot further up and minimum temperature to come down in the cardamom tract of Kerala. The mean summer maximum temperature never crossed 30.0°C across the highranges while the range in temperature was high over Ambalavayal during summer (10.9°C) when compared to that of Pampadumpara (9.9°C). At the same time, the temperature range at Ambalavayal during summer was low (10.9°C) when compared to winter (12.1°C). Increase in maximum temperature and decline in minimum temperature resulted to increase in temperature range. It is a concern to the plantation crops grown across the highranges of Kerala in the context of projected climate change scenario.

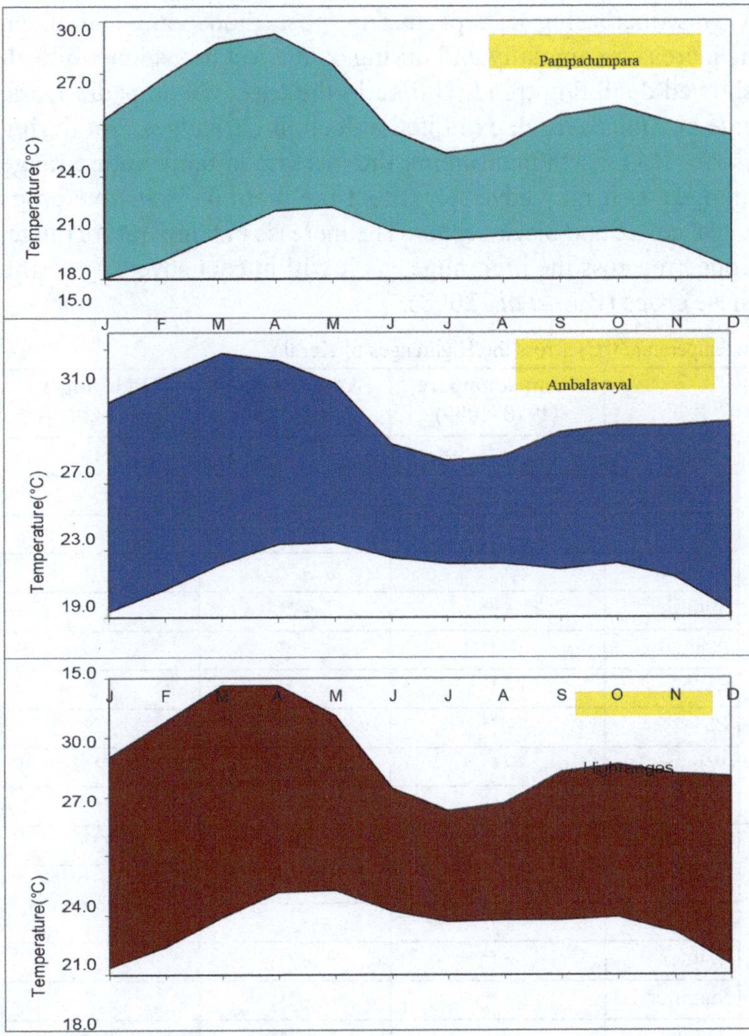

Fig. 7.6: Thermal regime of black pepper across the Highranges of Kerala

7.4.1 Temperature Trends

The maximum temperature across the highranges was increasing irrespective of the seasons while decline in minimum temperature (Table 7.5). The mean temperature also showed an increasing trend except during summer. It is obvious that the range of temperature (the difference between the day maximum and the night minimum temperature) is increasing. Similar was the trend in the case of Pampadumpara. In the case of Ambalavayal (Wayanad), increase in maximum temperature is not seen as noted in the case of Pampadumpara, except a marginal increase in southwest monsoon. In the case of minimum temperature, it was increasing except during post monsoon. The mean temperature was increasing annually and during southwest monsoon while all other seasons showed declining trend. Unlike in the case of Pampadumpara, temperature range at Ambalavayal, exhibited a declining trend, except during southwest monsoon. It is worth mentioning the increase in temperature range across the highranges as it may adversely affect the thermo - sensitive crops viz., cardamom, tea, coffee and black pepper. The increase in temperature range, therefore, is a concern across the highranges as it will impact adversely on the production of these crops (Rao *et al.*, 2008).

Table 7.5: Trend in temperature (°C) across the Highranges of Kerala

Maximum temperature (°C)		**Pampadumpara (1978–2009)**	**Ambalavayal (1984–2009)**	**Highranges (1984–2009)**
	Annual	+ **	- **	+ **
	SWM	+ **	+ **	+ **
	PM	+ **	- **	+ **
	Winter	+ **	- **	+ **
	Summer	+ **	- **	+ **
Minimum temperature(°C)	Annual	- **	+ **	- **
	SWM	- **	+ **	- **
	PM	- **	- **	- **
	Winter	- **	+ **	- **
	Summer	- **	+ **	- **
Mean temperature (°C)	Annual	+ **	+ **	+ **
	SWM	+ **	+ **	+ **
	PM	+ **	- **	+ **
	Winter	+ **	- **	+ **
	Summer	- **	- **	- **
Temperature range (°C)	Annual	+ *	- **	+ **
	SWM	+ **	+ **	+ **
	PM	+ **	- **	+ **
	Winter	+ **	- **	+ **
	Summer	+ **	- **	+ **

*- Significant at 5% level ** -Significant at 1% level

7.4.2 Temperature and Black Pepper

It revealed that when the annual maximum temperature was high, the yield was low. For example, the annual maximum temperature was high during

1985 (0.16 t/ha), 1987 (0.24 t/ha), 1991 (0.28 t/ha), 1998 (0.26 t/ha) and 2002 (0.29 t/ha), the yields were low. In contrast, when the mean annual maximum temperature was low, the yields were high. It was evident in 1990 (0.323 t/ha), 1993 (0.32 t/ha), 1994 (0.32 t/ha), 1995 (0.38 t/ha), 1996 (0.36 t/ha), 1999 (0.38 t/ha) and 2006 (0.31 t/ha). It revealed that mean annul maximum temperature above 27.4°C may adversely impact the black pepper yield. As in the case of maximum temperature, when the minimum temperature was high, the black pepper yield was low. It was evident in 1985 (0.16 t/ha), 1997 (0.31 t/ha) and 1998 (0.26 t/ha). In contrast, when the mean annual minimum temperature was low, as noticed in 1990 (0.32 t/ha), 1993 (0.32 t/ha), 1994 (0.332 t/ha), 1999 (0.38 t/ha), 2005 (0.32 t/ha) and 2001 (0.31 t/ha), the yields were high (Fig. 7.7). The highest yield was obtained in 1995 when the mean annual minimum temperature was 16.9°C (lowest).

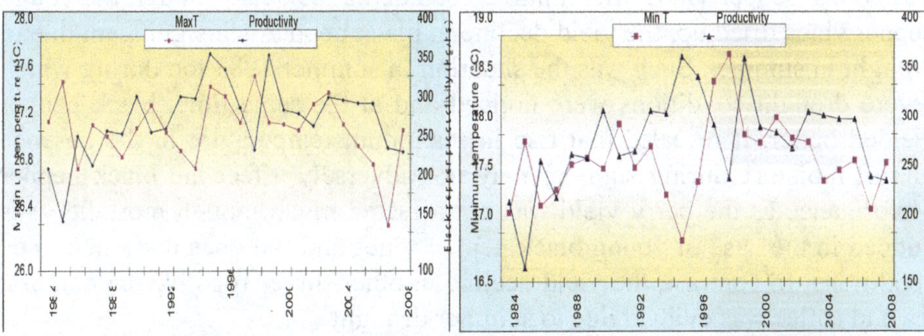

Fig. 7.7: Temperature and black pepper yield

7.4.3 Maximum Temperature During January-March, 2004 in Wayanad and Its Impact on Black Pepper

The State has witnessed severe water shortages, extreme temperatures and drying up of surface wells and ponds during summer 2004. Drinking water problem was severe. Most of the plantation crops were affected due to the severe water stress and extremely high surface air temperatures. The black pepper yields were not seen affected much due to summer drought in 2004. It is true that productive pepper plants can withstand continuous dry spells and produce better harvest if moderate summer showers during April-May are received before flowering process initiates in June after the onset of monsoon rains. But, black pepper vines, especially young and senile ones and vines planted in open spaces may not be able to tolerate extreme high temperatures as experienced in summer 2004. Black pepper vines planted in shade may be able tolerate such extreme temperatures to a certain extent. It was reported that pepper vines in some of the pepper plantations of Wayanad District were wiped out due to the extreme high day temperatures experienced throughout the dry period from January to March in the absence of soil moisture. Hence, an investigation was made to examine the march of day temperatures from January to March 2004. It revealed that 2/3rd of the total days experienced above normal temperatures (Fig. 7.8). Increase in day maximum temperature was noticed from the beginning of January to 21st March, 2004. The day maximum temperature shot above normal and reached as high as 34.7°C on 20th March, 2004 registering an increase of 3.6°C above normal followed by 34.6°C on 19th March, 2004, registering an increase of 3.0°C. On an average, the increase in maximum temperature over normal was 0.6°C. The decline in rainfall continuously from 1999 to 2003 led to hydrological drought in monsoon 2004 and several irrigation channels dried up. In the absence of soil moisture, increase in maximum temperature affected the black pepper gardens and young pepper vines dried up to a considerable extent. Though the young pepper vines dried up, the yield decline in black pepper is insignificant due to drought in summer. Such was the situation in summer 1983 too during which severe drought conditions were noticed and at the same time, black pepper yielded better. It revealed that rise in maximum temperature in the absence of soil moisture during summer may not adversely affect the black pepper performance as the berry yield was not affected much though mortality was noticed in the case of young black pepper vines and the ones under the open conditions. In contrast, the yield decline in other three crops like cardamom, tea and coffee was evident due to summer drought.

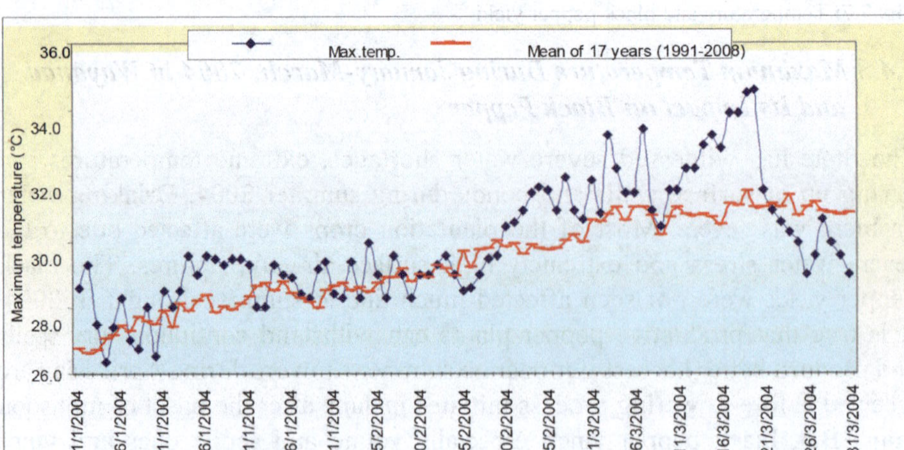

Fig. 7.8: Maximum surface air temperature from January- March, 2004 in ayanad district

7.4.4 Heat Units and Black Pepper

The heat unit concept was taken in to account to explain the relationship between temperature and crop yields of all the two crops tested as the heat units depend on the mean and base temperature. The base temperature was considered as 13°C in the case of plantation crops below which the crop growth ceases. The heat units were worked out based on the formula given below:

$$HU = \left(\frac{T_{max} + T_{min}}{2} \right) - T_b$$

Where, Tmax is the maximum temperature during the day

T_{min} is the minimum temperature of the day

T_b is the base temperature (13°C) below which no growth takes place.

Black pepper yield responded negatively with the heat units (Fig. 7.9). It indicated that whenever the heat units were more, the yields were low and vice versa as in the case of maximum and minimum temperature. Of course, the relationship may not be true always as the yield is a complex one and depends on several other environments factors. Interestingly, the yield in black pepper is low while coffee yield is high with the same amount of heat units. It indicated that the thermal environment is more suited to coffee when compared to that of black pepper. It appears that the plantations viz., cardamom, tea, coffee and black pepper may respond better to thermal regimes as they are thermo-sensitive and predominantly growing across the highranges. The relationship between the GDD and black pepper yields was the same as in the case of maximum and minimum temperatures. Higher yields were obtained when GDD was low and vice versa. For example, GDD was high during 1985 (3566°C) and 1999 (3737°C) and the yield was low (164 Kg/ha and 255 Kg/ha, respectively). In contrast, high yields were obtained when the growing degree days or heat units were low. For example, yield was 378 Kg/ha in 1995 while the heat unit was low (3479°C). Similarly, yields during 1986, 1987, 1988, 1989, 1990, 1991, 1992 and 1993 also showed the same trend. In majority of the years (92 per cent) the above pattern was observed.

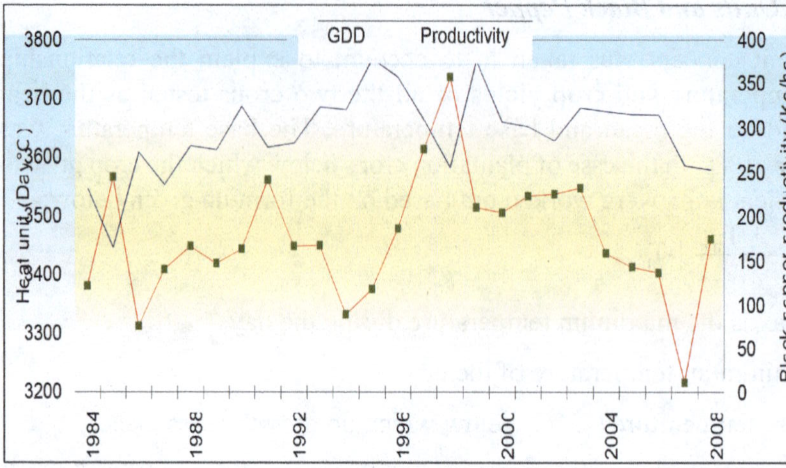

Fig. 7.9: Heat unit versus black pepper productivity

7.5 Moisture Regime

The monthly moisture index at Pampadumpara varied from 388 per cent in July to -9.5 per cent in May. During the rainy months from June to November, the moisture index was always above 100 per cent (Per humid A type climate). Soon after the rainy period, the moisture index all on a sudden fell down and the climate shifted to B1 humid type (Im=28.5 per cent) in December. In all the summer months from January to May, the moisture index was always below zero, ranging between -9.5 per cent in May to -41.3 in March, indicating that the moisture index during the above period were under the dry sub humid to arid type (Fig. 7.10). April and May fell under the dry sub humid as the moisture index was between -19 % to zero. January and February fell under Semiarid as the index was between -20 to -37.4 per cent. Only in March arid type climate was noticed, during which the moisture index was -41.3 per cent. In short, January to May, Idukki district experiences dry weather due to the non receipt of sufficient rainfall and hence the climate shifted from Perhumid during the rainy season to humid during the northeast monsoon. Thereafter the climate shifted from dry subhumid to arid during summer months as rainfall received was insignificant, if pre monsoon showers fail. Therefore, the monthly and seasonal climate shifts vary insignificantly. Similar was the trend in monthly climate fluctuations at Ambalavayal also. The moisture index varied between 420.6 per cent in July to -12 per cent in April. June, July, August and October only fell under the 'per humid A' type climate while September fell under 'B4 humid type' and November in 'B1 humid' climates. It may be noted that though September and October coming under the rainy months in Kerala, in the Wayanad region these months are shifting towards drier climates within the humid B type. In contrast to Pampadumpara, at Ambalavayal, December fell under even lesser type humid regime (C1 dry sub humid). January, February and March clearly showed a tendency to move towards 'Arid type (E)' climate as the moisture index was always more than -40 per cent during the above months. As a whole, the highranges experience from 'A' perhumid climate to arid climate 'E' depending upon the rainfall distribution. Since the moisture index is predominantly influenced by rainfall, the index follows a similar pattern to that of rainfall. During April and May, the moisture index fell under 'C1 dry subhumid' and 'C2 moist subhumid' type of climates. The fluctuations in moisture index at Ambalaval are more pronounced when compared to that of Pampadumpara as is evident from the analysis. Even during rainy months, moisture index showed a tendency to move towards lesser moisture regimes within the humid type climates (from B4 to B3 climate). Three months from January to March experience arid climate at Ambalavayal as against one (March) at Pampadumpara (Table 7.6). It reveals that Wayanad region of the Western Ghats in the highranges experience more dryspells during summer when compared to that of Pampadumpara. The trend analysis indicated that the annual moisture index across the highranges was declining. The annual moisture index at Pampadumpara was declining at the rate of 0.79 per cent per year while the declining rate was 0.67 per cent per year at Ambalavayal. As a whole, the decline of moisture index is at the rate of 0.65 per cent per year across the Highranges.

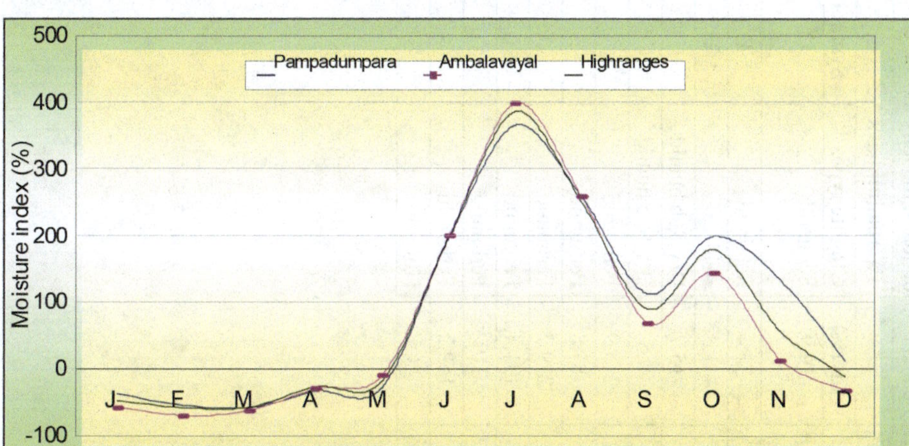

Fig. 7.10: Monthly moisture Index (%) across the highranges

Table 7.6: Spatial distribution of monthly moisture index (Im in %) across the highranges

Month	Pampadumpara			Ambalavayal			Highranges		
	Im (%)	**Climate types**	**Symbol**	**Im (%)**	**Climate types**	**Symbol**	**Im (%)**	**Climate types**	**Symbol**
Jan	-19.9	Semiarid	D	-40.8	Arid	E	-30.6	Semi Arid	D
Feb	-37.4	Semiarid	D	-53.1	Arid	E	-41.2	Arid	E
Mar	-41.3	Arid	E	-45.1	Arid	E	-40.2	Arid	E
Apr	-15.7	Dry Sub humid	C1	-12.0	Dry Sub humid	C1	-9.6	Dry Sub humid	C1
May	-9.5	Dry Sub humid	C1	8.9	Moist Sub humid	C2	0.0	Moist Sub humid	C2
Jun	215.9	Perhumid	A	220.4	Perhumid	A	217.9	Per humid	A
Jul	387.8	Perhumid	A	420.6	Perhumid	A	408.3	Per humid	A
Aug	279.7	Perhumid	A	279.6	Perhumid	A	271.6	Per humid	A
Sept.	131.4	Perhumid	A	87.0	Humid	B4	107.9	Per humid	A
Oct	219	Perhumid	A	164.3	Perhumid	A	198.4	Per humid	A
Nov	152.2	Perhumid	A	30.9	Humid	B1	73.5	Humid	B3
Dec	28.5	Humid	B1	-15.1	Dry Subhumid	C1	6.0	Moist Sub humid	C2

Summary

Black pepper requires a long dryspell during summer (February-April). Summer showers during the critical periods of black pepper (February –April) will adversely affect the crop yield. The ups and downs in the production are to a large extent determined by the occurrence of summer showers during February-April. The ever lowest production in 1984 was due to the receipt of continuous summer showers received through January to May. While black pepper is favoured with the continuous dry spell during the dormant phase of the crop, coffee requires summer showers during February-March (blossom showers) and March-April (backing showers) for better blooming and fruit set. Therefore, it is advisable to cultivate coffee and black pepper as a mixed cropping system in coffee growing tracts of Kerala in order to assure maximum returns from unit cropped area. Temperature increase during summer may adversely affect black pepper vines as seen in Wayanad District during summer 2004, during which high maximum temperature adversely affected young pepper vines to a considerable extent. In recent years, warm nights are noticed over the high ranges and the crop matures early. The berry weight also appears to be declining in recent years if temperature is high. These observations needs further study in detail to understand the effect of temperature on black pepper.

Suggested Readings

Anonymous, 1954. Annu.Rep.1953–54. Pepper Research Scheme. Madras State. Directorate of Agriculture, Madras.

Anonymous, 1984 Annu. Report 1984. Pepper Research Station , Panniyur. Economic Review. 2003. Chapter 4. Agriculture and Allied Sectors. Kerala State Planning Board, Govt. of Kerala, Thiruvananthapuram.

Lazarus, T.P Aipe, K.C., Sudheesh, M.V and Rao, GSLHVP. 2005. An analysis of yield and weather relationship on coffee and pepper in Wayanad. In: Proc. of National Workshop on Drought Management in Plantation Crops. (Eds. Nybe, E.V. and G.S.L.H.V. Prasada Rao).Kerala Agricultural University. Thrissur.176p

Mathai, C.K.1983. Growth and yield analysis in black pepper varieties (Piper nigrum L.) under different light conditions. Ph.D Thesis, Division of Crop Physiology- University of Agricultural Sciences, Bangalore.

Menon, R 1981. Growth, Flowering, Floral Biology and spike shedding in Pepper (Piper nigrum L.) M.Sc (Hort) Thesis, Kerala Agricultural University, Vellanikkara.

Nalini, P.V. 1983 Flower bud differentiation in pepper (Piper nigrum L)- M.Sc (Hort)Thesis, Kerala Agricultural University, Vellnaikkara.

Neema, V.P., Mammootty, K.P., Sudheesh, M.V., Vanaja, T and Rajagoapalan, A. 2005. Effect of drip irrigation on Panniyur varieties of black pepper. In proceedings of National Workshop on Drought Management in Plantation Crops. (Eds. Nybe, E.V. and G.S.L.H.V. Prasada Rao). Kerala Agricultural University. Thrissur.176p.

Pillay, V.S., Sasikumaran, S., and Ibrahim, K.K, 1988. Effect of rainfall pattern on yield of pepper (*Piper nigrum* L.) In: Proc. of the National Seminar on Agro-meteorology of Plantation

Crops (Eds. G.S.L.H.V. Prasada Rao and R.R. Nair), held at RARS, Pilicode during 12–13,. March 1987. Kerala Agricultural University. Mannuthy, Thrissur, kerala . 147–151pp.

Rao, B.V.R. and Rao, V.U.M. 2024. Agricultural Meteorology. Jain Brothers, New Delhi. 154p

Rao, G.S.L.H.V.P. 2021. Agricultural Meteorology (5th Reprint). PHI Learning Private Limited, New Delhi, India. 364p.

Rao, G.S.L.H.V.P. and Krishnakumar, K.N. 2024. Impact of weather and climate on plantation crops in the Humid Tropics. Chapter 21 (417-432pp). Agrometeorological Applications for Climate Resilient Agriculture (Eds.): Chattopadhyay, N., Stefanski, R., Attiri, S.D. and Rathore, L.S. Springer International Publishing AG. Singapore.445p

Sharangi, A.B., Rao, G.S.L.H.V.P., Das, S., Krishnamurthy, K.S.Upadhyay, T.K., Gopakumar, C.S. and Acharya. S.K. 2024. Brunt of Climate Change and Spice Crops: Scenario, Response, and Resilience.755-812pp (Eds: Ravindran, P.V., Sivaraman, K., Devasahayam,S., Nirmal Babu, K.). Handbook of Spices in India: 75 Years of Research and Development. Springer Nature Singapore Pte. Ltd.535p

Sujatha, V.S., Manikandan, N, Nybe, E.V and Rao, G.S.L.H.V.P. 2005. Relationship of weather with spiking and yield in black pepper. In: Proc. of National Workshop on

Drought Management in Plantation Crops (Eds. Nybe, E.V. and G.S.L.H.V. Prasada Rao). Kerala Agricultural University. Thrissur.176p.

Questions

I. Elucidate on the Following

1. Weather and black pepper
2. Crop growing environment of black pepper
3. Rainfall Regime of black pepper
4. Thermal Regime of black pepper

II. Write Short Notes on

1. Weather abnormalities and black pepper
2. Coffee and Black pepper as mixed crops
3. Rainfall and black pepper yields
4. Temperature and black pepper yields
5. HU and black pepper yields
6. Water deficit during summer and black pepper yields

III. Fill in the Blanks

1. ------------ is the relationship between summer soil moisture and black pepper yields
2. Black pepper is known as the ---------- of spices
3. Black pepper is also known as ----------
4. Black pepper vines wiped out in Wayanad during the year ---------
5. Black pepper yield in summer drought years 1983 and 2004 ------------
6. On an average --------- was the increase in maximum temperature during summer 2004
7. ----------^{0}C was considered as the base temperature for calculating HU in black pepper
8. Black pepper yield is affected if the mean annual maximum temperature goes beyond -------
9. Black pepper yield is poor if the summer rains are-----------
10. ----------------crop harvesting calendar of black pepper
11. ------------percentage is the black pepper contribution from India

IV. State Frue or False

1. Black pepper is a crop of Wayanad
2. Black pepper likes bimodal rainfall
3. Soil moisture during summer may be negative in black pepper yields
4. Temperature range is widening in both the seasons (SWM and PM)
5. Continuous rains require till berry formation in black pepper
6. Rains during November is good to black pepper duing the next crop
7. Black pepper can be an intercrop in coconut plantations
8. Rainfall pattern is similar across the western ghats where black pepper is cultivated
9. Negative relationship exists between rainfall during dry months and black pepper yield
10. Negative relationship exists between HU and black pepper yield

8

Tea

Tea (*Camellia sinensis*) is one of the most important beverage crops in the World and the second most commonly drank liquid on earth after water. It has numerous medicinal benefits mainly due to its antibacterial and antioxidant properties. The major tea growing regions of the world are South-East Asia (India and Sri Lanka), Eastern and South Africa consisting of Kenya, Malawi, Tanzania, Uganda and Mozambique. All these countries are producers of black tea. China and Japan are the major producers of green tea. These countries experience cool, tropical humid climate where this crop is being cultivated. However, at present, the crop is also being cultivated in wide range of climates from Mediterranean to warm, humid tropics to include countries such as Russia and Georgia in the northern latitudes and Argentina and Australia in the southern latitudes (Plate 8. 1). The region to which the crop is grown is closely linked with the climate. The crop is grown across wide altitudes up to 2200 m a.m.s.l. Sri Lanka, Kenya, China and Vietnam are the major competitors to India in tea exporting field. China is the major producer of green tea while Sri Lanka and Indonesia are producing mainly orthodox varieties of tea. Tea industry in India occupies an important place in the national economy since last 172 years. It is an essential item of domestic consumption. Tea is also considered as the cheapest beverage amongst the beverages available in India. In India, major tea growing areas are Assam, West Bengal, Tamil Nadu and Kerala. To a limited extent it is cultivated in Karnataka, Tripura, Himachal Pradesh, Uttaranchal, Arunachal

Pradesh, Manipur, Sikkim, Nagaland, Meghalaya, Mizoram and Bihar (Plate 8.2). Unlike most other tea producing and exporting countries, India produces green tea as well, apart from CTC (Crush, Tear and Curl) and Orthodox tea. Tea export has been an important foreign exchange earner for the Country. Tea industry provides direct employment to more than a million workers mainly from backward and socially weaker sections of the society.

Plate 8.1 Tea producing countries in the World

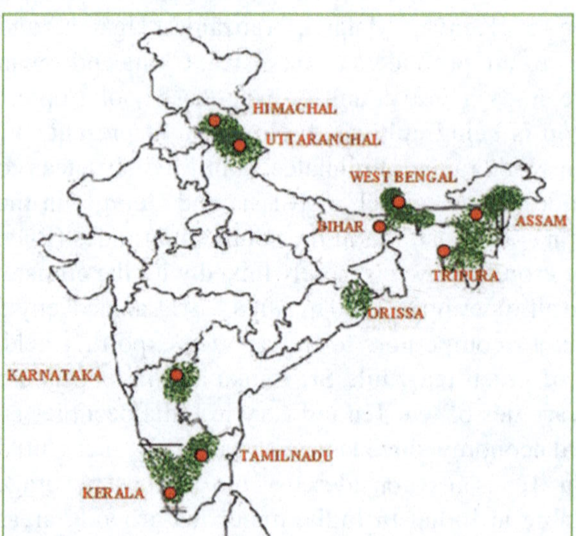

Plate 8.3 Tea producing States in India *Source*: Tea Board of India

8.1 Tea in Kerala

In Kerala, tea is cultivated mainly in Idukki and Wayanad Districts. To a limited extent it is cultivated in Palakkad District also. Being a crop of highranges it is subjected to vagaries in atmospheric temperature rather than rainfall fluctuations. High rainfall may not be constraint in the form of waterlogging and floods since it is mostly grown in terrains. As the carbon dioxide concentrations in the atmosphere have registered an increase from 230 ppm to the present level of 420 ppm, warming of the tea growing tract has become a reality. Modified rainfall patterns and temperature rise are projected in such an event. In Idukki, one of the major tea producing regions of Kerala, rise in maximum temperature and fall in minimum temperature is noticed. The temperature range is increasing. Monsoon rainfall is showing a declining trend. Changes in thermal regime are not at all beneficial to tea as it is a thermo- sensitive crop. Annual as well as monsoon rainfall is declining in the Wayanad district, where tea is cultivated extensively after Idukki. A fall in maximum and a rise in minimum temperature is observed across the highranges of Wayanad. Long term climate changes are likely to affect the developmental process that determine yield and yield attributes of tea. In a climate change scenario, the agro-climatic conditions prevailing in the tea producing areas are facing threat. Studies on crop weather relations of tea are limited and not much literature is available. It is high time to generate potential crop growing environments which can thrive well even under the projected rise in temperature with unpredicted rainfall patterns.

8.2 Area and Production of Tea

The area under tea has declined over the years. It has come down to 36400 ha in 2001–08 from 40400 ha in 1952–60. The decline in area was gradual over the years. It revealed that the area under tea cultivation was the lowest in 1981–90 (34.8 ha). It may be interesting to note that the 1981–90 was the warmest and driest decade over Kerala. Replacing the senile and unproductive plantations with young one, conversion of unproductive and uneconomical plantation area for non-agricultural purposes are some of the major reasons for decline in area under tea cultivation across the State. At the same time, tea production in the State has shown an increasing trend. A steady and gradual increase was seen over the years. It indicates that production has registered an increase of 67 per cent over the six decades.

8.3 Rainfall

The mean annual rainfall across the highranges of Kerala is 1901 mm, varying between 1865 mm at Pampadumpara and 1939 mm at Ambalavayal. The annual number of rainy days is 117. It is interesting to note that the number of rainy days did not show any significant difference between Pampadumpara and Ambalavayal. The monthly rainfall pattern at Pampadumpara (Idukki) and Ambalavayal (Wayanad) follows the same pattern. The distribution pattern of rainfall at Pampadumpara showed that the highest rainfall was received during July (365.3 mm), followed by June (295.6 mm), August (271.1 mm) and October (262.2 mm). The least was in February (14.8 mm), followed by January (18.7 mm). In the case of Ambalavayal, similar pattern was evident. The lowest rainfall at Ambalavayal was noticed during January (12.1 mm), followed by February (16.6mm) and December (24.3 mm). Both locations receive the highest rainfall in June, July, August and October due to the influence of both southwest and northeast monsoons (Fig. 8.1). Pamapdumpara receives relatively better rainfall from October to December when compared to that of Ambalavayal. It was attributed to the influence of northeast monsoon towards south of Kerala. It reveals that rainfall distribution across the cardamom tract of Kerala is bi-model, having a peak in July and another peak in October.

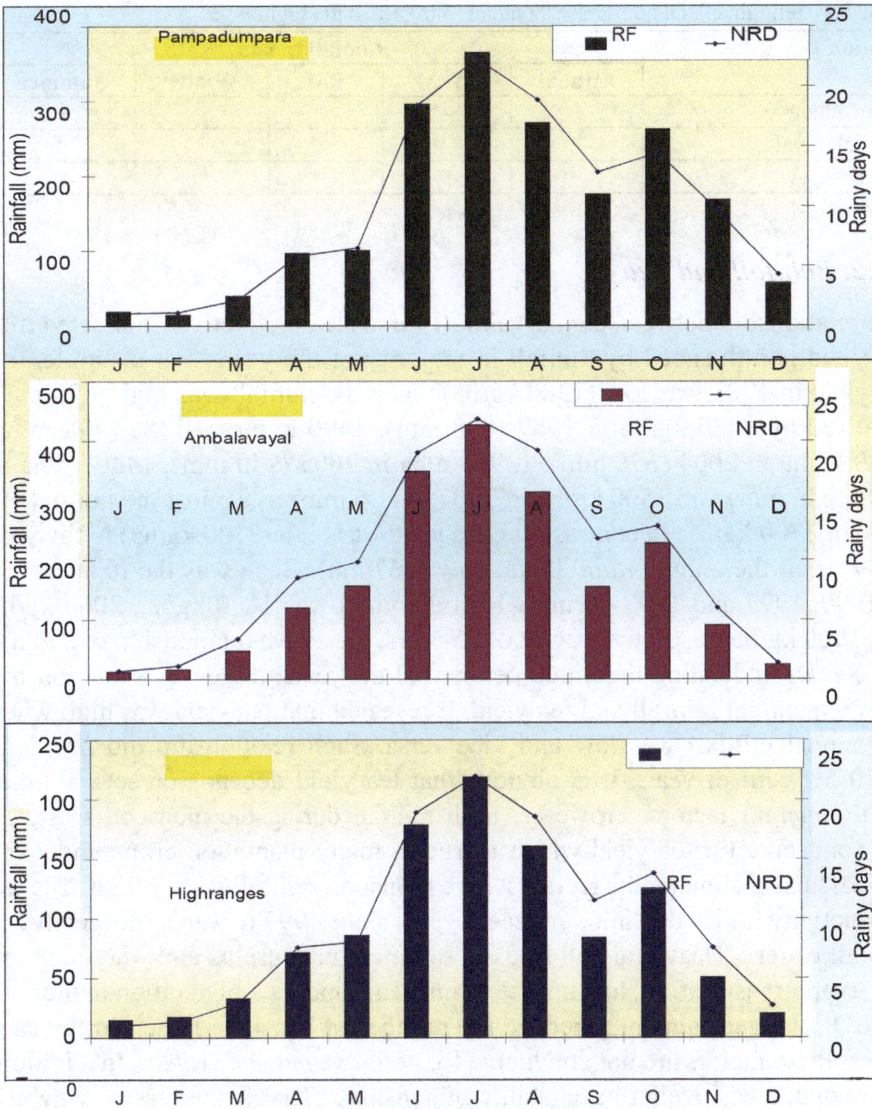

Fig. 8.1: Mean monthly rainfall and rainy days across the highranges

8.3.1 Rainfall Trends

Rainfall across the highranges of Kerala is declining during the southwest monsoon while rainfall during the post monsoon is increasing across the highranges. While Pampadumpara showed an increase in rainfall except during southwest monsoon, rainfall across Ambalavayal is declining (Table 8.1). The decline is more so since the last one and-a-half-decades.

Table 8.1 Trend in annual and seasonal rainfall (mm) across the highranges

Station	Rainfall trends				
	Annual	SWM	PM	Winter	Summer
Pampadumpara	+	-	+	+	+*
Ambalavayal	- **	- **	+ **	- **	- **
Highranges	- **	-**	+**	-**	-**

*- Significant at 5% level, ** - Significant at 1% level

8.3.2. Rainfall and Tea

The relationship between annual rainfall versus tea productivity indicated that the yield is influenced by rainfall in majority of the years. For example, the tea yield in 1985 was low (1380 kg/ha) when the rainfall was high (625mm), followed by 1450 kg/ha in 1987 (838 mm), 1490 kg/ha in 1988 (798 mm), 1500 kg/ha in 2004 (656 mm), 1390 kg/ha in 2005 (740 mm), 1440 kg/ha in 2006 (646 mm) and 1500 kg/ha in 2007 (811.6 mm) as against the normal tea yield of 1700 kg/ha. In contrast, the tea yield was high (2100 kg/ha) in the year 1989 when the annual rainfall was low (667mm). Such was the relationship in 1990, 1997 and 1999 during which the yield was 1870 kg/ha, 2000 kg/ha and 1900 kg/ha, respectively. Out of 25 years, the above relationship was found in 15 years, indicating that in 60 per cent of the years, there was a relationship between annual rainfall and tea yield. It revealed that tea yield was high when the annual rainfall was low and vice versa. Such relationship did not exist in 40 per cent of years. It is obvious that tea yield depends on several other environmental factors. However, high rainfall during the monsoon season is not conducive for tea yield which is true in many plantation crops under the topical humid climate like Kerala where monsoon rainfall is very high. The tea productivity is also declining in recent years since 1997 onwards irrespective of rainfall pattern. Heavy rainfall leads to soil moisture surplus and waterlogging, low evapotranspiration, low uptake of nutrients and lack of aeration in the root zone. Prolonged rains may reduce the number of sunshine hours. In the case of tea, these factors are not conducive for better vegetative growth. In addition, it responded positively to availability of sunshine. That is the reason, why, the tea yield was low wherever continuous rainfall with the overcast sky is noticed.

8.4 Thermal Regime

The mean annual maximum temperature varied between 25.2°C and 27.2°C, and the annual mean minimum between 17.3°C and 17.7°C, respectively at Pampadumpara and Ambalavayal. Irrespective of the seasons, variations of minimum temperature at both the locations across the cardamom tract are relatively less when compared to that of maximum temperature. However, the diurnal changes in minimum temperature during winter are predominant and in the case of maximum temperature, the diurnal changes are predominant during summer. The mean annual maximum temperature at Pampadumpara is high (29.3°C) while March (30.3°C) over Ambalavayal. The annual minimum temperature during December-January is low (15–16°C) across the highranges of Kerala. The annual range in temperature is high (9.5°C) over Ambalavayal when compared to that of Pampadumpara (7.7°C). The maximum temperature during the rainy season (southwest as well as post monsoon) varies between 23 and 27°C while the minimum between 17 and 18°C across the two locations. The range in temperature varies between 6 and 9°C. A gradual decline in temperature is noticed during winter from that of rainy season. The minimum temperature during this period across the cardamom tract is around 15–16°C and the temperature range shoots up to 9–13°C (Fig. 8.2). It is due to the fact that the maximum temperature showing a trend to shoot further up and minimum temperature to come down in the cardamom tract of Kerala. The mean summer maximum temperature never crossed 30.0°C across the highranges while the range in temperature was high over Ambalavayal during summer (10.9°C) when compared to that of Pampadumpara (9.9°C). At the same time, the temperature range at Ambalavayal during summer was low (10.9°C) when compared to winter (12.1°C). Increase in maximum temperature and decline in minimum temperature resulted to increase in temperature range. It is a concern to the plantation crops grown across the highranges of Kerala in the context of projected climate change scenario.

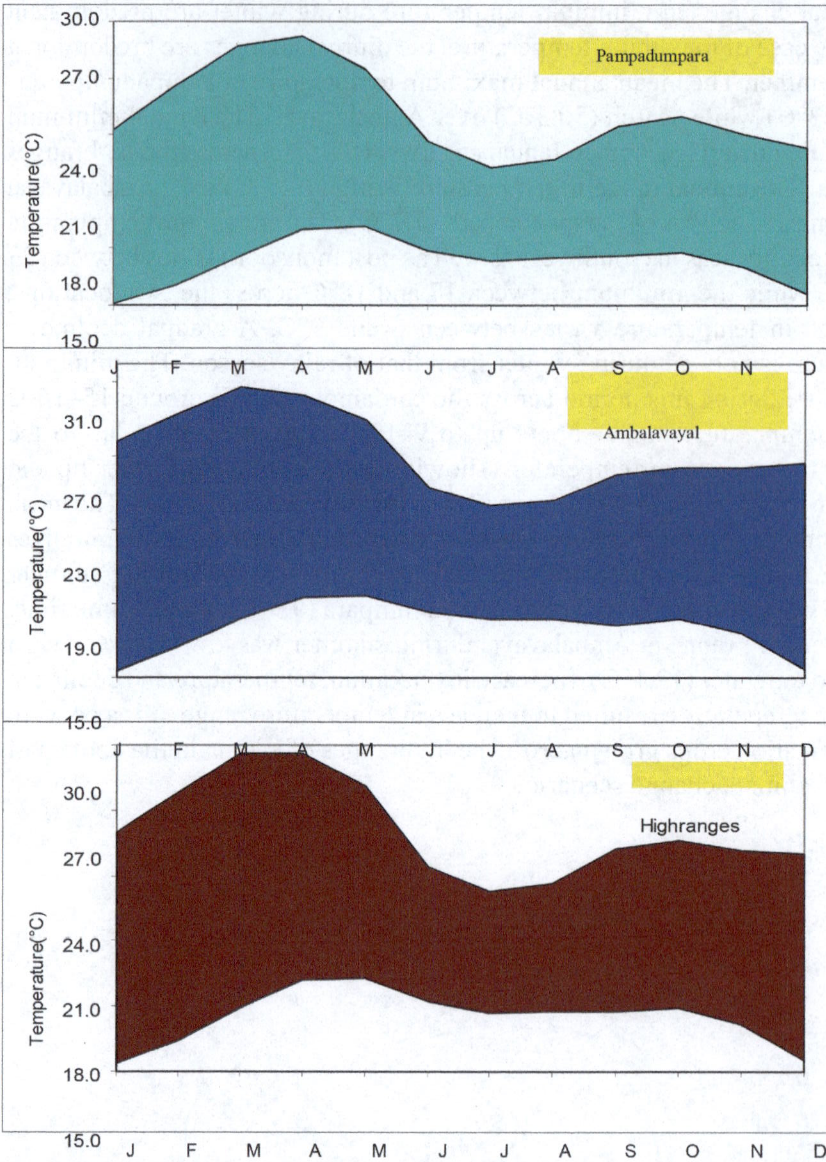

Fig. 8.2: Thermal regime across the highranges

8.4.1 Temperature Trends

The maximum temperature across the highranges was increasing irrespective of the seasons while the minimum temperature declined (**Table 8.2**). The mean temperature also showed an increasing trend except during summer. It is obvious that the range of temperature (the difference between the day maximum and the night minimum temperature) is increasing. Similar was the trend in the case of Pampadumpara. In the case of Ambalavayal (Wayanad), increase in maximum temperature is not seen as noted in the case of Pampadumpara, except a marginal increase in southwest monsoon. In the case of minimum temperature, it was increasing except during post monsoon. The mean temperature was increasing annually and during southwest monsoon while all other seasons showed declining trend. Unlike in the case of Pampadumpara, temperature range at Ambalavayal, exhibited a declining trend, except during southwest monsoon. It is worth mentioning the increase in temperature range across the highranges as it may adversely affect the thermo - sensitive crops viz., cardamom, tea, coffee and black pepper. The increase in temperature range, therefore, is a concern across the highranges as it will impact adversely on the production of these crops (Rao *et al.*, 2008).

Table 8.2: Trend in temperature (°C) across the highranges of Kerala

Maximum temperature (°C)		**Pampadumpara (1978-2009)**	**Ambalavayal (1984-2009)**	**Highranges (1984-2009)**
	Annual	+ **	- **	+ **
	SWM	+ **	+ **	+ **
	PM	+ **	- **	+ **
	Winter	+ **	- **	+ **
	Summer	+ **	- **	+ **
Minimum temperature (°C)	Annual	- **	+ **	- **
	SWM	- **	+ **	- **
	PM	- **	- **	- **
	Winter	- **	+ **	- **
	Summer	- **	+ **	- **
Mean temperature (°C)	Annual	+ **	+ **	+ **
	SWM	+ **	+ **	+ **
	PM	+ **	- **	+ **
	Winter	+ **	- **	+ **
	Summer	- **	- **	- **
Temperature range (°C)	Annual	+ *	- **	+ **
	SWM	+ **	+ **	+ **
	PM	+ **	- **	+ **
	Winter	+ **	- **	+ **
	Summer	+ **	- **	+ **

*- Significant at 5% level ** -Significant at 1% level

'+' - Increasing '–' Deceasing

8.4.2 Temperature and Tea

The tea yield was better when the maximum temperature was below 26.5°C and minimum temperature was below 17.5°C. The upper limit of maximum temperature and minimum temperature appears to be 26.5°C and 17.5°C, respectively. If the maximum temperature was more than 26.5°C, the tea yield was low. The maximum temperature was high during 2003 (28.0°C), 2005 (27.5°C) and 2008 (27.6°C) during which the tea yield was low 1470 kg/ha, 1390 kg/ha and 1430 kg/ha, respectively. Therefore, increase in temperature beyond 26.5°C is likely to affect the tea yield adversely as noticed. The tea yield was also adversely affected since 2002 onwards due to high variability in maximum temperature. The maximum temperature and productivity trends also indicated that the maximum temperature was in increasing trend while the yield declining. It revealed that the maximum temperature had a negative correlation with productivity (r = -0.564) and production (r= -0.45). It is significant at 1% level and 5% level in the case of tea productivity and production, respectively. It indicated that there exists an inverse relationship between the annual minimum temperature and tea yield as in the case of rainfall. When the minimum temperature was high (17.7°C) in 1985, the tea yield was low (1380 kg/ha). Similar was the case in 1987 during which the tea yield was low (1450 kg/ha), followed by 1988 (1510 kg/ha), 1991 (1750kg/ha), 1993 (1580 kg/ha), 1998 (1880 kg/ha), 2006 (1450 kg /ha) and 2007 (1520kg/ha). In contrast, tea yield was comparatively high when the annual minimum temperature was low. It was evident in 1980 (2050 kg/ha), 1990 (1910 kg/ha), 1992 (1910 kg/ha), 1994 (1950 kg/ha), 1995 (1750 kg/ha) and 1996 (1760 kg/ha). Such was the case in the remaining years also. It reveals that the minimum temperature and tea yield was inversely related. It was observed that the above relationship holds good in majority of years (60 per cent) during the study period. A significant negative relationship was obtained between the minimum temperature and tea productivity (r= -0.58) and production (r=–0.51). It is significant at 1% level and 5% level in the case of tea productivity and production, respectively with minimum temperature. The study revealed that the optimum temperature for tea productivity appears to be 26.5°C in terms of maximum temperature while 17.5°C in the case of minimum temperature. Based on the experimental data (Kumar and Kumar,2011) it was explained that the cumulative effect of maximum temperature, after noon relative humidity and rainfall contributed 82 per cent variation in tea yield. The contribution of sunshine hours in tea yield variation was 8.25 per cent while other variables which include minimum temperature, relative humidity in the morning and number of rainy and sunny days collectively contributed to 8.61 per cent variation. The relationship between Growing Degree Days (GDD) and yield also exhibited similar trend as in the case of annul minimum temperature. It was an inverse relationship. For example, GDD was high in 1985 (4442.3°C),1998 (4511°C), 2003 (4562.7°C), 2004 (4333.3°C), 2005 (4481°C), 2006 (4299°C), 2007 (4325°C) and 2008 (4402°C) and the yield recorded was poor (1380 kg/ha, 1880 kg/ha, 1480 kg/ha, 1500kg/ha, 1390 kg/ha, 1440 kg/ha,1510 kg/ha and 1430 kg/ha, respectively). On the other hand, yield recorded was comparatively better when the GDD was low. It was evident that yields were better when the GDD was low as seen in 1989, 1990, 1992, 1994, 1996,1997, 1999, 2000, 2001, 2002. In majority of the years, inverse relationship holds good in the case of GDD and tea yield since the GDD is a function of mean and base temperature.

8.5 Moisture Regime

The monthly moisture index at Pampadumpara varied from 388 per cent in July to -9.5 per cent in May. During the rainy months from June to November, the moisture index was always above 100 per cent (Per humid A type climate). Soon after the rainy period, the moisture index all on a sudden fell down and the climate shifted to B_1 humid type (Im=28.5 per cent) in December. In all the summer months from January to May, the moisture index was always below zero, ranging between -9.5 per cent in May to -41.3 in March, indicating that the moisture index during the above period were under the dry sub humid to arid type (Fig. 8.3). April and May fell under the dry sub humid as the moisture index was between -19 % to zero. January and February fell under Semiarid as the index was between -20 to -37.4 per cent. Only in March arid type climate was noticed, during which the moisture index was -41.3 per cent. In short, January to May, Idukki district experiences dry weather due to the non receipt of sufficient rainfall and hence the climate shifted from Perhumid during the rainy season to humid during the northeast monsoon. Thereafter the climate shifted from dry subhumid to arid during summer months as rainfall received was insignificant if pre monsoon showers fail. Therefore, the monthly and seasonal climate shifts vary insignificantly. Similar was the trend in monthly climate fluctuations at Ambalavayal also. The moisture index varied between 420.6 per cent in July to -12 per cent in April. June, July, August and October only fell under the 'perhumid A' type climate while September fell under 'B_4 humid type' and November in 'B_1 humid' climates. It may be noted that though September and October coming under the rainy months in Kerala, in the Wayanad region these months are shifting towards drier climates within the humid B type. In contrast to Pampadumpara, at Ambalavayal, December fell under even lesser type humid regime (C_1 dry sub humid). January, February and March clearly showed a tendency to move towards 'Arid type (E)'climate as the moisture index was always more than -40 per cent during the above months. As a whole, the highranges experience from 'A' perhumid climate to arid climate 'E' depending upon the rainfall distribution. Since the moisture index is predominantly influenced by rainfall, the index follows a similar pattern to that of rainfall. During April and May, the moisture index fell under 'C_1 dry subhumid' and 'C_2 moist subhumid' type of climates. The fluctuations in moisture index at Ambalavayal are more pronounced when compared to that of Pampadumpara as is evident from the analysis. Even during rainy months, moisture index showed a tendency to move towards lesser moisture regimes within the humid type climates (from B_4 to B_3 climate). Three months from January to March experience arid climate at Ambalavayal as against one (March) at Pampadumpara (Table 8.3). It reveals that Wayanad region of the Western Ghats in the highranges experience more dryspells during summer when compared to that of Pampadumpara.

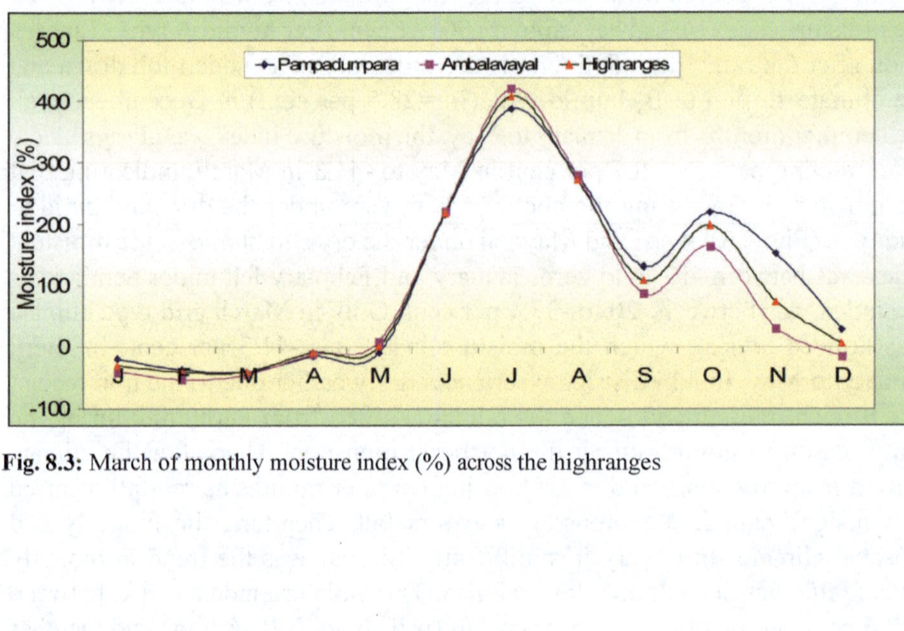

Fig. 8.3: March of monthly moisture index (%) across the highranges

Table 8.3 Mean monthly moisture index (MI) across the highranges

Month	Pampadumpara			Ambalavayal			Highranges		
	MI (%)	Climate types	Symbol	MI (%)	Climate types	Symbol	MI (%)	Climate types	Symbol
Jan	-19.9	Semiarid	D	-40.8	Arid	E	-30.6	Semi Arid	D
Feb	-37.4	Semiarid	D	-53.1	Arid	E	-41.2	Arid	E
Mar	-41.3	Arid	E	-45.1	Arid	E	-40.2	Arid	E
Apr	-15.7	Dry Sub humid	C1	-12.0	Dry Sub humid	C1	-9.6	Dry Sub humid	C1
May	-9.5	Dry Sub humid	C1	8.9	Moist Sub humid	C2	0.0	Moist Sub humid	C2
Jun	215.9	Perhumid	A	220.4	Perhumid	A	217.9	Per humid	A
Jul	387.8	Perhumid	A	420.6	Perhumid	A	408.3	Per humid	A
Aug	279.7	Perhumid	A	279.6	Perhumid	A	271.6	Per humid	A
Sept.	131.4	Perhumid	A	87.0	Humid	B4	107.9	Per humid	A
Oct	219	Perhumid	A	164.3	Perhumid	A	198.4	Per humid	A
Nov	152.2	Perhumid	A	30.9	Humid	B1	73.5	Humid	B3
Dec	28.5	Humid	B1	-15.1	Dry Subhumid	C1	6.0	Moist Sub humid	C2

8.6 Trend in Climate Shifts

The trend analysis indicated that the annual moisture index across the highranges was declining. At Pampadumpara, itwas declining at the rate of 0.79 per cent per year while 0.67 per cent per year at Ambalavayal. As a whole, the decline of moisture index is at the rate of 0.65 per cent per year across the highranges (Fig.8.4).

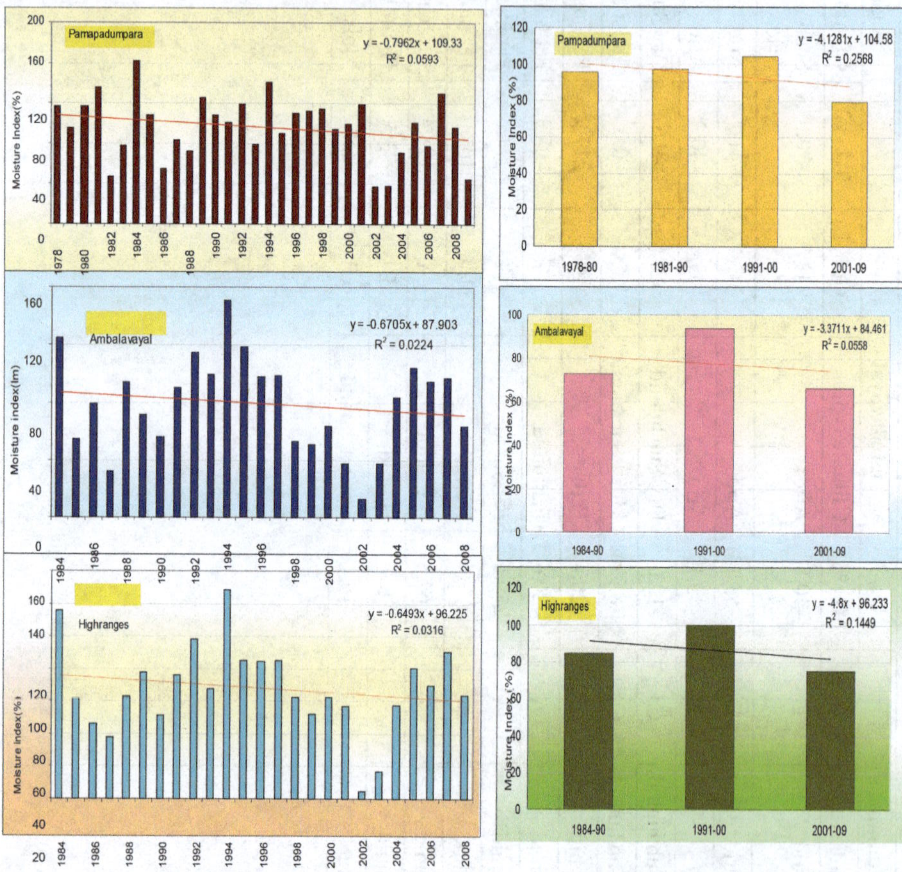

Fig. 8.4: Annual and decadal trend of moisture index across the highranges

8.7 Water Deficit and Tea Yield

The effect of soil moisture status during summer could be felt to some extent in the case of tea. The study on the effect of water deficit during summer on tea across the highranges revealed that if the water deficit during summer is low(less than 200mm) the tea production is high and vice versa **(Fig. 8.5)**

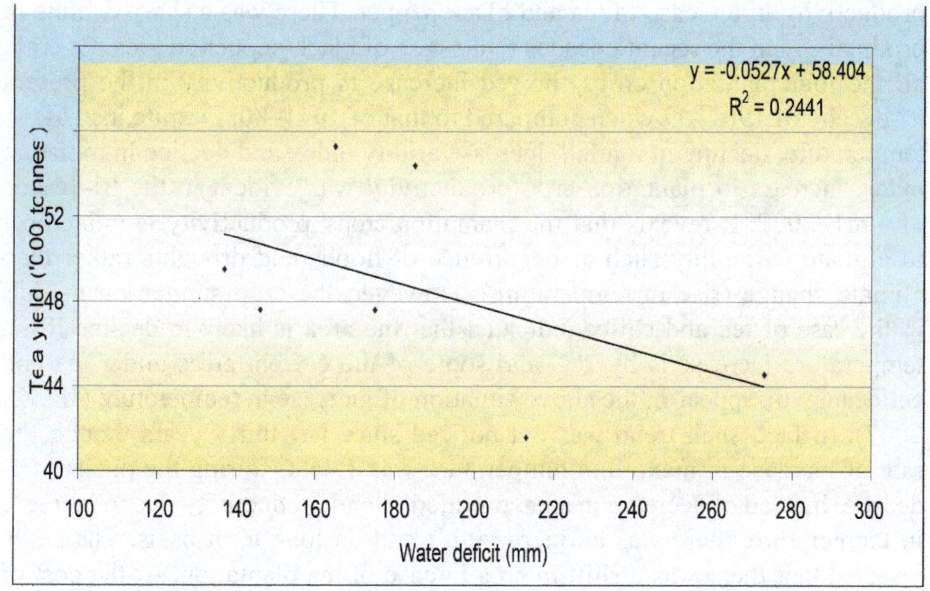

Fig. 8.5: Water deficit versus yield relationship in tea

8.8 Climate Variability and Climate Change

There was a decline in annual rainfall during the study period while increase in surface air temperature. The annual rainfall was significantly low in the decade 1981–90 and led to several summer droughts. As a result, increase in temperature was noticed in the above decade. The rate of increase in temperature was more evident since 1980s. Decline in rainfall and increase in temperature led to decline in moisture index and increase in aridity index, which is an indicator for dryspell. It led to more number of droughts (nine) in the present tri-decade of 1981–2009 when compared to that of 1951–80. The tri-decadal analysis also indicated that the decline in rainfall and moisture index, increase in temperature and aridity index were evident in the tri-decade of 1981–2009 (Figs. 8.6 and 8.7). All these factors reveal that the global warming and climate change are real in Kerala. Therefore, warming and dryness within the humid climate in the State of Kerala is a concern in the plantation sector of the State. In case of tea, the area decline was evident since 1950s and it was the lowest in the decade 1981–90. A marginal increase was noticed thereafter in the area under tea. However, increase in area was evident in the case of coffee and black pepper during the study period while decline in area under tea and cardamom. Interestingly, the production of all the plantation crops (coffee, cardamom, black pepper and tea) was in tune with the area, it was so in the case of cardamom and tea. There was a decline in productivity in the case of tea and black pepper. There was a sharp decline in productivity in the recent decades in the case of black pepper and tea. Overall, all the four plantation crops showed increase in productivity in the present tri-decade of 1981–09 when compared to that of 1951–80. Despite increase in temperature, decline in rainfall, increase aridity index and decline in moisture index, increase in plantation crop productivity was evident in the tri-decade of 1981–2009. It reveals that the plantation crops' productivity is vulnerable to climate variability such as occurrence of floods and droughts rather than climate change (rise in temperature). However, the crop simulation models in the case of tea and coffee indicated that the area is likely to decline if the temperature increase is by 2°C and some of the current areas under tea and coffee may disappear in the above situation of increase in temperature (IPCC, 2007). In fact, such trend was not noticed since last thirty years though the rate of increase in maximum temperature was 1.46°C during the present tri-decade. Instead of decrease in area, production and productivity due to increase in temperature, there was an increasing trend on long term basis. The study revealed that there was a shift in area in case of tea plantation. At the cost of tea and cardamom area, the area under coffee and black pepper was increasing and it was phenomenal since last two decades. There may be several reasons for shift in area among the four plantation crops. From farmers' point of view, a crop mix with combination of coffee and black pepper may be beneficial under the current climate variability scenario as one crop like coffee suffers, another crop like black pepper gains in a same weather situation. It may not be so in the case of monocrop like tea and cardamom. It may be probably, one of the reasons why, shift was noticed in the cropped area of these four plantations. However, it needs further probe in depth from all the angles in view of the climate change scenario. In the case of tea, area was the least during 1981–90. It was the warmest and driest decade over Kerala. Replacing of the senile and unproductive plantations with young ones, conversion of unproductive and uneconomical plantation area for non-agricultural purposes were some of the major reasons for decline in area under cultivation across the State. It is true that 1981–90 was the warmest and driest decade as far as Kerala is concerned. Tea production was adversely affected during 1983, 1985 and 1987 as evident from the data. 1987 was the warmest year on record, followed by 1983. Also, 1983 was the warmest summer across Kerala. 1982–83 was an unprecedented summer drought year that adversely affected most of the plantation crops of the State. Tea production was also affected during that year. The tea leaf production was 44600 tonnes during 1982–83 while it was 58200 tonnes in 1983–84 when there was no summer drought. Similarly tea yield was poor in 1987 (5034660 tonnes), when compared to preceeding and proceeding years.

In terms of productivity also, the productivity was poor during the drought year 1982–83 (1266 kg/ha). It showed a decline of 3.98 per cent when compared to the preceeding year 1981–82 (1318 kg/ha). During the warmest year of 1987 also, the tea productivity was less (1453 kg/ha). It was less by 4.03 per cent when compared to the previous year 1985–86 (1514 kg/ha). It indicates that tea is more vulnerable to abrupt changes in weather during the peak crop growth period. It was one of the reasons for decline in the yield since last one decade. Change in cropping pattern, variations in leaf quality and increased pest attacks triggered by transformed climatic conditions have put the tea industry in Kerala under tremendous pressure. Pest attacks, especially in the Vandiperiyar-Peermedu belt in Kerala, had resulted in doubling of expenses on pest control and fighting diseases over the past five years. It appears that there was also a noticeable increase in atmospheric temperature in the recent years. Increased temperature has resulted in the outbreaks of various diseases and pest attacks, especially in the belt where these have traditionally been a problem. Tea leaves needed a lot of sunshine. Slow growth induced by poor sunshine resulted in leaves becoming hard and leathery, which may result in poor quality. The tea leaf production had been significantly hit by heavy rain during the September-December phase in 2010, which used to be the high cropping period. Reports indicate that there have also been significant swings in relative humidity, rainfall and sunshine hours per day when the decennial mean is compared with conditions in 2009. It was observed that climate variability appeared to favour the building of sucking pests like red spider mite. Hence, the weather aberrations not only adversely affect tea productivity directly but also indirectly through pest and disease intensity. Temperature (both maximum and minimum) had a cumulative impact on crop productivity irrespective of the topography. Each weather variable either alone or in combination imposed stress on the plants and the compounded effect of climate variability not only affected the plant metabolism but also influenced the crop productivity. Therefore, a gradual increase in temperature may not impose any deleterious impact on the crop productivity, but a sudden change may put the plants under stress, leading to low yields and trigger a sudden outbreak of pest and pathogens as well.

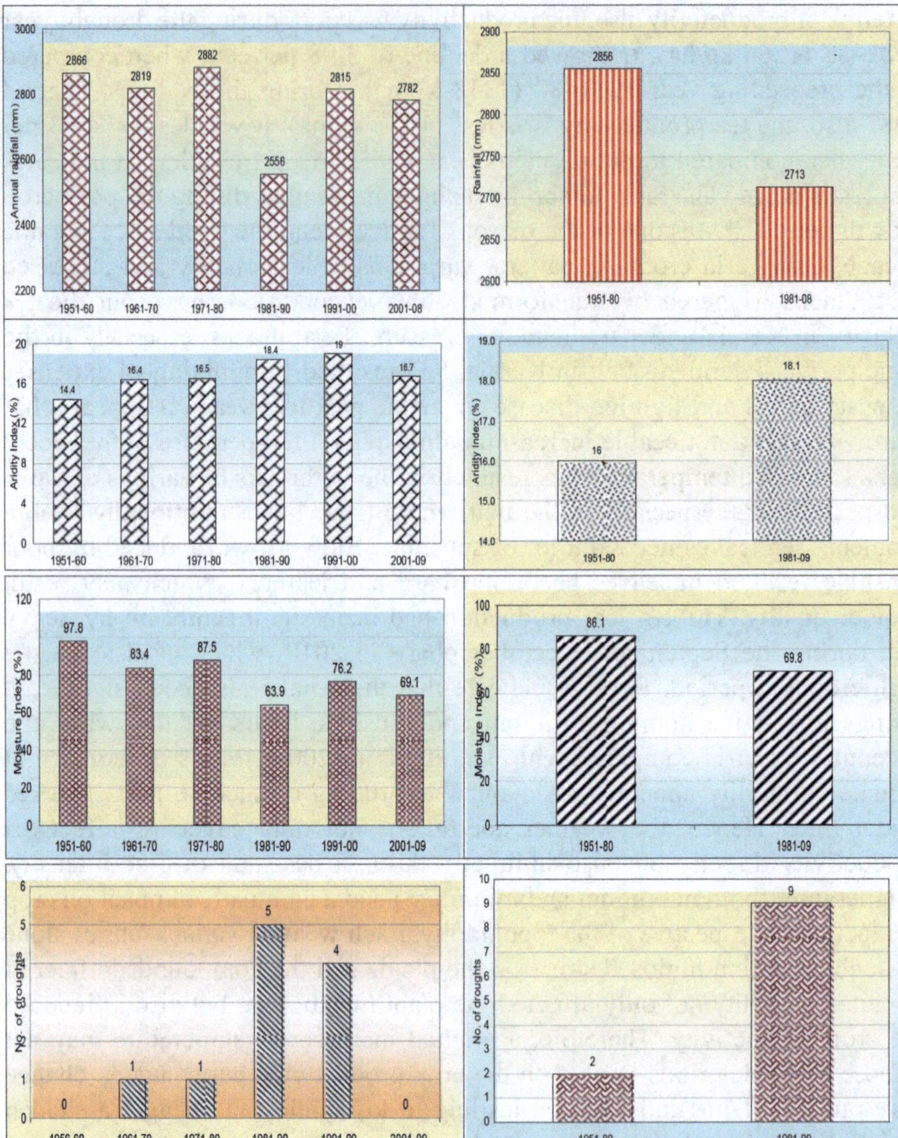

Fig. 8.6: Decadal and tri-decadal changes in annual rainfall, aridity index, moisture index and droughts over Kerala

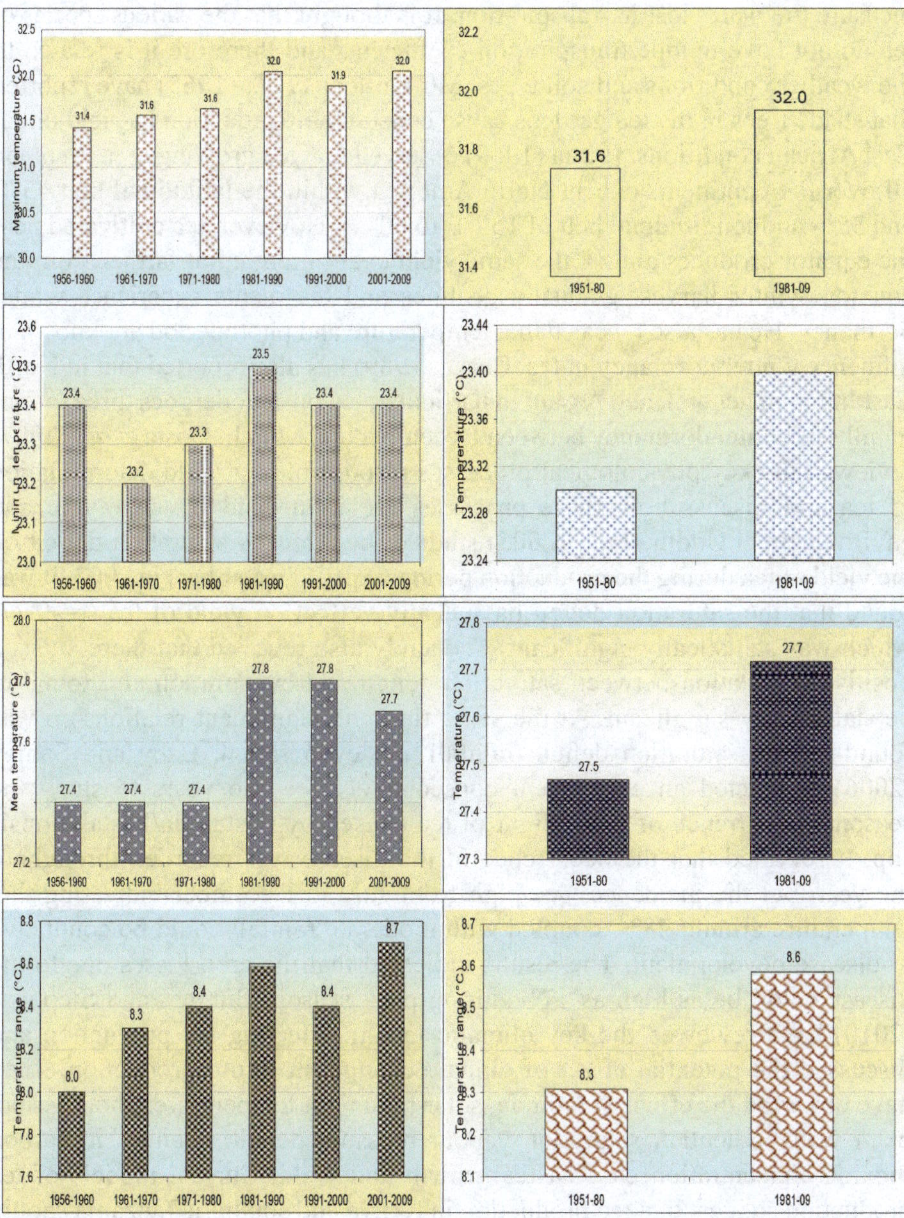

Fig. 8.7: Decadal and tri-decadal changes in temperature across the State of Kerala

Summary

It is difficult to specify the ideal climate requires for good growth especially with regards to rainfall. Attempts have been made using many species to measure the water lost in transpiration. It is thought that the various species of tea do not have unique transpiration coefficients and therefore it is related to the weather conditions and soil types. MC Culloch *et al.,* (1965) have reported that shade trees in the tea gardens cause considerable reduction in yield under East African conditions. Barua (1969) observed that tea growing area is spread all over the continents except North America within the latitudinal belt 45°C and 35°S and longitudinal belt of 150°E to 60°W. However, tea cultivated near the equator produces almost the same yield every month, but farther from the equator, winter harvest gradually declines and tea plants experience winter dormancy. He has also revealed that temperature and photoperiod are known to influence winter dormancy in tea. Barua (1989) has also reported that in North East India, in general, and Assam in particular, tea bush undergoes three to four months of winter dormancy between December and March. Costa *et al.,* (2007) reviewed the key physiological processes responsible for yield determination of tea and discuss how these processes are influenced by genotypic and environment. J Uddin *et al.,* (2005) studied the effect of saturation deficit on the yield of tea during the production period (April- December) in 2003. It was found that the saturation deficit has negative effect on yield of tea (r=-0.77) which was statistically significant. The study also revealed that there exists a positive correlation between saturation deficit and evaporation and found to be statistically significant. At the same time, no significant relationship was found among saturation deficit, rainfall and evaporation. Baby and Sanjay (2006) conducted an experiment consecutively for two years to study the seasonal occurrence of grey blight of tea caused by Pestalotia/Pestalotiopsis spp. It revealed that the occurrence of the disease was recorded throughout the year, but the incidence was high from July to December indicating that temperature around 28°C coupled with moderate rainfall could be conducive to disease development. The results indicated that the crop losses due to the disease could be as high as 17% during peak season. Burgess and Stephens (2010) have reviewed the key climatic factors affecting tea production and discussed the potential effect of climate change on future production. They have observed the effect of latitude, solar radiation, temperature, photoperiod, plant water potential, saturation vapour pressure deficit, changes in carbon dioxide concentration etc., on tea growth and yield. Current models of tea production suggest that tea production in Kenya and southern Tanzania should be well placed to increased concentrations of CO_2.

In southern Malawi, where mean temperatures are higher, further increase in temperature is generally expected to have a negative effect on production. The net effect in both the places will depend on the climatic response of pests. Mohotti and Lawlor (2002) related the diurnal changes in the rate of photosynthesis of mature tea bushes at high elevations in the field in Sri Lanka in relation to environmental conditions. Interactions between factors regulating photosynthesis in tea are also discussed. Mathews and Stephens (1998) described a model predicting biomass production and water use in tea plantations. Tripathi, et al., (2004) attempted to generate tea yield prediction model using Leaf Area Index (LAI) derived from IRS-IC LISS-III sensor records. LAI measurements were made in the fields and an empirical model was developed to find the suitability of using field LAI for annual tea yields. Field LAI values were related to satellite based LAI estimates. They found a strong relationship with a correlation coefficient of 0.76. Thus satellite based LAI values can be effectively used for tea yield prediction model. Ranganathan and Natesan (1987) developed an equation for agro-climatic potential applicable to South Indian conditions based on growth of tea plant as related to photosynthetic efficiency and weather parameters. A prediction value of 50 per cent was achieved by using agro-climatic potential of the previous month and that of the current month. The rest of the prediction should come from crop husbandry and manuring practices which affect the per cent of total dry matter harvested as crop and availability of minerals. Pandey (2009) conducted an experiment and demonstrated that biochemical could be used effectively to enhance early bud break in pruned tea bushes and /or reducing the winterdormancy and thereby achieve a higher level of productivity without affecting the quality. Certain chemicals like cycocel when used judiciously can be applied to regulate rush season crop in Northeast India. Squire et al., (1993) studied the shoot production for two clones at four sites in western Kenya differing by 300 m in altitude and 16.4°C to 18.1 °C in mean air temperature. It was observed that the duration of shoot growth cycle increased with altitude and the rate of shoot extension decreased due to temperature differences. Wijeratne (1996) discusses the various aspects of the adverse effects of climate change on Sri Lanka's tea industry. He has observed that tea yield is greatly influenced by weather. Drought causes irreparable losses to tea gardens as irrigation is seldom practiced in tea gardens. In the projected climate change scenario, Sri Lanka is likely to experience heavy rainfall events and warmers temperatures. He noted that increase in temperature, soil moisture deficit and saturation vapour pressure deficit in the low elevations will adversely affect growth and yield of tea. Rao et al., (2008) have opined that the thermo-sensitive crops like cocoa, black pepper, coffee, cardamom and tea may need attention as temperature range is likely to increase and rainfall is likely to decline in addition to deforestation as these crops are grown under the influence of typical forest-agro-ecosystems. Rao, et al., (2009) while highlighting the climate change/variability studies in the State, have observed that climate variability in the highranges of Kerala is likely to adversely affect the thermo-sensitive crops like tea, coffee, black pepper and cardamom.

Suggested Readings

Barua, 1969. Seasonal dormancy in Tea. (Camellia sinensis L). Nature 224:514.

Barua, 1989. Science and practice in Tea culture. Pp. 91–363. Tea Research Association, Culcutta - Jorhat.

Burgess, P.J. and Stephens, W 2010. Climate, Climate Change and Tea Production in Kenya, Tanzania and Malawi. Infini-Tea: Ensuring the Future of Tea 10–11 May 2010. New Delhi, India.

Costa, W.A.D.J.,Janaka, M and Madawala A. Wijeratna. 2007 Ecophyisiology of Tea.Braz. J. Plant Physiol. Vol.19. No.4 Londrina Oct - Dec. 2007.

Matthews and Stephens, W. 1998. Cuppa-Tea: A Simulation Model Describing Seasonal Yield Variation Simulation Model Describing Seasonal Yield Variation and Potential production of Tea 2. Biomass Production And Water Use. Expl.Agric. 1998. 34: 369–389.

MCCulloch, J.S.G., Pereira, H.C., Kerfoot, O and Goodchild, N.A. 1965. Effect of shade trees on tea yields. Agric.l Meteorol. 2(6) December 1965. pp.385–399

Rao, G.S.L.H.V.P., Mohan, H.S.R., Gopakumar, C.S and Krishnakumar, K.N. 2008. Climate change and cropping systems over Kerala in the humid tropics. J. of Agrometeorol. (Special issue-Part 2): 286-291. pp. 286-291

Ranganathan,V and Natesan, S. 1987. Agrometeorology of productivity in Tea in South India. In: Proc. of National seminar on Agrometeorology of Plantation Crops. KAU Press, Mannuthy. Pp.187.

Squire, G.R., Obaga, S.M.O and Othieno, C.O. 1993. Altitude, temperature and shoot production of tea in the Kenyan highlands. J. of Experimental Agri., (29): 107–120. Cambridge University Press.

Subrahmanyam, V.P. 1982. Water balance and applications with special reference to India. Andhra University, Waltair, No.173: 102.

Tripathi, K.N., Rajapakse, S.S and Honda, K. 2004. Tea yield modeling based on satellite derived LAI. Geocarto International, 19(3):51–54.

Uddin, J.Md., Rafiqul Hoque, Md., Ahmed, M and Saha, J.K. 2005. Studies on the effect of saturation deficit on the yield of Tea. Pakistan J.of Meteorol. 12 (4): 3–8.

Wijeratne, M.A. 1996. Vulnerability of Sri Lanka Tea production to Global Climate Change. J. of Water, Air, and Soil Pollution Kluwer Academic Publishers 92: 87–94.

Questions

I. Elucidate on the Following

1. Rainfall Regime and tea yield
2. Temperature Regime and tea yield
3. Moisture Regime and tea yield

II. Write Short notes on

1. Water deficit during summer versus tea yield
2. Rainfall and tea yield
3. Temperature and tea yield
4. Heat Units and tea yield
5. Thermal Regime across the highranges
6. Rainfall Regime across the highranges
7. Tea producing States in India

III. Fill in the Blanks

1. The ----------- beverage among beverages in India
2. The ----------- largest drink after water
3. ----------, ------------ and --------- reduce tea growth and development in low elevations
4. Tea is grown across the wide altitudes up to ---------------AMSL
5. Tea is cultivated in ----------and ----------Districts in Kerala
6. The temperature range across the tea tract is ----------
7. Tea is a ----------- crop
8. Increase in temperature trends is not ---------- for tea
9. Tea responds-----------to sunshine while ------------to rainfall
10. The moisture index declines from July and increases again in the month of -----------

IV. State True or False

1. Relationship exists between water deficit during summer and tea yield
2. Winter dormancy in tea production is noticed in NER
3. Round-the-year tea harvest is noticed very near the equator
4. Relative humidity is the most important factor for tea production
5. Two peaks in tea harvest periods are noticed across the Western Ghats
6. Tea productivity is declining in Kerala since 1997 onwards
7. The number of rainy days is almost the same between Ambalavayal and Pumpadumpara
8. Rainfall during the post monsoon season is declining

9. Variations in minimum temperature are minimum when compared to maximum temperature across the tea tract over Kerala
10. The maximum temperature across the high ranges was increasing in all the seasons

9

Rubber

The natural habitat of rubber (*Heavea brasiliensis*) is rain forests of the Amazon basin, situated within 5° North and South at altitudes below 200m. The climate of this region is equatorial monsoon type characterized by mean monthly temperature by 25 to 28°C, well distributed rainfall and no marked dry weather. Though it is originated in the Amazon basin, it is now predominantly grown in the tropics where an equatorial monsoon type climate prevails. Regions between 8°N and 10°S which includes Indonesian archipelago, Malaysia, southern part of Sri Lanka and some other Islands, are better suited to rubber cultivation. The production of natural rubber in the country was 8.65 lakh MT in 2008–09, registering a 4.74 per cent growth compared to the previous year. India is the fourth largest producer of natural rubber with a share of eight per cent in the World after Thailand, Indonesia and Malaysia and the fourth largest consumer of natural rubber after China, USA and Japan. Among the rubber growing countries in the World, India occupies a pride place with the top most position in productivity (Table 9.1). The productivity of natural rubber (NR) in India in 2009–10 was 1784 kg/ha. In 1950–51, it was 284 kg/ha. The growth achieved in terms of the productivity of natural rubber is 628 per cent. This achievement is rare in comparison with other agricultural and plantation crops (Table 9.2). Kerala accounts for 81 per cent of the area under rubber in the Country. The coverage under the crop in 2008–09 was 5.17 lakh ha, higher by 5500 ha over the previous year. The production of natural rubber in Kerala during the year was 7.83 lakh tonnes indicating a four per cent increase over the previous year.

The increasing trend in productivity continued during 2008–09. It was 1190 kg/ha in 1998–99, which rose to 1514 kg/ha during 2008–09. In terms of tapping area, the productivity of natural rubber recorded was 1514 kg/ha during the year 2008–09. Though the production and productivity of rubber is showing an uptrend since long, the crop production is to a certain extent determined by the weather conditions. Rubber requires an equally distributed monthly rainfall of 125 mm which is equivalent to the potential evapotranspiration loss from the tree under tropical monsoon climates. In the tropical monsoon climate, the potential evapoptranspiratrion rate is about 4mm per day (Monteith, 1977). Rubber trees are generally grown under rainfed conditions except in nurseries where essential irrigation is provided. In traditional rubber growing areas, the total rainfall ranges between 2000 to 4000 mm spread in 140–200 rainy days. The rainfall is evenly distributed throughout the year. In heavy soils, excessive rainfall may lead to waterlogging. Heavy rainfall also causes nutrient loss by run off and leaching. It also causes soil erosion. Regional yield fluctuations in south India were related to the soil moisture availability (Rao *et al.,* 1990). By increasing soil moisture availability to the plant by providing irrigation, adopting soil moisture conservation techniques, yield loss can be avoided to some extent (Haridas *et al.,* 1987; Vijay Kumar *et al.*, 1988). According to Liyanage *et al.,* (1984), Haridas and Subramaniam (1955) rainfall exceeding 9 to 11 mm per day is not congenial as it hinder harvesting and other operations.

Table 9.1: Production of natural rubber in major rubber producing countries in the world

Country	**Production ('000 tonnes)**	
	2009	**2010***
Thailand	3164	3072
Indonesia	2843	2938
Malaysia	970	1050
India	845	890
Vietnam	750	780
China	647	690
Sri Lanka	148	153
Philippines	102	107
Cambodia	45	63

*Anticipating figures

Table 9.2: Growth in area, production and productivity of NR from 1950-51 to 2009-10 in India

Item	1950–51	2009–10	Growth (% increase)
Area (ha)	74915	686882	917
Production (MT)	15800	931400	5895
Productivity (Kg/ha)	284	1784	628

(After Mohanan, 2010)

Satheesan (1987) has reviewed the work done in characterizing the radiation climate in rubber plantations in determining the evapotranspiration and canopy photosynthesis. The evapotranspiration rate of rubber plantations ranges from 4 to 6 mm a day after refoliation with plenty of soil water. Unlike other plantation crops, overall production of natural rubber does not seem to be influenced by drought of a particular year. At micro level drought may influence the production of natural rubber.Yield of rubber is influenced by the climatic conditions prevailing over a place and crop management practices adopted rather than the fluctuations in weather (Sailajadevi, *et al.*, 2005). Temperature is another environmental factor which influences plant growth. The plant gets affected by extreme temperatures. Mean monthly temperature 20–25°C has been found to be optimum. High temperature conditions result in higher rates of evapotranspiration leading to severe soil moisture stress in the absence of rainfall. High temperatures above 37°C, coupled with soil moisture stress, result to injury to leaf and drying of leaf margins (Chandrasekhar *et al.*, 1990 and Vijayakumar *et al.,* 1998).Thermal injury coupled with water deficit results in increased tree loss. In areas experiencing low temperature, the growth rate increases with increase in temperature. Under the projected climate change scenario, rubber production also poses threat.

Studies conducted at the Rubber Research Institute of India (RRII) have also point towards an increased NR productivity due to warming in the North-Eastern States, where low temperature is currently a limiting factor for cultivation. This aspect would also promote expansion of rubber cultivation to more areas in such regions. Models developed at RRII clearly indicate that if both maximum and minimum temperatures rose concurrently by 1°C, productivity could be inhibited by 9–16 per cent in the agro-climatic conditions of Kerala. Studies conducted using genetically-modified (GM) rubber plants in the laboratories of the RRII indicate that such plants may be able to tolerate drought better. The approach holds great potential for developing location-specific clones that can tolerate the adverse effects of climate change such as drought and attack of insect, pest and diseases.

9.1 Trend in Area, Production and Productivity

Area under rubber was just 62.6 thousand ha in 1952–53. It has increased by 727 per cent and reached 517.5 thousand ha in 2008–09. The area under rubber has been steadily increasing over the years (Fig.9.1). Decadal area indicated that it was in increasing over the decades. During 1952–60, area under rubber was 83 thousand ha which has increased to 490 thousand ha in 2001–09. The percentage increase in area was more (79%) from the decade 1952–60 to 1961–70, followed by 1971–80 to 1981–90 (56 %) and 1981–90 to 1991–00 (41 per cent). The increase in area was the least in the recent decade 2001–09 (9.6 per cent). Tri-decadal area has increased significantly by 179 per cent. Rubber Production also has been in upward trend since 1952. Increase in area was tremendous over a period of six decades. The increase in production was 3960 thousand tonnes from 1952 to 2009. Decadal production was in increasing trend. The lowest production (21 thousand tonnes) was in the decade 1952–60. The production during the decade 2001–09 was 684 thousand tonnes. The percentage increase in decadal production was the highest in 1971–80 (155 per cent), followed by 1991–90 (141 per cent) and 1961–70 (114 per cent). The least increase in production was recorded in 2001–09 (51 per cent), followed by 1981–90 (63 per cent). The tri-decadal production of rubber has registered an increase of 589 per cent. As in the case of production, productivity of rubber has been increasing over the decades. The average productivity of rubber during 1952–60 was only 271 kg/ha. It has reached 1393 kg/ha in 2001–09. Tri-decadal productivity also has registered an increase of 155 per cent (Fig. 9.2).

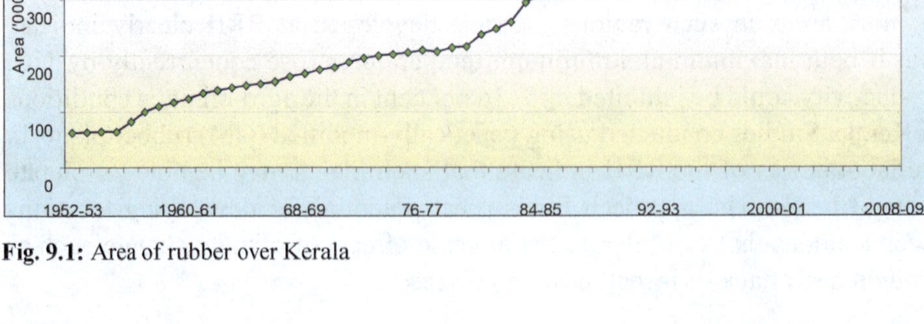

Fig. 9.1: Area of rubber over Kerala

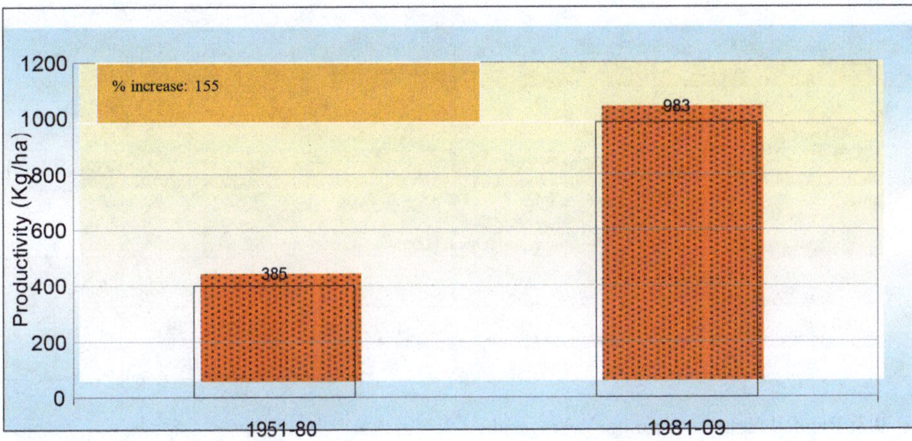

Fig. 9.2: Tri-decadal change in productivity of rubber over Kerala

9.2 Rainfall

Rubber plants are traditionally grown under rainfed conditions. Irrigation is practiced only in nurseries. Under rainfed conditions, monthly rainfall should be sufficient to meet the water requirement of the crop. In the tropical monsoon climate, the potential evapotranspiration rate is about 4 mm per day (Monteith, 1977) and therefore, an amount of 125 mm per month is considered essential to maintain optimum growth of plants. In traditional rubber growing areas like Kerala, annual rainfall ranges from 1000 to 4000 mm. The distribution pattern of rainfall in Kerala is such that it is having two peaks one in southwest monsoon season and another one in post monsoon period. However, the bimodel pattern of distribution of rainfall is restricted from central and southern districts where the influences of both the monsoons are felt. Rubber being a crop requires well distributed rainfall during its yielding phase may not perform well under a unimodel rainfall regime. That is why, rubber production is comparatively low in areas where unimodel rainfall is seen such as northern districts of Kerala. Heavy rainfall from June to September followed by weak rainfall in October - November is the rainfall scenario of Kannur and Kasaragod districts. If the pre-monsoon showers fail most of the plantation crops are subjected to soil moisture stress under rainfed conditions. However, such long dryspells are not conducive to rubber as it yields poor under dryspell/ drought situation. This is more so in northern Kerala consisting of Kannur and Kasaragod districts where the rubber area is comparatively poor (Fig. 9.3).

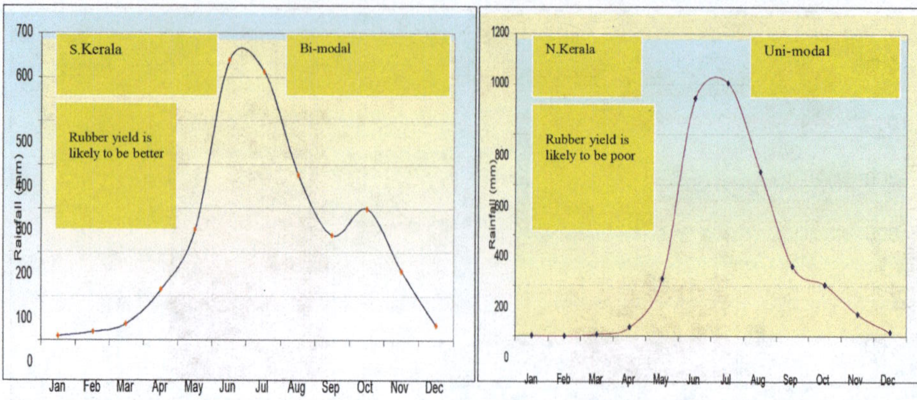

Fig. 9.3: Rainfall regimes of rubber over Kerala

9.3 Thermal Regimes

Temperature is one of the key environmental factors which influence plant growth. Rubber, being a crop adapted to moderate temperatures, gets affected by extremes in temperatures. Higher temperature results in higher rates of evapotranspiration and thereby to severe soil moisture stress in the absence of rainfall. Rubber is grown extensively across the midlands of Kerala. It can thrive well and produce better in the midlands of Kerala where the thermal environment is conducive to its growth. The mean annual maximum temperature across the midlands is

32.6°C while it is 22.8°C in the case of mean annual minimum temperature. The mean maximum temperature in the midlands of Kerala where the crop is grown extensively varies between 29.2°C in July and 36.7°C in March. In the case of minimum temperature it varied between 21.3°C in January and 24.4°C in April (Fig. 9.4). The thermal environment in Kerala in general, midlands of the State in particular is conducive for the growth and development of rubber.

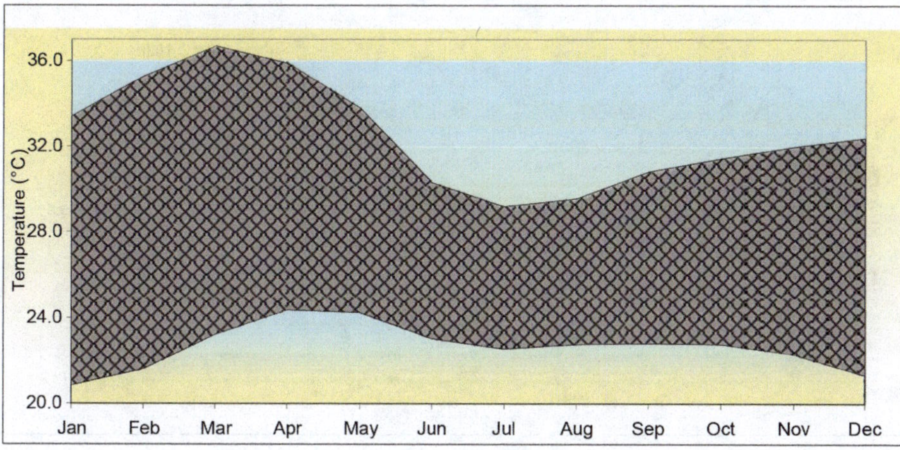

Fig. 9.4: Thermal regime of rubber in Kerala

9.4 Climate Variability and Rubber

The rate of increase in rubber productivity was affected during 1959–63 (-24.3%), followed by 1979–83 (-8.3%), 1984–88 (2.2%), 1999–03 (13.5%) and 2004–08 (18.9%). The percentage decline in productivity of rubber could be attributed to the adverse weather situations prevailed during that period. 1961 was severe flood year as far as Kerala is concerned. Severe flood due to continuous rainfall and more number of rainy days led to low productivity, though it was not very evident from the productivity chart. Continuous rainfall adversely affects rubber production. Tapping is adversely affected in the event of heavy rainfall and extended number of rainy days. Similarly, during 1979–83 and 1984–88 there was a decline in the rate of increase of yield (Fig. 9.5).

1981–90 was the warmest and hottest decade in Kerala, which might have influenced adversely the rubber yield during that decade. The productivity of rubber showed a different pattern when the increase in productivity was considered on decade-wise. It was the lowest (6.7%) during 1981–90 followed by 31.7% in the recent decade 2001–09 (Fig. 9.6). It is interesting to note that the above pentads fell under these decades.

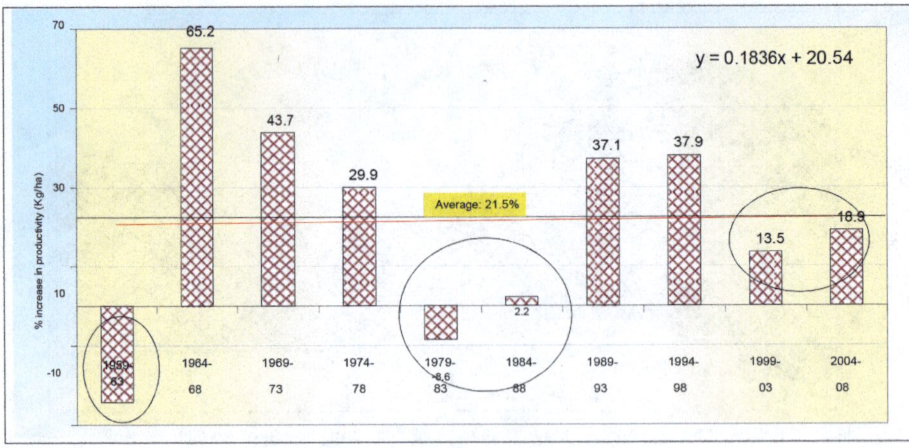

Fig. 9.5: Pentad-wise rate of increase in rubber productivity in Kerala

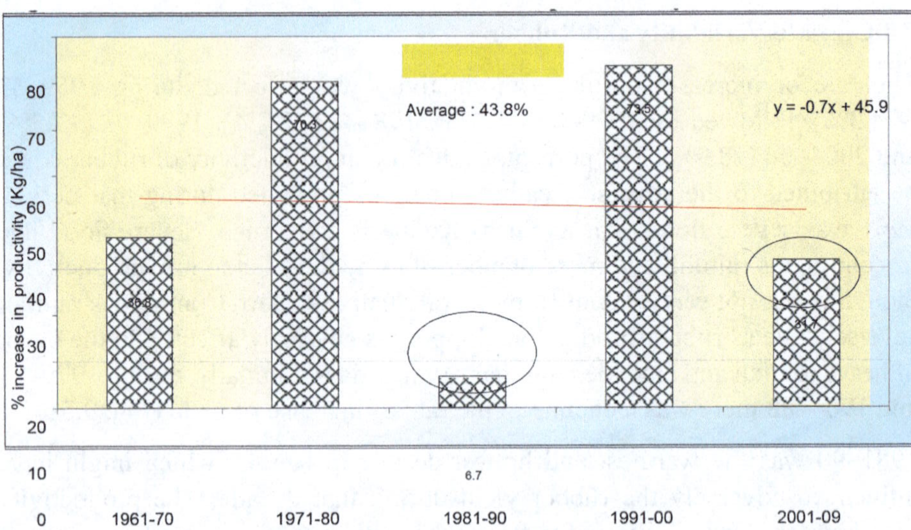

Fig. 9.6: Decadal rate of increase in rubber productivity in Kerala

9.5 Climate Change and Rubber

Area, production and productivity of rubber have been increasing over a period of time. Rubber cultivating area since 1952–53 has increased by 727 per cent while it was 3960 per cent in the case of production. The increase in the case of productivity was 392 per cent. Tri-decadal area under rubber cultivation has registered an increase of 179 per cent while it was 589 per cent in the case of production. The increase was 155 per cent in the case of productivity. Overall, area, production and productivity of rubber have been increasing over the decades. Despite increase in temperature, aridity index, number of droughts, decline in rainfall and moisture index (index of climate shifts), the rubber productivity was in increasing trend. It indicated that the long term increase in temperature is unlikely to affect rubber yield adversely. At the same time, the prolonged dryspells during summer due to absence of rainfall during northeast monsoon and failure of summer showers is likely to affect rubber productivity adversely as seen in recent years. Abrupt changes in weather may affect the productivity due to continuous rainfall and more number of rainy days when compared to that of previous year 2006–07 (1554 kg/ha). Climate change in the form of climate variability, especially unpredictable, irregular and deficient rainfall pattern adversely affects the growth and productivity of natural rubber. Changes in rainfall pattern triggers fungal infection of leaves, which adversely affects productivity. The Country's production has declined to 8, 25,000 tonnes in 2007–08 from 8,52,000 tonnes in the previous year, according to official sources. Kerala, the main rubber growing state, witnessed about 30,000 tonnes fall in production at 7.53 lakh tonnes from previous year. Heavy and extended rainfall with good number of rainy days hampered the tapping in rubber. Similarly, the yield in rubber is likely to be declined in 2010–11 due to the extended rains in post monsoon. Of course, soil erosion, soil degradation and insect, pest and diseases affect the yield of rubber which is again determined by the weather vagaries. Rise in temperature may not have much role in determining rubber productivity as the thermal regime of rubber is not subjected to frequent fluctuations even during summer. It is rainfall- its quantum, distribution and number of rainy days which determine the yield decline in rubber. Rubber growing areas in south Kerala suffered due to heavy rainfall which extended up to December 2010. Good quality planting material having tolerance to drought and pest and disease attack are to be used extensively to combat the changes in weather in the contest of climate change and global warming to sustain rubber production in the country. In short, it revealed that rubber yield is not affected due to long term changes in temperature, but is likely to be affected due to short term monsoon uncertainties like prolonged rain during monsoon, followed by extended rain during post monsoon as seen in 2007 and 2010. Similarly, prolonged dryspells during summer may adversely affect rubber yield under rainfed conditions. Therefore, detailed studies in these directions are to be taken up with experimental data systematically collected for this purpose.

Summary

The area, production and productivity of rubber have been increasing over a period of last four decades. Rubber plants are traditionally grown under rainfed conditions. Irrigation is practiced only in nurseries. Under rainfed conditions, monthly rainfall should be sufficient to meet the water requirement of the crop. In the tropical monsoon climate, the potential evapotranspiration rate is about 4-5 mm per day and therefore, an amount of 120-150 mm per month is considered essential to maintain optimum growth of plants. In traditional rubber growing areas like Kerala, annual rainfall ranges from 1000 to 4000 mm. The distribution pattern of rainfall in Kerala is such that it is having two peaks one in southwest monsoon season and another one in post monsoon period. However, the bimodal pattern of distribution of rainfall is restricted to central and southern districts where the influences of both the monsoons are felt. Rubber being a crop requires well distributed rainfall during its yielding phase may not perform well under a unimodal rainfall regime. That is why, rubber production is comparatively low in areas where unimodel rainfall is seen such as northern districts of Kerala. Heavy rainfall from June to September followed by weak rainfall in October - November is the rainfall scenario of Kannur and Kasaragod districts. If the pre-monsoon showers fail most of the plantation crops are subjected to soil moisture stress under rainfed conditions. However, such long dry spells are not conducive to rubber as it yields poor under dryspell / drought situation. This is more so in northern Kerala consisting of Kannur and Kasaragod districts where the rubber area is comparatively less. Rubber, being a crop adapted to moderate temperatures, gets affected by extremes in temperatures. Higher temperature results in higher rates of evapotranspiration and thereby to severe soil moisture stress in the absence of rainfall. Rubber is grown extensively across the midlands of Kerala. It can thrive well and produce better in the midlands of Kerala where the thermal environment is conducive to its growth. Rubber yield is affected due to short term monsoon uncertainties like prolonged rain during monsoon, followed by extended rain during post monsoon as seen in 2007 and 2010. Similarly, prolonged dry spells during summer may adversely affect rubber yield under rainfed conditions.

Rubber and cashew compete in terms of area over a period of time depending upon market price. Cashew is always replaced by rubber, sometimes vice versa. These two crops are mostly seen in mid lands of Kerala. However, the crop comes up well in southern districts of Kerala due to the influence of southwest and northeast monsoon rainfall as indicated in the preceding paragraph. Cashew possesses a special place in Kannur and Kasaragod Districts while rubber in Kottayam District. It indicates that the uni-modal rainfall is favourable to cashew while bimodal rainfall to rubber. It is one of the reasons, why, rubber yield is relatively in northern districts. Delay in commencement of rubber taping is also noticed in northern districts of Kerala due to less girth even after 6-8 years of its growth and development on rubber planting and number of tapings is also less. Prolonged summer droughts adversely affect rubber yield under rain fed conditions.

Suggested Readings

Chandrasekhar,T.R., Jana, M.K., Thomas, J., Vijayakumar, K.R. and Sethuraj, M.R.1990. Seasonal changes in physiological characteristics and yield in newly opened trees of Hevea Brasiliensis In North Konkan. Indian J. of Nat. Rubber Res.3(2): 88–97.

Gopakumar,C.S. 2011. Impacts of Climate Variability on Agriculture in Kerala. Ph.D. Thesis. Cochin University of Science and Technology. Kochi, Kerala.286p

Haridas, G., Eusof, Z., Sudin, M.N., Bachick, A.T and Young R.Y.N. 1987. Soil moisture conservation in rubber. In Proc. Rubber Research institute of Malaysia, Rubber Growers' Conference. Johore, Malaysia. Pp.154–166.

Haridas, G. and Subramanian, T. 1985. A critical study of the hydrological cycle in mature stand of rubber (Hevea brasiliensis Muell. Arg.). J. of the Rubber Research Institute of Malaysis. 33: 70–82.

Monteith, J.L. 1977. Climate. In: Ecophysiology of Tropical Crops. (Eds. P.de T. Alvin and T.T.Kozlowski). Academic Press. New York, pp.1–27.

Nybe, E.V. and Rao, G.S.L.H.V.P. 2005. Drought management in plantation crops. Proceedings of the National Workshop. Kerala Agricultural University. Thrissur, Kerala, India. 176p.

Rao, B.V.R. and Rao, V.U.M. 2024. Agricultural Meteorology. Jain Brothers, New Delhi. 154p

Rao, G.S.L.H.V.P. 2021. Agricultural Meteorology (5th Reprint). PHI Learning Private Limited, New Delhi, India. 364p.

Rao, G.S.L.H.V.P. and Krishnakumar, K.N. 2024. Impact of weather and climate on plantation crops in the Humid Tropics. Chapter 21 (417-432pp). Agrometeorological Applications for Climate Resilient Agriculture (Eds.): Chattopadhyay, N., Stefanski, R., Attiri, S.D. and Rathore, L.S. Springer International Publishing AG. Singapore.445p

Rao, P.S., Jayarathnam, K. and Sethuraj, M.R.1990. Water balance studies of the rubber growing regions of south India. J. of Applied Hydrology. 3(2): 23–30.

Sailajadevi, T., Nair, B.R., Jacab, C.K and Vijayakumar, K.R. 2005. Does drought influence the overall production of natural rubber? Pp 99–104. In: Proc. of the National workshop on Drought Management in Plantation Crops.176p. KAU, Thrissur.

Sharangi, A.B., Rao, G.S.L.H.V.P., Das, S., Krishnamurthy, K.S.Upadhyay, T.K., Gopakumar, C.S. and Acharya. S.K. 2024. Brunt of Climate Change and Spice Crops: Scenario, Response, and Resilience. 755-812pp. Eds: Ravindran, P.V., Sivaraman, K.,Devasahayam,S., Nirmal Babu, K. Handbook of Spices in India: 75 Years of Research and Development. Springer Nature Singapore Pte. Ltd.535p

Vijayakumar, K.R., Dey, S.K.,Chandrasekhar, T.R., Devakumar, A.S., Mohanakrishna, T., Rao, P.S and Sethuraj, M.R. 1998. Irrigation requirement of rubber trees (Hevea Brasiliensis) in the subhumid tropics. Agrl. Water Management, 35: 245–259.

Questions

I. Elucidate on the following

1. Monsooned Malabar Coffee
2. Weather over Coffee Processing
3. Weather and nut size in coconut
4. Weather and kernel weight in cashew

II. Write Short notes on

1. Monsooned Malabar Coffee
2. Rainfall Regime of coffee processing
3. Temperature Regime over coffee processing
4. Seasonal nut size in coconut
5. Kernel weight in cashew

III. Fill in the Blanks

1. -------------- Coast where monsoon coffee is processed
2. ------------ happens when coffee beans are spread on the floor during the monsoon season
3. --------- is reduced in coffee beans after processing
4. ----------- increases in coffee beans after processing
5. ------------ plays a major role to change the taste of coffee
6. ---------- number of pods harvested during November is equal to ---------number of pods harvested during summer months
7. -----------, -----------, and ---------- are tested to know the taste of coffee
8. The maximum temperature during the monsooning varies between ---------and ----------^{0} C
9. The minimum temperature during the monsooning varies between ---------and ----------0 C
10. The moisture content in coffee beans was estimated based on ---------- and ---------
11. ---------------variety is tastier as it contains less acidity and high sugars
12. -------------- unit is a function of temperature
13. ------------ is the critical phase in coconut development
14. The final nut size in coconut is influenced by ----------

15. Copra content is maximum in -----------season
16. Husked nut weight is maximum in -------------season

IV. State True or False

1. The final nut size in coconut is dependent on the exposure to temperature at female flower stage
2. Coffee comes up well where coconut cultivation is uneconomic altitudinally
3. Coffee requires open conditions for its better performance
4. Coffee is processed where profuse rainfall is noticed during the rainy season
5. The coffee beans don't bulge during the process of monsooning
6. The acidity and sugars are low in Robusta when compared to that of Arabica
7. Relative moisture content is more with increase in minimum temperature
8. Relative moisture content is more with increase in vapour pressure deficit
9. Dry spell is required during nut set and development in cashew
10. It is easy to dehusk the kernels if it is in better size
11. Three varieties were tested to understand the quality in cashew
12. Three hybrids were analysed to test the copra content in coconut
13. Copra content is less in WCTxCDO when compared to other tested hybrids
14. Coconut hybrids record less copra content in southwest monsoon
15. Three phases are seen in coconut development from female flower fertilization to harvest
16. Five cocoa beans in summer are equivalent to seven cocoa beans in post monsoon season in terms of weight
17. The growth of nut size in coconut is slow during the first phase
18. The total heat units below 2100 day^0 is not congenial for coconut development
19. Madakkathara records high temperatures when compared to Pilicode during the reproductive phase of cashew
20. Monsoon adds flavour to Malabar coffee

10

Weather and Quality

Coffee is grown at higher elevation across the Western Ghats between 750 and 1200 m amsl. It requires a certain degree of shade for its better performance. The annual rainfall over the coffee growing tracts varies between 1400 and 2600mm, having uni/bi-model rainfall with prolonged dryspell from November to May. It is more so over North of the Western Ghats. The mean annual maximum temperature varies between 24 and 30°C while the minimum between 15 and 18°C. The difference between maximum and minimum temperatures (temperature range) varies between 5 and 16°C across the coffee tract. The 'Monsooned Malabar Coffee' is a specialty coffee produced from the erstwhile 'Malabar Region'. It undergoes certain changes in physical and biochemical characters when the coffee beans are processed during the summer monsoon (June to September). It is processed at Mangalore, located in Coastal Karnataka. The processing location experiences heavy rainfall of more than 1000 mm in July, followed by 950 mm in June and 600 mm in August. Out of 3264 mm of annual rainfall, 2852 mm is received during the monsoon period. It accounts more than 80 per cent of annual rainfall. The maximum temperature during the coffee monsooning period varies between 28 and 30°C while night temperature revolves between 22 and 24°C. Being located in the tropics along the West Coast, the crop processing environment in terms of surface air temperature is relatively uniform and moderate with a characteristic feature of heavy clouds and continuous rainfall during the monsoon period. These monsoon features are more predominant from the middle of June to August during which the coffee beans are subjected to monsooning. The coffee beans bulge in the process of monsooning and the volume increases significantly due to absorption of atmospheric moisture. It is directly related to the monsoonal wind during which coffee beans are processed. The bulging was less in all the samples processed during monsoon 2003, which varied from 2.1 to 2.7 ml/100 number, while it ranged between 2.7 and 5.8 ml/100 number in the samples processed during monsoon 2004. The density of coffee beans declines as the change in weight of coffee beans is marginal and the change in volume is large. The acidity and polyphenols are low and sugars are high in Arabica types to that of Robusta while it possesses more carbohydrates when compared to Arabica. The starch and amino acids appear to be similar in both the types. Probably, it might be the reason for better aroma in Arabica as well in Robusta though Arabica is superior in terms of taste as it contains less acidity and high sugars. The moisture content of Arabica varied from 12.2 to 14.8% while 12.8 to 15.7% in the case of Robusta. The moisture content, accrued during monsooning, leads to certain physical and biochemical changes in coffee beans. Robusta responds better to monsooning in terms of body appearance like volume, weight, density and moisture content due to its initial better bean size though Arabica attains superiority in quality. Arabica also responds slowly to monsooning as it takes more time to complete its processing cycle. Contrary to the above, the Arabica pea berry responds better to monsooning when compared to that of Robusta pea berry. The superiority of arabica types may be due to inherent property in addition to monsooning. The moisture content of coffee beans was high in 2005. It resulted in high sugars, starch content and amino acids. Such trend was not seen in other biochemical properties. It appears that continuous clouds with heavy rainfall and its better distribution, high atmospheric water vapour content, moderate surface air temperature and gentle sea breeze would lead to better moisture content in coffee beans, if processed during the monsoon period. Probably, active monsoon period within the monsoon season may be more conducive for such processing of coffee beans. It is one of the factors leading to superior quality in addition to other genetical and environmental factors. Vapour pressure deficit and night temperature had significant relationship with moisture content of coffee beans.

10.1 Quality of Coffee: Monsooned Malabar Coffee

Monsooned Malabar Coffee is a speciality coffee processed during the monsoon season along the West Coast. It experiences heavy rainfall of more than 1000 mm in July, followed by 950 mm in June and 600 mm in August. It accounts about 80 per cent of annual rainfall, where coffee beans are processed. Maximum temperature dips to 28–29°C during the processing period due to heavy monsoon rains and difference between day-and-night surface air temperatures is around 6–8°C. Occasionally, it shoots up to 10–12°C depending upon variations in maximum and minimum temperatures during the monsoon season (June- September) if no rains are received. It appears that continuous rainfall together with moderate day-and-night temperatures (22–30°C) during the processing period may be conducive for coffee bean processing as the quality of coffee beans is superior. It is believed that when the coffee beans are subjected to monsooning, certain biochemical changes take place through physical characters. These physical and biochemical changes appear to vary with monsoon features and poor rainfall with its ill distribution may adversely affect the quality of coffee beans. For the first time, the monsoon rainfall was very poor (less by 33 %) and it was deficit by more than 58 % in July along the West Coast in 2002 and its distribution was erratic. Process of coffee beans commences normally in the end of June or July with receipt of good rains during which rainfall was less in 2002. It might have affected the quality of coffee beans adversely though it depends on genotype, time of harvest, nutrient status of soils, crop management and cultural practices. Studies on "Monsoon influence on quality of coffee beans" are not seen in literature or scarce. Keeping this in view, a study was undertaken in collaboration with M/s Aspinwall to understand the influence of monsoon on quality of coffee beans in terms of physical and biochemical properties for a period of three years, commencing from 2003 to 2005. The popular cultivars tested for the study were Arabica and Robusta. The results emerged out of this project are projected for the benefit of coffee industry like in previous chapters.

10.1.1 Rainfall

The number of rainy days during the monsoon season (June-September) in 2003, 2004 and 2005 was 87, 82 and 88 days, respectively. The total amount of rainfall in that order was 2858 mm, 1758 mm and 2184 mm. It reveals that rainfall spread was relatively uniform in terms of number of rainy days during the study period while its daily distribution and the amount were not the same and it was more so in 2004 (Fig. 10.1). The processing of coffee beans was taken up in July and August in 2003 and 2005 while mid-June to first week of August in 2004 in the case of Robusta. Similar was the case in Arabica also except in 2005 during which, the spread was between 14th June and 9th August. Arabica took 57–59 days while 50–55 days in the case of Robusta for completion of its processing cycle. It is evident that the response is relatively slow in Arabica to monsooning when compared to that of Robusta. Both the cultivars exposed to relatively high rainfall and number of rainy days in 2003 (1210 mm in 48 days) and 2005 (1227 mm in 47 days) while low rainfall and number of rainy days in 2004 (825 mm in 42 days) as noticed similarly during the monsoon season (Table 10.1).

Table 10.1 Mean weather during monsooning of coffee

Year	Mean weather during monsooning of coffee										
	Max. T (°C)	**Min. T (°C)**	**Mean T (°C)**	**Rainfall (mm)**	**No.of rainy days**	**Wind speed (Km/h)**	**Sunshine hours (hrs/day)**	**R.H (%)**	**Vapour pressure (mm of Hg)**	**Saturation vapour pressure (mm of Hg)**	**Vapour pressure deficit (mm of Hg)**
2003	29.2	22.9	26.0	1210	48	2.7	3.5	91	22.6	25.7	2.7
2004	28.8	22.1	25.4	825	42	3.0	4.9	89	23.0	26.2	3.2
2005	28.8	21.6	25.2	1227	47	2.7	3.0	90	23.2	25.7	2.7

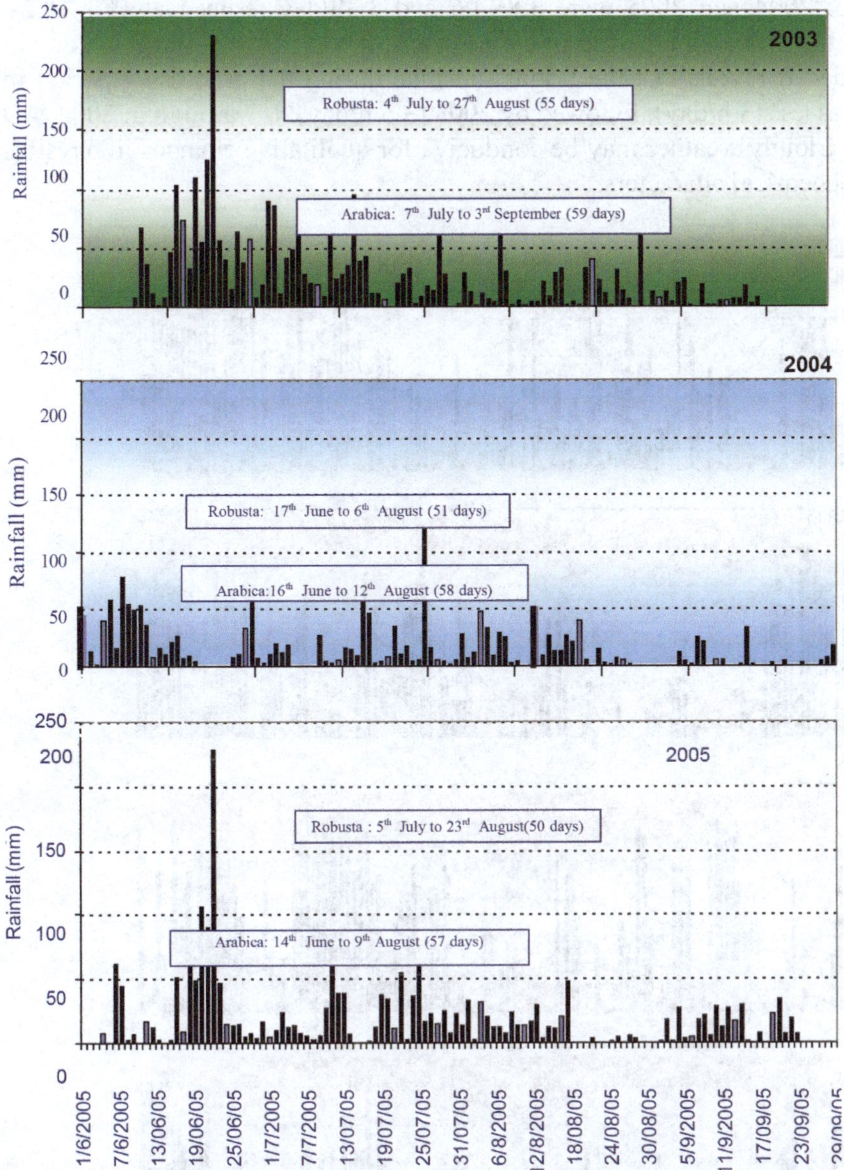

Fig. 10.1: Daily rainfall during monsoon season over coffee processing units

10.1.2 Bright Sunshine

The bright sunshine is an indication of cloud amount. If the cloud amount is more the number of bright sunshine hours is less and vice-versa during the monsoon period.The average bright sunshine hours during the monsoon season in 2003, 2004 and 2005 were 3.8, 4.6 and 3.3h/day respectively**(Fig.10.2).** During the processing period of coffee beans, it was 3.5, 4.9 and 3.0 h/day in that order. It reveals that the bright sunshine during the processing period in 2005 was less (3 h/day), followed by 2003 (3.5 h/day). It was high in 2004 (4.9 h/day). Cloudy weather may be conducive for qualitative changes in physical and biochemical characters.

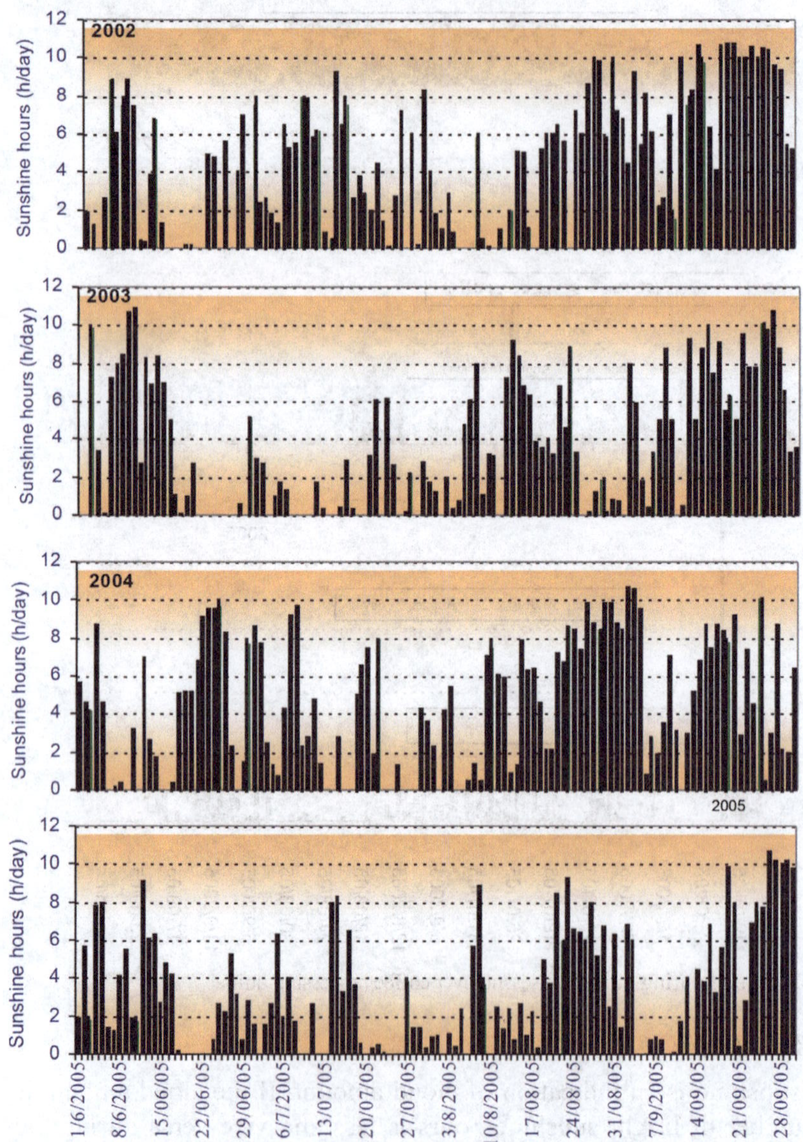

Fig. 10.2: Daily bright sunshine (h/day) during monsoon season over coffee processing units

10.1.3 Wind

The average wind speed varied between one and three km/hr during the monsoon period on majority of days, reaching above 6 km/hr occasionally in June, 2003. On an average, it was 2.7 km/hr during the processing period of coffee beans. In 2004, it was 3 km/hr and reached up to 6–9 km/hr in the initial period of processing (Fig.10.3). The wind speed in 2005 was 2.7 km/hr during the processing period, reaching up to 6–9 km/hr during the initial period of processing like in 2004. It revealed that moderate wind was noticed in 2004 and 2005 when compared to that of 2003. Monsoon wind carries lot of atmospheric moisture and hit the West Coast, where the processing industries are located. It results in cloudy and rainy weather on majority of days during monsoon period unlike in other parts of the Country.

10.1.4 Relative Humidity and Vapour Pressure

The actual vapour pressure was 22.1, 22.0 and 22.7 mm of Hg in monsoon 2003, 2004 and 2005, respectively. The mean saturation vapour pressure deficit was 3.4, 3.5 and 3.3 in the above order while it was 5.3, 5.1 and 4.8 in the afternoon hours (Fig. 10.4). It reveals that the saturation vapour pressure deficit was relatively low in 2005 as the actual vapour pressure was high, followed by 2004 and 2003. There was only a marginal difference (89–90%) in relative humidity among the years under study (Fig. 10.5). However, the difference between forenoon and afternoon was marginally low (9%) in 2005 as against 11% in 2003 and 2004 during the monsoon period. Relative humidity is the ratio between actual and saturation vapor pressure expressed in percentage while the saturation vapour pressure deficit is the difference between the saturation vapour pressure and actual vapour pressure. The saturation vapour pressure is a function of atmospheric temperature. It revealed that the saturation vapour pressure deficit during the processing of coffee beans was less in 2005 (4.8 mm of Hg), followed by 2003 (5.4 mm of Hg) and 2004 (5.5 mm of Hg) as seen similarly for the monsoon period.

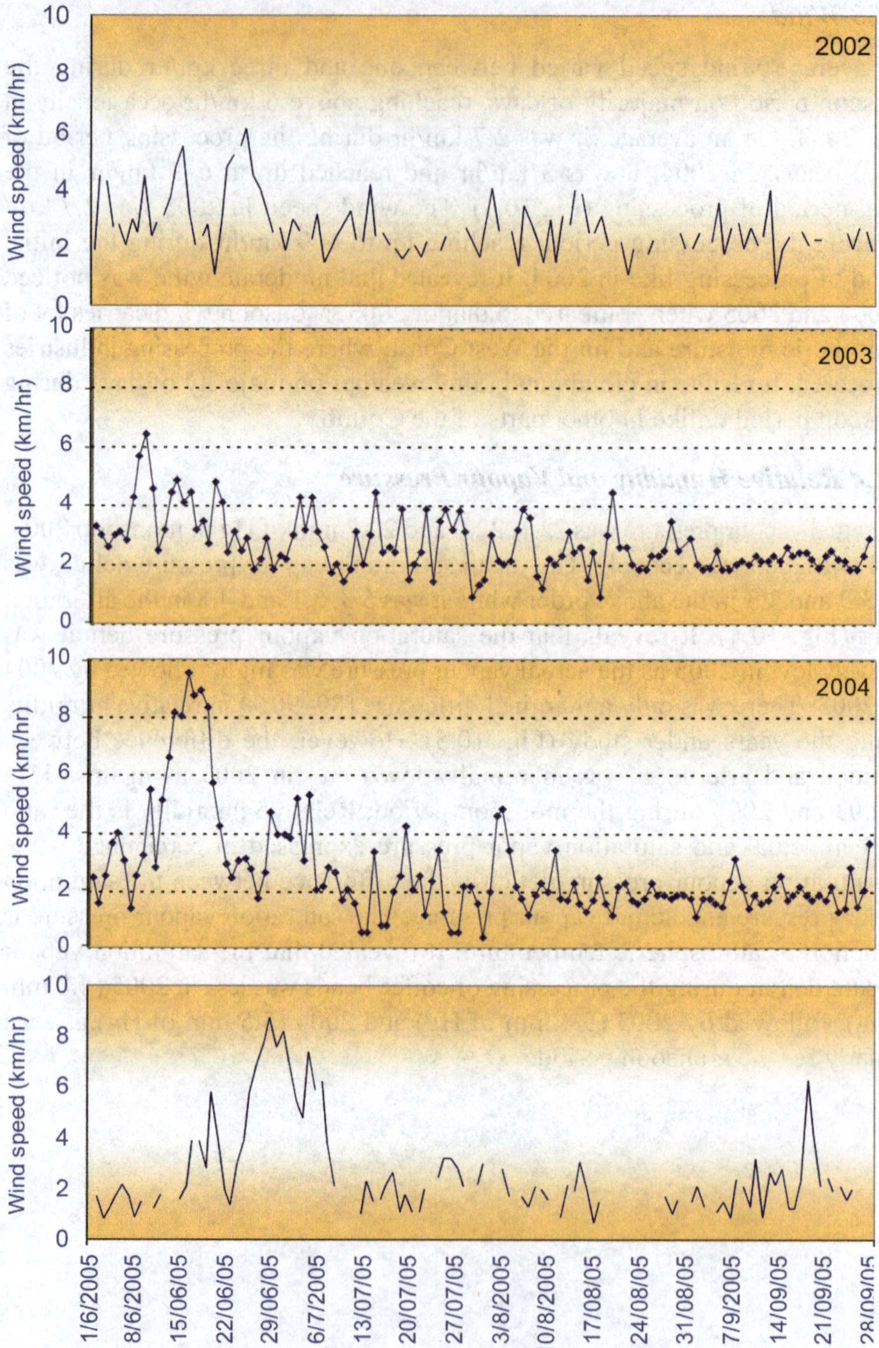

Fig. 10.3: Daily wind speed (km/hour) during monsoon season over coffee

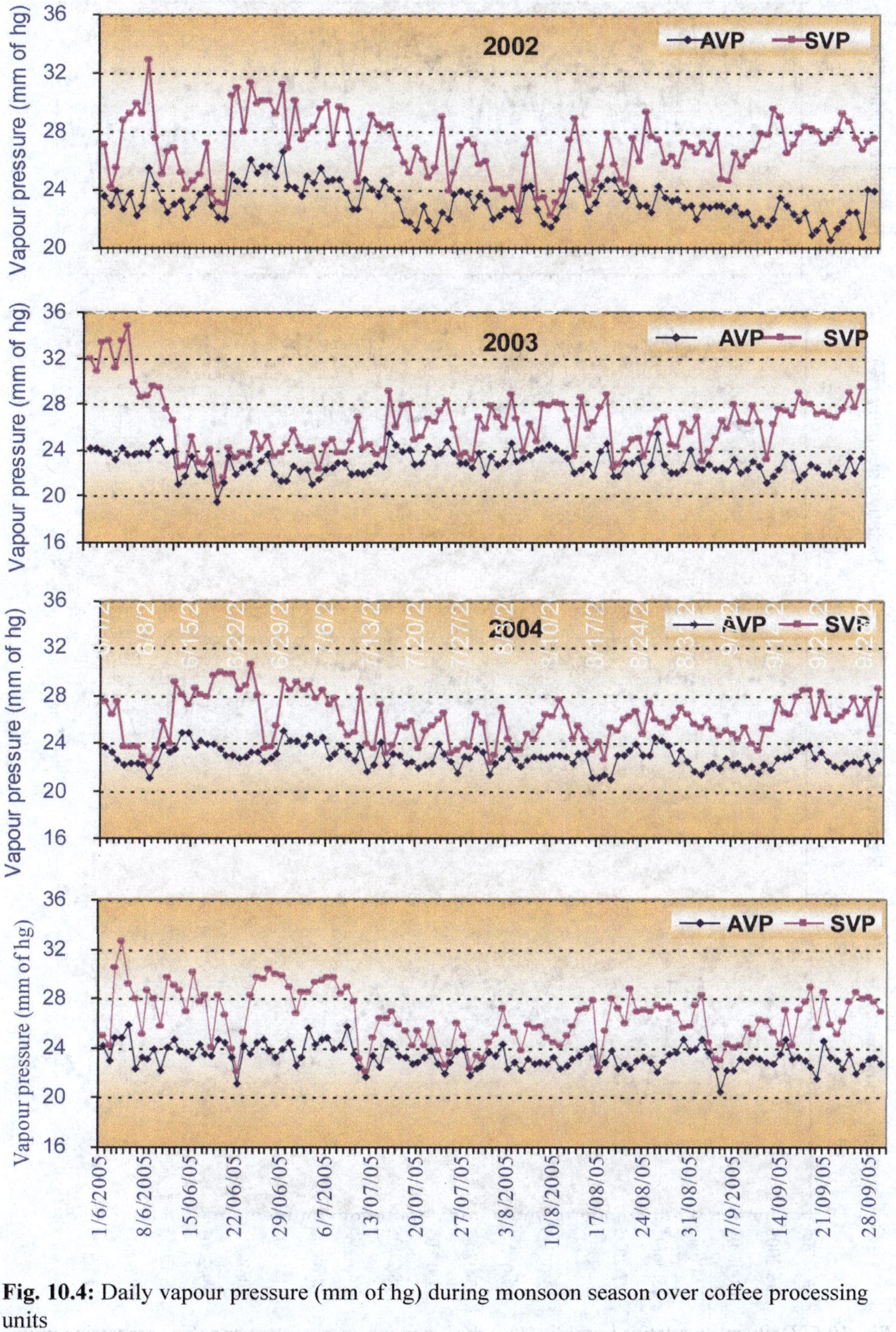

Fig. 10.4: Daily vapour pressure (mm of hg) during monsoon season over coffee processing units

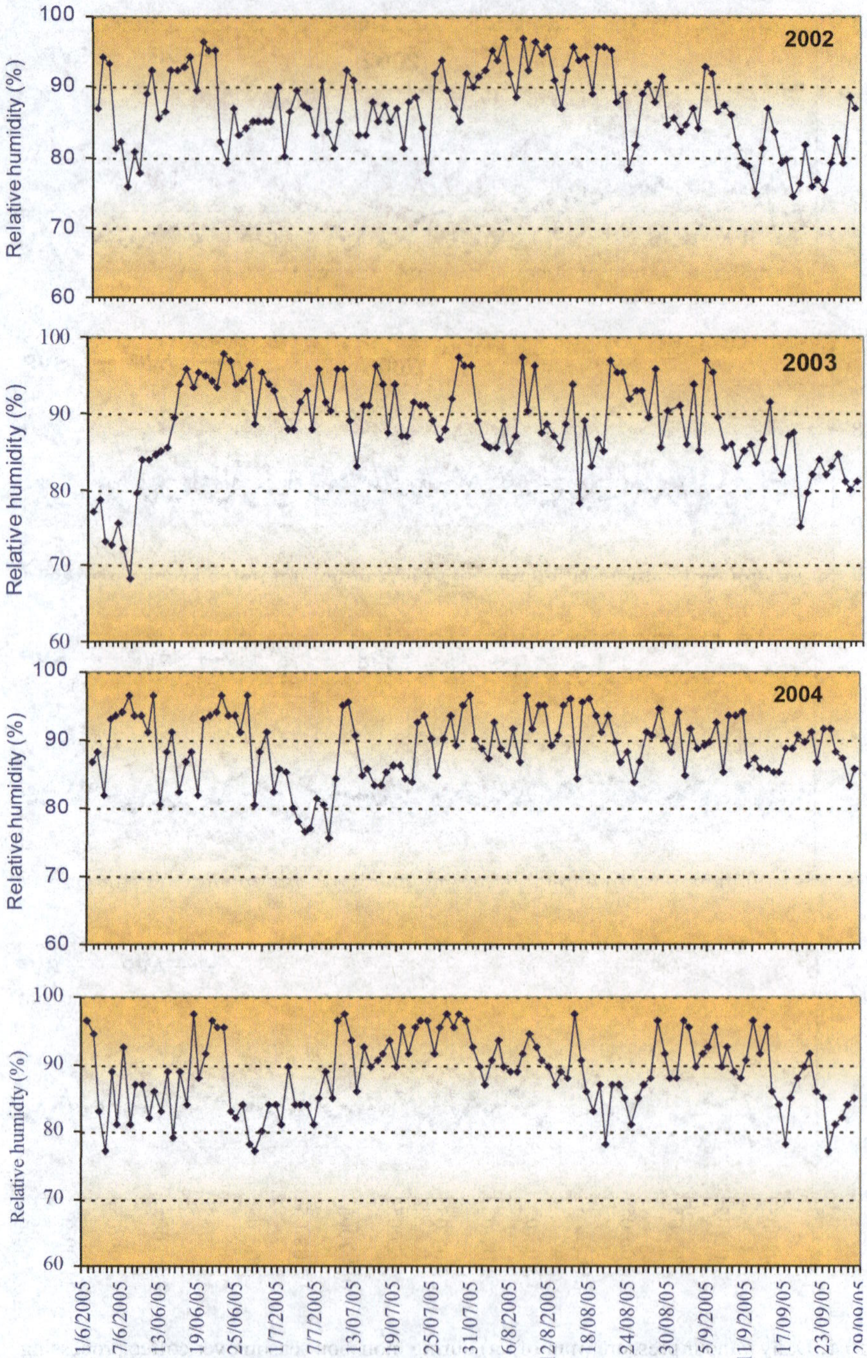

Fig. 10.5: Daily mean relative humidity (%) during monsoon season over coffee processing units

10.1.5 Temperature and its Range

The mean maximum temperature was 29.7°C, 29.1°C and 29.2°C in 2003, 2004 and 2005, respectively. In that order, the minimum temperature was 22.8°C, 21.9°C and 21.8°C. It reveals that the temperature range, the difference between maximum and minimum temperatures, was 6.9°C, 7.2°C and 7.4°C in monsoon period of 2003, 2004 and 2005, respectively. The maximum temperature shooted up to 35°C in June 2003 while it was only up to 32°C in 2004 and 2005 (Fig. 10.6). Though the temperature range was less in 2003, the mean temperature was relatively high (26.0°C) when compared to that of 2004 (25.4°C) and 2005 (25.2°C) during the processing of coffee beans. It indicated that the maximum and minimum temperatures were relatively low in 2005, followed by 2004 and 2003 (Table 10.2). The study on meteorological variables during the processing of coffee beans reveal that continuous clouds with heavy rainfall spread in more number of rainy days together with moderate winds and temperature were noticed in 2005 while similar weather conditions prevailed except in temperature and vapour pressure deficit in 2003 during the processing of coffee beans. In 2004, relatively low rainfall with less number of rainy days and cloudiness was noticed. The saturation vapour pressure deficit was relatively high in 2004 when compared to that of 2005 and 2003.

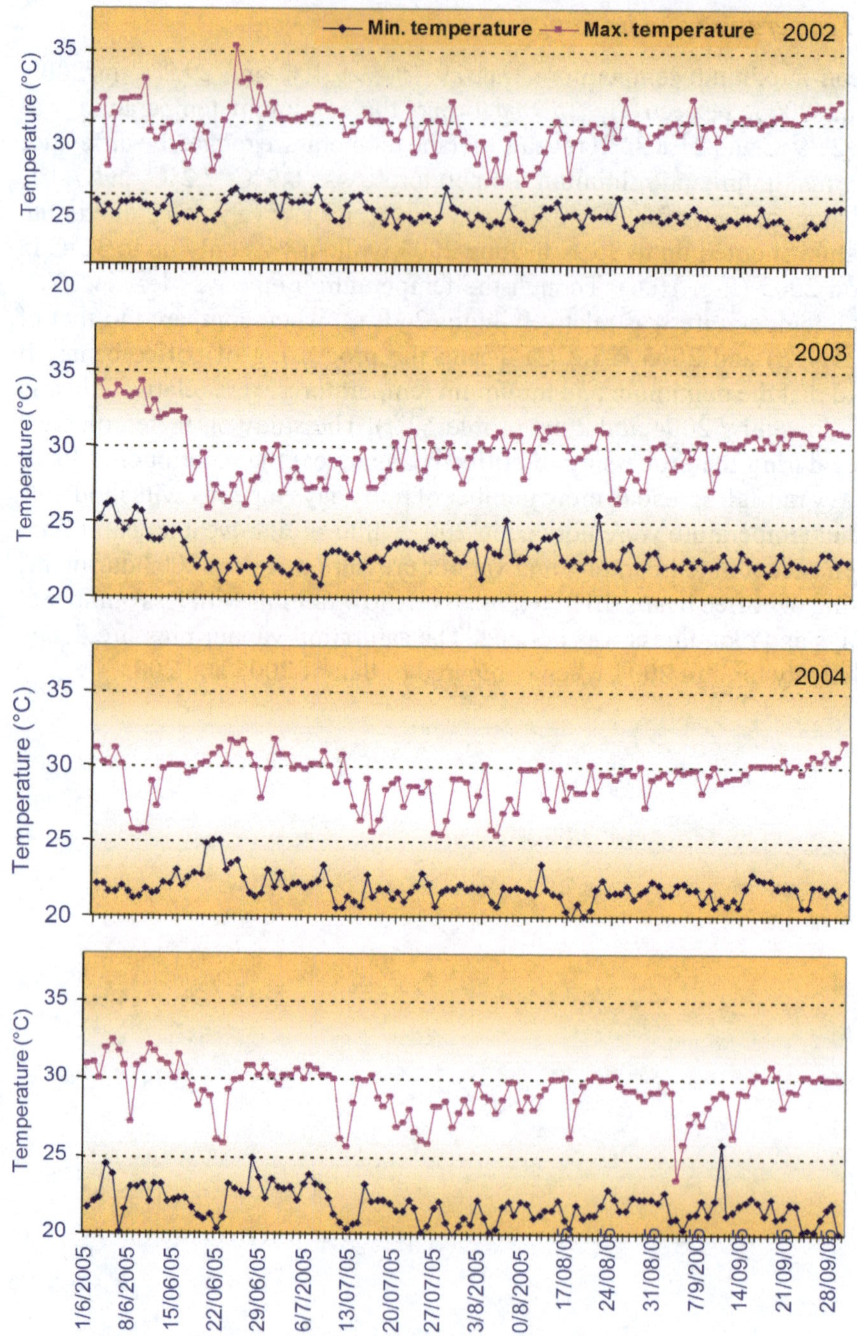

Fig. 10.6: Daily maximum and minimum surface temperatures (^{0}C) during monsoon season over coffee processing units

Table 10.2: Mean meteorological parameters during the process of coffee bean from 2003 to 2005

Robusta	Year	Weather variables										
		Max.T (°C)	Min.T (°C)	Mean T (°C)	RF(mm)	NRD	WS (Km/ hr)	SSH (hrs/ day)	RH (%)	VP (mm of Hg)	SVP (mm of Hg)	VPD (mm of Hg)
	2003	29.1	22.9	25.9	1232	46	2.7	3.5	91	23.0	25.6	2.7
	2004	28.7	22.1	25.4	779	39	2.9	4.9	86	23.0	26.2	3.2
	2005	28.7	21.4	25.0	927	42	2.2	3.2	91	23.1	25.6	2.8

2003: (4th July to 27th August : 55days), 2004: (17th June to 6th August : 51 days), 2005: (5th July to 23rd August : 50 days)

Arabica	Year	Weather variables										
		Max.T (°C)	Min.T (°C)	Mean T (°C)	RF(mm)	NRD	WS (Km/ hr)	SSH (hrs/ day)	RH (%)	VP (mm of Hg)	SVP (mm of Hg)	VPD (mm of Hg)
	2003	29.2	22.9	26.0	1188	49	2.7	3.4	91	23.1	25.7	2.7
	2004	28.8	22.1	25.4	870	44	3.0	4.9	86	22.9	26.2.	3.2
	2005	28.9	21.7	25.4	1526	51	3.1	2.7	90	23.3	25.7	2.6

2003: (7th July to 3rd September : 59 days), 2004: (16th June to 12th August : 58 days), 2005: (14th June to 9th August : 57 days)

10.1.6 Moisture Content of Coffee Beans

Increase in moisture content on an average in Arabica beans was 17.5 % while 18.6 % in Robusta. It was high in 2005, varying between 22.8 % in Arabica and 25.7 % in Robusta when compared to that of 2003 (14.8 % in Arabica and 13.5 % in Robusta) and 2004 (14.6 % in Arabica and 15.9 % in Robusta). It also revealed that Robusta responds to monsooning much better when compared to that of Arabica due to its bigger bean size. Arabica also responds slowly to monsooning as it takes more time for completion of its processing cycle. Irrespective of types, the average moisture content of coffee beans increased by 19.0% due to monsooning and it was high (22.4%) in 2005 (Table 10.3), followed by 2004 (17.9%) and 2003 (16.8 %). Increase in volume and bulging of coffee beans might be due to the absorption of atmospheric moisture. Absorption of atmospheric moisture and resultant swelling of coffee beans may contribute to changes in biochemical reactions. On an average, increase in sugars was 15.8 % in Arabica while 21.7 % in Robusta. The percentage increase in sugars was more in Robusta types than that of Arabica. However, Arabica coffee is considered superior in quality than Robusta in international market due to inherent low level of acidity and high level of sugars. Increase in sugars and decrease in acidity and polyphenols might be contributing to a favourable sugars acid blend in coffee, which is responsible in upgradation of quality. Increase in sugars was high (20.6%) in 2005 when compared to that of 2004 (17.7 %) and 2003 (17.5 %). Similar was the trend in starch and to some extent in amino acids. Such trends were missing in other biochemical reactions. It might be due to various other factors. Significant negative correlations were noticed between surface air temperature and vapour pressure deficit and moisture content of coffee beans. Relatively low moisture content of coffee beans in 2003 and 2004 could be attributed to high average surface air temperature and vapour pressure deficit. It was night temperature to which the moisture content of coffee beans responded negatively, but not the day-maxima temperature. Though moderate wind, cloudy weather and continuous rains play pivotal role in controlling surface air temperature and atmospheric water vapour content and vice-versa, there was no significant correlations between the above causal parameters and moisture content of coffee beans (Table 10.4). It might be due to the monsoon environment that prevails during processing of coffee beans. Moist laden winds with overcast sky and continuously heavy rains are the monsoon features in the humid tropics. Therefore, it appears that continuous clouds with heavy rainfall and its better distribution, high atmospheric water vapour content, moderate temperature and gentle sea breeze would lead to better moisture content in coffee beans if processed during the monsoon season.

Table 10.3: Average moisture content of coffee beans (%) during monsooning

Year	Moisture content (%)			
	I	F	D	Relative increase
2003	12.5	14.5	2.1	16.8
2004	12.8	15.2	2.3	17.9
2005	12.5	15.3	2.8	22.4
Mean	12.6	15.0	2.4	19.0

Table 10.4: Correlation between meteorological variables and moisture content of coffee beans

Pooled data	Rainfall	Rainy days	Bright sunshine	Wind speed	Relative humidity	Vapour pressure deficit	Surface air temperature		
							Max.	Min.	Mean
2003-05	-0.27	-0.23	-0.37	-0.43	0.40	-0.87*	-0.50	-0.86*	-0.79*

* Significant at 0.01 level

Probably, active monsoon period within the monsoon season is conducive for such coffee processing to obtain superior quality in "Monsooned Malabar Coffee". A simple regression equation was worked out based on vapour pressure deficit and night temperature for estimating moisture content of coffee beans. The observed and estimated moisture content of coffee beans are in agreement with 87 % accountability (Fig. 10.7). The regression equation developed for estimating moisture content of coffee beans is as follows:

$$Y = 183.32 - 4.81662\, X_1 - 6.486X2 \; (r^2 = 0.87)$$

Where,

Y is moisture content of coffee beans

X_1 is vapour pressure deficit in the afternoon

X_2 is night temperature

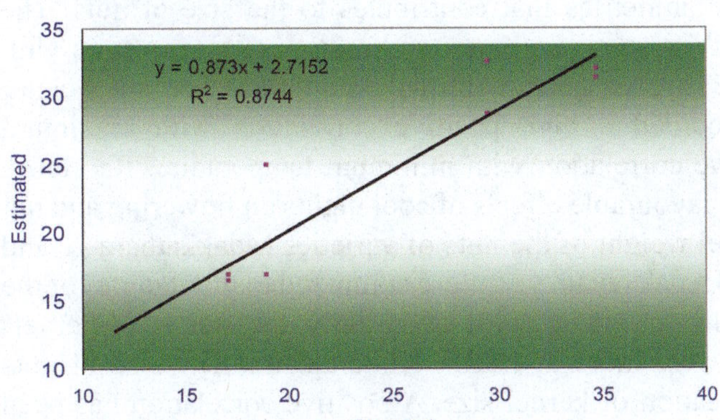

Fig. 10.7 Observed and estimated moisture content (%) in coffee beans

10.2 Quality of Cashew Nut

It is a fact that the performance of varieties and hybrids are found region specific and they are well suited only for the regions for which they are recommended. The suitability is dependent on factors like soil type and weather parameters especially temperature, humidity, rainfall and solar radiation. This aspect remains to be unstudied one when the whole cashew trade depends on quality of kernels. An attempt was made to analyze the effect of weather parameters on quality of cashewnut. Fully matured nuts of the varieties Madakkathara -1, Kananka, K-22-1 and Dhana were collocated from the CRS, Madakkathara (Central zone), FSRS, Kottarakkara(SZ), CRS, Anakkayam(NZ) and RARS, Pilicode (NZ) representing different climatic zones. The cashew nuts were analyzed for different physical biochemical characters viz., nut weight, length, breadth, shelling percentage, kernel weight and white whole recovery percentage of kernels. The characters recorded for a variety at particular region was correlated with the average maximum and minimum temperature, total rainfall(mm) and relative humidity experienced at that region from August to May , which is the period of flushing, flowering and fruiting of cashew . Based on this the weather inflicted qualitative changes in cashew was brought out. The results revealed that the characters of nuts of all the varieties differ with respect to the region I which they are grown. The nuts belonging to FSRS, Kottarakkara and CRS, Anakkayam recorded more length, breadth and weight compared to nuts of the same variety belonging to other regions. This may be due to the influence of temperature prevailed during nut set and development at those two regions. The day temperature and nut set and development period (December and January) prevailed at FSRS, Kottarakkara and CRS, Anakkayam was more whereas the night temperature was comparatively cool (Fig. 10.8). This might have favoured production of big nuts. The length and breadth of nuts are the biometrics that contributes to the size of nuts. The correlations worked out between maximum and minimum temperature with size of nuts (Table 10.5) are supportive to this assumption. The nut weight, length and breadth recorded a high positive correlation with maximum temperature and negative correlation with minimum temperature. Rao *et al.* (2000) also reported the favourable effects of cool nights on flowering and nut set of cashew. The kernel weight of the nuts of varieties Mdakkathara -1 and K-22-1 collected at CRS Anakkayam was more compared to the samples of the same varieties collected from other regions. Similarly, the weight of kernels of Kanaka and Dhana were more at FSRS, Kottarakkara. The size of nuts can be taken as an indication of kernel size. A positive correlation has been reported between nut size and kernel size (Swami *et al.*, 1998). In this present study also the nut weight and kernel weight recorded high positive correlation.

Recovery of kernels with better size will positively reflect the cashew trade in the international market, where big kernels will be graded higher which will fetch premium size. The kernel weight recorded high positive correlation with maximum temperature, negative correlation with minimum temperature, rainfall and humidity. This emphasizes the need for dryspell during nut set and development phase. All the varieties were found to be performing better where such conditions prevail. The kernel size and while whole recovery percentage recorded a high positive relation. It is easy to dehusk the kernels having better size and chance of getting pieces and rejects are less.

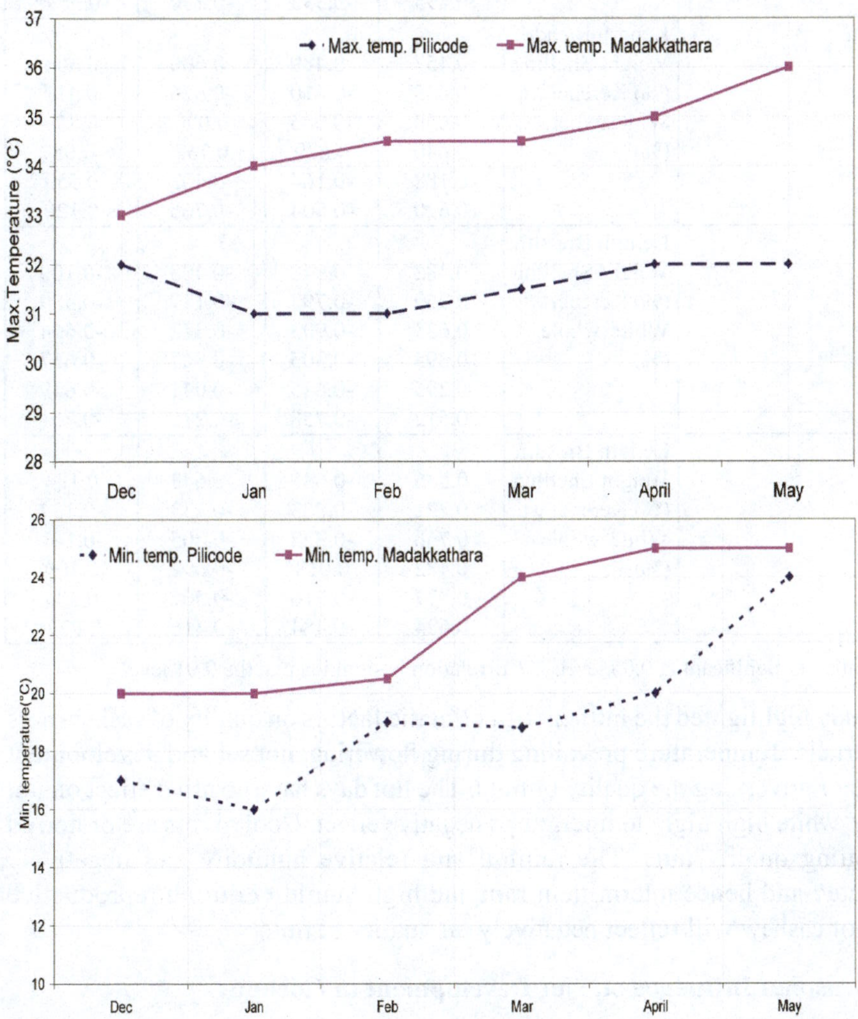

Fig.10.8: Maximum and minimum temperatures (^{O}C) at two locations from December to May

Table 10.5: Correlation coefficients of quality parameters of cashew with weather parameters

Variety		Weather parameters			
		T max (°C)	T min (°C)	Rainfall (mm)	RH(%)
Madakkarathara-1	Length Breadth				
	Weight Shelling	0.671	-0.999*	-0.559	-0.376
	(%) Kerenel wt.	0.980*	-0.680	-0.992**	-0.552
	White whole	0.327	-0.817	-0.321	0.192
	(%)	-0.709	0.981*	0.559	0.544
		0.853	-0.699	-0.690	-0.935
		0.495	-0.333	-0.280	-0.959*
Kanaka	Length Breadth				
	Weight Shelling	0.457	-0.489	-0.609	0.306
	(%) Kerenel wt.	0.455	-0.410	-0.626	0.311
	White whole	0.658	-0.085	0.090	0.850
	(%)	0.640	0.629	0.752	-0.087
		0.188	-0.164	-0.402	0.559
		0.620	-0.504	-0.765	0.120
K-22-1	Length Breadth				
	Weight Shelling	0.482	-0.948	-0.405	-0.100
	(%) Kerenel wt.	0.369	-0.793	-0.137	-0.559
	White whole	0.623	-0.903	-0.422	-0.664
	(%)	0.895	-0.305	-0.927	-0.657
		0.295	-0.645	-0.041	-0.649
		0.512	-0.758	-0.275	-0.751
Dhana	Length Breadth				
	Weight Shelling	0.536	-0.648	-0.643	0.194
	(%) Kerenel wt.	0.721	-0.883	-0.733	-0.134
	White whole	0.766	-0.822	-0.805	-0.141
	(%)	0.482	0.018	-0.692	0.109
		0.377	-0.316	-0.567	0.389
		0.624	-0.484	-0.416	-0.977*

*Correlation is significant at 0.05 level., ** Correlation is significant at the 0.01 level

The study highlighted the influence of climatic factors on quality of cashew nuts and kernels. Temperature prevailing during flowering, nut set and development is a factor governing the quality of nuts. The hot days have positive effect on nut quality, while high night temperatures negative effect. Cool nights are preferred for getting quality nuts. The rainfall and relative humidity and negatively correlated and hence intermittent rain and high humidity during reproductive phase of cashew will reflect negatively on quality of nuts.

10.3 Seasonal Influence on Nut Development in Coconut

The monthly variation in nut development of hybrid coconuts was studied at the Regional Agricultural Research Station, Pilicode during the year 1985.

Five palms each of the three coconut hybrids viz., Chowghat Dwarf Orange x West Coast Tall (CDO x WCT), West Coast Tall x Chowghat Dwarf Orange (WCT x CDO) and Chowghat Dwarf Orange x Lakshadweep Ordinary (CDO x LO) formed the material for the study. The hybrid CDO x WCT registered higher (538 g per nut) nut weight than WCT x CDO (498 g/nut) and CDO x LO (504 g/nut). In contrast, WCT x CDO (Table 10.6).

Table 10.6: Monthly husked nut weight (g/nut), and copra content (g/nut) of coconut hybrids during January, 1985 to January 1986.

Month	CDO x WCT		WCT x CDO		CDO x LO	
	Husked nut weight (g/nut)	Copra content (g/nut)	Husked nut weight (g/nut)	Copra content (g/nut)	Husked nut weight (g/nut)	Copra content (g/nut)
January	1042	-	611	-	820	-
February	696	155	663	190	720	163
March	733	200	700	206	708	183
April	650	200	673	183	675	166
May	700	111	708	201	530	134
June	-	-	-	-	-	-
July	437	121	433	130	280	76
August	366	66	366	100	300	62
September	244	93	303	122	325	100
October	304	91	320	108	350	87
November	330	110	333	113	385	100
December	410	126	373	133	466	133
January	581	172	420	146	500	154
Mean	538	131	498	149	504	123

recorded relatively more copra content (149 g/nut) than CDO x WCT (131 g/nut) and CDO x LO (123 g/nut). All the three hybrids recorded higher nut weight and copra content in the summer months of March, April and May. The fall in nut weight and copra yield was conspicuous in the rainy and post rainy periods (Table 10.7). In the light of the observations made by early researchers that any abnormal weather factor coinciding with the critical phases of nut development would adversely affect the rate of growth and final size of the nut and copra content, an attempt was made at the RARS, Pilicode to work out the relationship between the total heat units during the course of nut development and nut weight. The results indicated a negative relationship between the heat units during the second phase of nut development (4–7 months after fertilization) and the husked nut weight (Table 10.8).

There was a marked decline in nut size from July to December in response to an increase in total heat units indicating that the total heat units above 2100 day° were not congenial for unit development. The following quadratic equations were developed for estimating the nut weight of the test varieties.

$$Y\ (WCT \times CDO) = 26493.2094 - 23.1107\ X + 0.0051\ X^2$$

$$Y\ (CDO \times WCT) = 47766.2579 - 42.2450\ X + 0.0094\ X^2$$

$$Y\ (CDO \times LO) = 55053.4040 - 49.7958\ X + 0.011\ X2$$

Where Y(WCT x CDO), Y (CDo x WCT) and Y (CDO x LO) are the estimated husked nut weight of the corresponding variety and X is the total heat unit during the second phase of the nut development. These equations has multiple correlation coefficients of 0.8689, 0.8443 and 0.7977 for the hybrids CDO x WCT, WCT x CDO and CDO x LO, respectively.

Table 10.7: Husked nut weight (g/nut) and copra content (g/nut) during different seasons, 1985

Hybrid	**Summer (March to May)**		**Southwest monsoon (June to Sept)**		**Post monsoon (October-November)**		**Winter (December to February)**	
	Husked nut weight (g/nut)	**Copra content (g/nut)**	**Husked nut weight (g/nut)**	**Copra content (g/nut)**	**Husked nut weight (g/nut)**	**Copra content (g/nut)**	**Husked nut weight (g/nut**	**Copra content (g/nut)**
CDO x T	694	170	339	94	317	100	526	155
T x CDO	693	197	394	120	327	110	469	152
CDO x LO	635	161	301	79	367	93	555	155

Table 10.8: Total heat units during the second phase of nut development and husked nut weight of coconut hybrids

Month	**Heat units (°C) During the second phase of nut development**	**Variety**		
		CDO x T (g/nut)	**T x CDO (g/nut)**	**CDO x LO (g/nut)**
January	2018	1942	611	820
February	2040	696	663	720
March	2040	733	700	700
April	2018	650	673	675
May	2040	700	708	530
June	-	-	-	-
July	2077	407	433	280
August	2184	366	366	300
September	2259	244	383	325
October	2285	304	320	350

Month	Heat units (°C) During the second phase of nut development	Variety		
		CDO x T (g/nut)	T x CDO (g/ nut)	CDO x LO (g/ nut)
November	2219	330	333	385
December	2219	410	373	466
January	2033	581	420	500
Correlation coefficient between heat units and husked nut weight (r)		-0.83*	-0.83*	-0.72**

*significant at 0.05 level ** significant at 0.01 level

The process of nut development commencing from fertilization of female flower ripening takes place in three distinct phases covering a total of ten to twelve months. During the first phase (0–4 months), the growth is slow. There is a decline in growth rate after the second phase. The nut development from fertilization to harvest takes place 10-12 months (around sixteen months at higher altitudes) normally. High ambient temperature during the active growth phase adversely affects the rate of growth of the nut with a consequent decline in nut weight under the rainfed conditions. Out of all the weather parameters studied, the heat unit , which is a function of temperature, has a distinct influence on the final nut size.

10.4 Seasonal Influence on Cocoa Pods

The cocoa pods harvested during October and November are superior in pod and bean weights when compared to that of remaining seasons though the number of pods harvested during summer are more. On an average, a harvest of five pods during the post monsoon season (October and November) is equal to nine pods harvested during summer (March to May). It reveals that the number of cocoa pods harvested is more in summer while the bean weight is more in post monsoon season.

Summary

The plantation crops like coconut, cashew, cardamom, coffee, cocoa, black pepper, tea, areca nut, rubber and oil palm play a major role in export economy at the national level. The plantation crops' production and its yearly fluctuations with reference to climate variability/change is not fully understood or studied to a limited extent only. At the same time, the quality aspects of the plantation crops' produce in relation to climate variability/change is yet to be attempted. In fact, the quality produce is the one which plays a pivotal role in plantation crops in deciding the marketability outside. For the first time in 2002 when the monsoon failed across the Western Ghats, the effect of monsoon and its behaviour on the quality of 'Monsoon Malabar Coffee' was highlighted in it's processing. If the coffee beans are processed in July when the active phase of monsoon is noticed along the West Coast, the quality of black coffee is superior in terms of less acidity and better sugar content. In the case of coconut also, coconut size is likely to be adversely affected when its active development phase (4–7 months) coincides with the hot summer during which the surface temperature is high. Probably, it is one of the reasons, why, nut size is relatively small in Tamil Nadu, Karnataka and Andhra Pradesh in non-traditional coconut growing areas though the number of nuts harvested per palm is high. The mean summer temperature across the coconut plantations is always between 40 and 45° C over non-traditional areas in the South while traditional coconut growing regions experience between 30 and 35° C. In the case of cashew also, nut size and kernel weight are much influenced by the maximum temperature during the nut development period. Also, a good summer shower during the nut development stage in sandy soils is enough to obtain better nut size in cashew. In contrast, unusual rains during the nut development stage may lead to immature fruit drop, thereby decline in nut quality along the West Coast. The unusual summer rains adversely affect the nut quality in cashew. The inferior nut quality of cashew due to unusual summer rains adversely affects the farmers' income and future market to a greater extent. The nut size and kernel weight of the same variety also vary significantly depending upon the agroclimatic zone. All these observations indicate that the climate variability is one of the factors deciding the quantity and quality produce of plantation crops to a greater extent. Therefore, further studies in this direction need to be taken up on priority basis in all the plantation crops under the projected climate change scenario.

Suggested Readings

Majumdas, S.K; Muthu, M;, Srinivasan, K.S; Natarajan, C.P., Bhatia, D.S and Subrahmania, V, 1969. Studies on storage of coffee beans, control and related storage experiments. Food Science 10: 332–338.

Manay,N.S and Swamy,S.M. 1987. Food Facts and principles. New age international (p) Ltd., NewDelhi,p.524.

Mc Cusker, R.R., Golelberger, b.A and Locone, E.J.2003. Coffeine content of speciality coffee. Journal of Analytical Toxicology. Vol.27(7) pp.520–522.

Menon, S. 2005 Quality coffee regions in India.- coffee Int. 177(6). Mobbet, T.1999. why worry. Indian Coffee XIII (8): 31–32.

Natarajan, C.P., Majumdas, S.K., Srinivasan, K.S; Balachandran, A, Bhatia, D.S and Subrahmania, V. 1961. Studies on the storage, physical, chemical and biological changes in coffee of high humid conditions. India coffee .pp.8–12.

Panse, V.G and Sukatme, P.V.1985. Statistical methods for agricultural workers. Indian Council of Agricultural research, New Delhi. P.339.

Rajagopal, H.B.1998. Specialty Coffees. Indian Coffee XIII (8): 37–39.

Rajeev Bhat, Sanju Francis, B.Y.K..Rao, Rashhed Ahmad and Lalit Varshney. Thermans. 2002. Thermal investigation on Indian Monsooned Coffee.

Rajesh, V.P.1999.Preparation of specialty coffee and marketing with our estate brand. Indian Coffee XIII (2): 29–30.

Rao, Dorajee A.V.D., Ravisankar, C. and Reddy, M.L.N. 2000. Influence of flowering phases on nut yield of different cashew clones. J. Plantation Crops 28(1):55–60.

Rao, G.S.L.H.V., Pushpalatha, P.B., Gopakumar, C.S and Mahesh Chandran. 2005. Influence of monsoon on quality of coffee beans in the humid tropics. In proceedings of National Workshop on Drought Management in Plantation Crops. Kerala Agricultural University, Thrissur, Kerala.176p

Rashhed Ahmad. Trivia and Science of Indian monsoon coffee. Indian Coffee. April 2000.

Salam, M.A. 1998. Processing characteristics and production potential of cashew (Anacardium occidentale L.) varieties. The cashew 12(1):27–29.

Sivetz, M and Foote, H.E.1963. coffee processing Technolfy. Vol.I. p 48–99. the Avi Publishing Company, Inc.Westport, Connectivity, Connecticut.

Swamy, K.R.M., Rao, E.V.V.B and Bhat, M.G. 1998 Catalogue of minimum descriptors cashew (Anacardium occidentale L.) Germplasm Accession – II. National Research Centre for Cashew, Puttur. Karnataka, India p.54.

Vijayan 1999. Preparation of speciality coffee and marketing with an estate brand. Indian coffee. Indian Coffee XIII (2): 29–30.

Questions

I. Elucidate on the following

1. Climate variability/change and rubber
2. Climate and rubber across the Country
3. Rainfall Regime and rubber yield in Kerala
4. Temperature Regime and rubber yield in Kerala

II. Write Short notes on

1. Water deficit during summer versus rubber yield
2. Rainfall and rubber yield
3. Temperature and rubber yield
4. Heat Units and rubber yield
5. Thermal Regime across the rubber tract in Kerala
6. Rainfall Regime across the rubber tract in Kerala
7. Rubber producing States in India

III. Fill in the Blanks

1. Increase in temperature trends is not ---------- for rubber
2. Rubber responds-----------to sunshine while ------------to rainfall
3. Monthly rainfall declines from July and increases again in --------- over Kerala
4. India is the ---------------largest producer of rubber
5. Evapotranspiration rate per day varies between ------ and -------mm in rubber plantations
6. The productivity of rubber in Kottayam is the ---------- in Kerala
7. Rubber is a crop of ------------District in Kerala
8. -----------------is the percentage increase in rubber productivity between 1951-80 and 1981-09
9. over Kerala
10. -------------- was the flood year affected rubber yield in Kerala
11. -------------and -----------crops compete across the midlands in Kerala

IV. State True or False

1. Relationship exists between water deficit during summer and rubber yield
2. Winter dormancy in rubber production is noticed in NER

3. True or False
4. Round-the-year rubber harvest is noticed in Kerala
5. Relative humidity is the most important factor for rubber production
6. Two peaks in rubber harvest periods are noticed across Kerala
7. Rubber productivity is declining in Kerala since 1997 onwards
8. The number of rainy days is almost the same across the rubber tract in Kerala
9. Rainfall during the post monsoon season is declining
10. Rubber is a crop of midlands
11. Rubber and cashew compete in the midlands of Kerala
12. Rubber likes uni-model rainfall
13. With the age the girth of a rubber tree is uniform across the rubber tract
14. Rubber is a crop of Idukki
15. Rainy season is detrimental for rubber tapping
16. Summer soil moisture deficit is detrimental to rubber productivity

11

Weather Risk Management

The global economy has adversely been affected to a considerable extent due to weather related disasters like droughts, floods, heat waves and cold waves which are not uncommon in the recent past due to global warming and climate change. For the first time, the year 1998 was declared as the weather related disastrous year globally. The year 2015 was the warmest year globally, again. Such weather phenomenon of warming more and more year after year was noticed in the current decade as well when we look at 2023 and 2024. The year 2010 was the warmest year ever recorded, followed by 2009 in India. The occurrence of floods and droughts, cold and heat waves adversely affected the Indian foodgrains production to a large extent across the Country in one or other region as seen in 2014, 2012, 2009, 2004, 2002 and 1987. The cold wave in winter 2003-04 across the Northeast region of the Country and the heat wave in summer 2004 across the same region are still in memory to understand the ill effects in agriculture/horticulture/poultry/inland fisheries sectors. Model simulations indicate that a marked increase in rainfall and temperature over India would be seen during the current century. The maximum expected increase in rainfall is likely to be 10–30% over the Central India. Temperatures are likely to increase between 2 and 3^{0}Cby the end of 2100 A.D. However, steps taken towards the green energy across the world may yield results not to allow atmospheric global temperature beyond 1.5 to 2^{0}C. Greater number of high surges and increased occurrence of cyclones in post monsoon period along with increased maximum wind speed are also expected along the East Coast of the Country. The summer droughts and heavy rains leading to waterlogging, floods and landslides during the monsoon period is a threat to plantation crops which are predominantly grown in the Humid Tropics like Kerala (8°15` N and 12°50`N latitudes and between 74°50`E and 77°30`E longitudes). Waterlogging and floods due to excess rains could be minimised by providing surface and subsurface drainage facilities in plantation gardens which will improve the soil health and nutritional status, in turn yields can be increased in plantation crops. Floods are mostly location /region specific due to heavy monsoon rains or cloud bursts during pre-and-post monsoon seasons while the effects of drought could be relatively seen in larger areas due to non-receipt of rains during the Northeast monsoon and summer showers in Kerala. In the case of occurrence of summer droughts, drought management techniques could be adopted with irrigation and without irrigation under the rain fed conditions. Irrigation may not be possible during summer due to scarcity of water, where plantation crops grow predominantly. However, the available water could be effectively used through various micro-irrigation systems and crop management and protection practices. At the same time, pro-active crop management practices are the only option to mitigate the ill effects of drought on plantation crops under the rain fed conditions. Therefore, a few of them are discussed in this chapter with case studies as part of weather risk management in plantation crops.

11.1 Drought Management

The State of Kerala always appears to be greenery due to heavy monsoon rainfall supported by rains during northeast monsoon. At the same time, the northern districts of Kerala and Palakkad experience prolonged dry spell from December to May, if pre-monsoon showers fail, leading to soil moisture stress and finally results to drought conditions. The occurrence of drought for the State as a whole was noticed during summer 1983, 2004 and 2013. There may be similar events sometimes even during the last one decade. It is the lack of emphasis on conservation measures which results in the recurring of drought situation. Rainwater during the monsoon should be harnessed with small investments. If careful arrangements for conserving of water are not done in time, it could result in the drying up of even bore wells and tube wells. In order to keep rivers and wells adequately replenished with water during the dry season, small sub surface dams and anicuts up to 40-odd rivers originating from the Western Ghats and flowing into the Arabian sea are necessary. Implementation of irrigation and hydel development projects in all the valleys would automatically take care of domestic water supply, sanitation and partial flood control. As even partial measures have not been undertaken in all the river systems, the water potential in the State remains poorly harnessed. Watershed based participatory natural resource development and management schemes such as tanks, vented cross bars, diversion weirs and lift irrigation can increase and sustain irrigation, ground water recharge, crop production and drinking water supply considerably. Even in the worst drought conditions, Kerala's rainfall statistics shows annual average as 1600 mm. This much of water if properly conserved and used would be plenty even to spare. The rivers of the State except one or two cease their flow with the cessation of the Northeast monsoon and thereafter practically all the rivers dry up and remain dry until till monsoon sets in May or June for six months. It is during this period that the State tends to become drought prone recurrently. In addition, drought management practices have to be followed under field conditions. Cultivation of drought tolerant crop varieties, proper soil and water conservation practices like ploughing before the receipt of southwest monsoon and immediately before the cessation of the northeast monsoon, opening basins of crops like coconut before the Northeast monsoon, organic mulching and husk burial can definitely conserve the available moisture in the soil and use water judiciously. Judicious use of water in day-to- day life and for agriculture purpose can reduce the impact of drought in Kerala. Awareness on drought management practices is the need of the hour to sustain plantation crop production in the State against the weather abnormalities. Drought may be a relative term used in case of deficit rainfall or soil moisture stress. Breaks or failure in monsoon may lead to droughts during *kharif* season across the Country. The *kharif* foodgrains production in India is dependent on the monsoon rainfall and its distribution under the rain fed conditions. At the same time, it is the northeast monsoon and pre-monsoon showers, which are important in the case of plantation crops. In he Tropical Monsoon Climates where monsoon is assured, it is the summerrainfall, which is also critical in the case of plantation crops as they are perennial in nature and the availability of soil moisture round the year is very important for survival of the crops. There are different types of drought viz., meteorological, hydrological, agricultural, physiological and atmospheric droughts (Fig. 11.1).

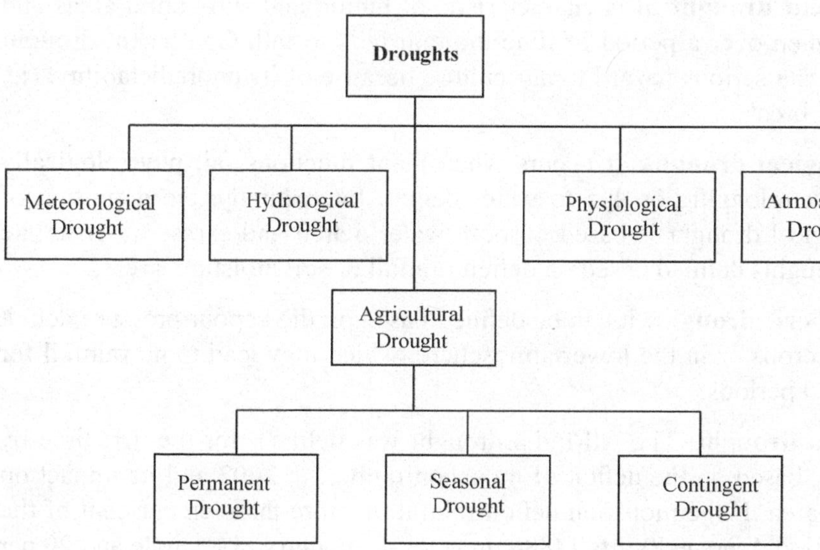

Fig. 11.1: Different types of drought

11.1.1 Types of Drought

Meteorological drought: If the rainfall decrease is more than 25 per cent of the normal over an area, it is known as the meteorological drought as per the India Meteorological Department.

Hydrological drought: Meteorological droughts, if prolonged, results in hydrological drought with marked depletion of surface water and consequent drying of reservoirs, lakes, streams and rivers, cessation of spring flows and fall in ground water table.

Agricultural drought: It occurs when soil moisture and rainfall are inadequate during the growing season to support healthy crop growth and maturity, causing extreme stress and wilting.

The agricultural droughts can further be classified as follows:

Permanent drought: It is found in arid areas where there is no enough precipitation to satisfy the water needs of plants. In such areas, agriculture is impossible without irrigation throughout the season.

Seasonal drought: It occurs in areas with well defined wet and dry seasons as in most of the tropics. Drought can be expected every year owing to seasonal changes in atmospheric circulation patterns. Agriculture is possible during rainy season or with the use of irrigation during dry season.

Contingent drought: It is characteristic of humid and sub-humid areas and occurs when over a period of time the rain fails to fall. Contingent drought constitutes a serious hazard to agriculture because of its unpredictability. (ie. monsoon breaks).

Physiological drought: It occurs when plant functions fail physiologically due to waterlogging or due to water deficit. Interestingly, the definition of physiological drought is based on both water deficit and excess water unlike other droughts defined based on deficit rainfall or soil moisture stress.

Atmospheric drought: It can be defined based on the vapour pressure deficit over the crops or in the lower atmosphere, which may lead to no rainfall for prolonged periods.

All-India drought: The All-India drought was defined, for the first time by the IMD, based on the deficit of monsoon rainfall in 2002 and its impact on cropped area. If the monsoon deficit rainfall is more than ten per cent of the Long Period Average Rainfall (880 mm) for the country as a whole and 20 per cent of the cropped area is affected without crop sown, it is known as the All India drought.

Rainfall Anomaly Index (RAI), Standardized Precipitation Index (SPI), Palmer Drought Index (PDI), Crop Moisture Index (CMI), Keetch-Byram Drought Index (KBDI), Reconnaissance Drought Index (RDI), Drought Severity Index (DSI), Evapotranspiration Deficit Index (ETDI), Soil Moisture Index (SMI), Y-Index and Remote Sensing Based Indices are some of the drought indices available to study the commencement and cessation of drought period with various intensities. These drought indices can be used with reference to crops for drought monitoring as well. Several workers have studied the incidence of drought over India using different criteria. Most of them classified the droughts based on the deficiency of rainfall during the monsoon period. Mooley (1984) studied the origin, incidence and impact of droughts over India. The years 1877, 1899, 1918, 1972,1979, 1987, 2002, 2004, 2009, 2012 and 2014 were identified as drought years based on deficit monsoon rainfall as against the normal rainfall of 880 mm from June to September. The effect of monsoon drought on the total food grain production of the country in 1987 was less by about 34 million tonnes (138.41 million tonnes in 1987–88 as against 172.18 million tonnes in 1988-89). Similar was the loss of food grains in 2002 due to deficit rainfall during monsoon 2002. However, the effect of drought on plantation crops in Kerala was not felt during 1987 and 2002 due to break in monsoon during July. At the same time, the total plantation crop production was severely affected due to unprecedented drought that occurred in Kerala during summer 1983. Identical summer droughts were noticed in summer 2004 and 2013. The low coconut production across the coconut growing regions in the South during 2013-14 and 2014-15 was attributed to the occurrence of summer drought. From the above, it is understood that the effect of drought with reference to plantation crops in the Humid Tropics should be studied based on the distribution of rainfall during summer, but not based on rainfall that is assured during the monsoon. Moreover, the plantation crops are never under soil moisture stress during the monsoon (June to September) as rainfall/ soil moisture is assured in the Humid Tropics in almost all the years due to continuous wet spells.

The prolonged summer droughts during 1983, 2004, 2013 and such similar events during the last one decade adversely affected plantation crops production to a large extent and the State's economy was adversely affected not only in the same year but also in the subsequent years depending on the nature of crops. Coconut palms in the State experienced drought wherever the plantations are maintained as the rain fed. Drying and breakage of lower whorl, button shedding and immature nut fall were noticed in coconut. The effect of drought on monthly nut yield was noticed in the following year and the yield decline was noticed up to 50% depending upon the type of crop management practices under the rain fed conditions. The decline in yield in the case of areca nut, cardamom, coffee and black pepper was also noticed. Black pepper vines wilted and dried up. Heavy loss was noticed in black pepper gardens. It was attributed to the lack of moisture in the surface soil layers coupled with high temperature and low atmospheric humidity. Growers were forced to harvest their crop due to wilting of the plant and subsequent dropping of th e berries and several black pepper gardens were wiped out in Wayanad District in 2004, which was one of the worst hit districts across the State. Yield reduction was more than 30% in cardamom tracts of Idukki district alone owing to severe moisture stress. Bean shedding in vanilla and immature fruit fall in nutmeg was also noticed even under irrigated conditions due to atmospheric dryness. Cocoa yield was also adversely affected due to high temperature in the absence of soil moisture. Therefore, there is need to mitigate the ill effects of weather abnormalities like floods and droughts and cold and heat waves through scarce water resources management and better crop management practices/pro-active measures at micro-level or at watershed level.

11.1.2 Cropwat

In Kerala, the distribution of rainfall is totally different from South to North of the State. In the South (Trivandrum), the distribution of rainfall is bi-model while uni-model in the North (Kannur and Kassragod districts). As a result, the crops that are grown in the South experience relatively less soil moisture stress when compared to that of northern districts of Kerala. For effectiveness of precipitation, the concept of evapotranspiration was introduced and defined the potential evapotranspiratlion (PET) by Thornthwaite (1948) and Penmann (1948). The PET was defined as the combination of evaporation and transpiration when the surface is completely covered with vegetation and there is an abundance of soil moisture. It depends upon a number of factors including temperature, incoming solar radiation, wind speed and saturation vapour pressure deficit. A number of empirical formulae are available in literature for computing potential evapotranspiration based on meteorological data. Penman's method of estimating potential evapotranspiration is widely used as the method considers both energy balance and aerodynamic components. The CROPWAT software developed by FAO in 1990 could be used for computing reference evapotranspiration (ETo) of crops based on surface air temperature, humidity, wind speed and sunshine hours. The following Table gives the computer output of reference evapotranspiration (ETo) according Penman-Monteith formula. The computed ETo underestimates from December to April while coincides in remaining months when compared to that of open pan evaporation (Fig. 11.2). This was due to abnormal seasonal fluctuations in vapour pressure deficit and the amount of solar radiation received. However, as a whole, the Penman is the best suit for computing reference evapotranspiration when compared to that of other formulae and it can be used under field conditions for irrigation requirement of crops. The underestimate of reference evapotranspiration of modified Penman-Monteith during winter can be attributed due to strong winds that blow through Palghat gap (Fig. 11. 3). Based on pan coefficient, the reference evapotranspiration could be worked out using the pan evaporation as well. The pan coefficient worked out as 0.83 for winter and 0.89 for summer under the weather conditions that prevail over Palghat and Thrissur regions. It was worked out using the regression equation Y=bX, where Y is reference evapotranspiration, X is pan evaporation and b is constant. The computational methods give the potential evapotranspiration while evapotranspiration can be worked out under direct measurements. The ratio between the evapotranspiration and the potential evapotranspiration gives the crop coefficient.

or

ET requirement of a crop = Kc x PET where, Kc is crop coefficient, ET - evapotranspiration and PET-potential evapotranspiration.

Table 11.1: Reference Evapotranspiration ETo according to Penman-Monteith during 2011

Country: India Altitude:25 meter				Meteo Station : Vellanikkara Co-ordinates: 10.00 N.L 76.00 E.L			
Month	**Max. Temp (°C)**	**Min. Temp (°C)**	**Humidity (%)**	**Wind (km/ day)**	**Sun-shine hours**	**Radiation (MJ/m2/day)**	**ETo-Penman (mm/day)**
January	32.7	22.2	45	151	8.5	10.2	4.6
February	33.7	22.0	42	127	8.5	11.3	5.3
March	34.8	23.9	48	99	8.7	13.0	5.4
April	34.3	24.5	62	78	6.6	12.4	4.7
May	33.0	24.9	66	74	6.8	12.4	4.5
June	29.3	23.6	84	62	2.5	8.8	2.9
July	29.1	22.9	83	52	1.6	8.1	2.6
August	29.4	22.9	81	58	2.2	8.7	2.9
September	30.0	23.1	77	54	4.4	10.5	3.4
October	32.1	23.5	68	74	6.1	11.1	4.0
November	31.4	22.9	60	119	6.3	9.9	4.1
December	31.9	22.6	52	151	7.3	9.6	4.6
Year	31.8	23.2	64.0	91.5	5.8	10.5	1487.0

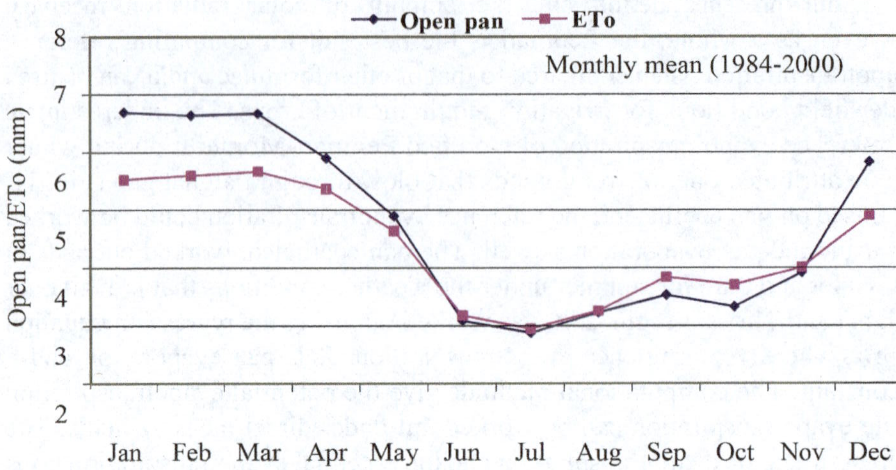

Fig. 11.2: Mean monthly Potential evapotranspiration (modified Penman- CROPWAT) versus open pan evaporation (1984–2000) at Vellanikkara

For many of the dry land annual crops, Kc from the time of sowing to establishment is 0.3 to 0.6 increasing to a maximum of 1.0 to 1.3 when the crop canopy covers most or all of the incoming solar radiation. When the crop attains physiological maturity, Kc declines to approximately 0.5. The values of Kc for number of crops are available in literature and vary from 0.6 to 1.2.

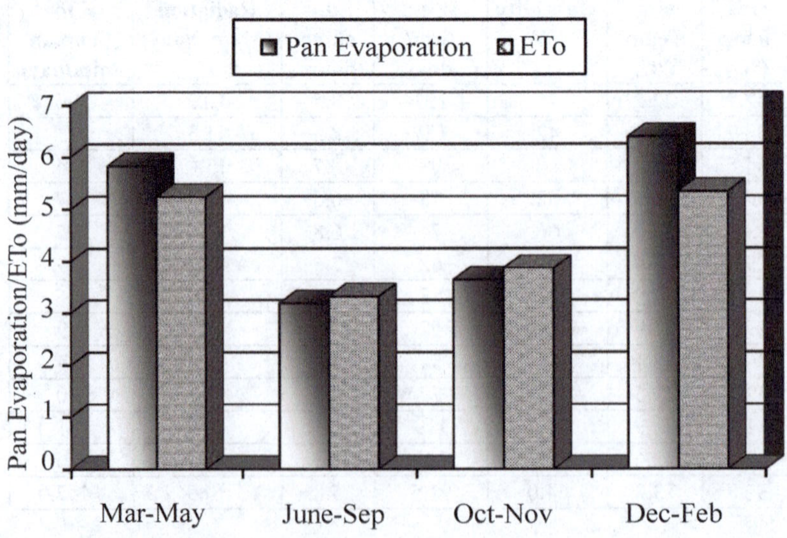

Fig. 11.3: Season-wise potential evapotranspiration (modified Penman- CROPWAT) versus open pan evaporation at Vellanikkara from 1984 to 2000

11.1.3 Rainfall and Pan Evaporation

Though rainfall is the main source for water supply of plants, the plant growth does not depend on rainfall alone, but it should balance the evapotranspiration of crops. Based on rainfall and potential evapotranspiration, crop growth periods can be adjusted. For most of the crops, the period during which rainfall exceeds the PET is the best suited for the maximum growth period. Of course, this is the period during which surplus rainfall leads to run off and in the absence of drainage facilities, the crops are subjected to waterlogging and it is detrimental to crops. The early vegetative phase including sowing period and the later phase that is ripening period can be adjusted when the rainfall is less than PET (the rainfall should be at least 50% of PET). The above conditions hold good in India wherever crops are grown only under the rain fed conditions. At the same time, it does not hold good under rain fed conditions of Kerala as rainfall is abundant during the monsoon season. The surplus water during the first crop season leads to waterlogging which is detrimental to crops. The second and third crops suffer due to soil moisture stress and crop failure is a common phenomenon during the above seasons if irrigation is not assured. The irrigation schedule or surplus water could be determined based on precipitation and potential evapotranspiratiaon of a particular location. The pan evaporation could be taken into account in place of potential evapotranspiration wherever pan evaporimeter data are available for operational use. The weekly rainfall and open pan evaporation were compared to explain water deficit and surplus periods with reference to crops (Fig. 11. 4).

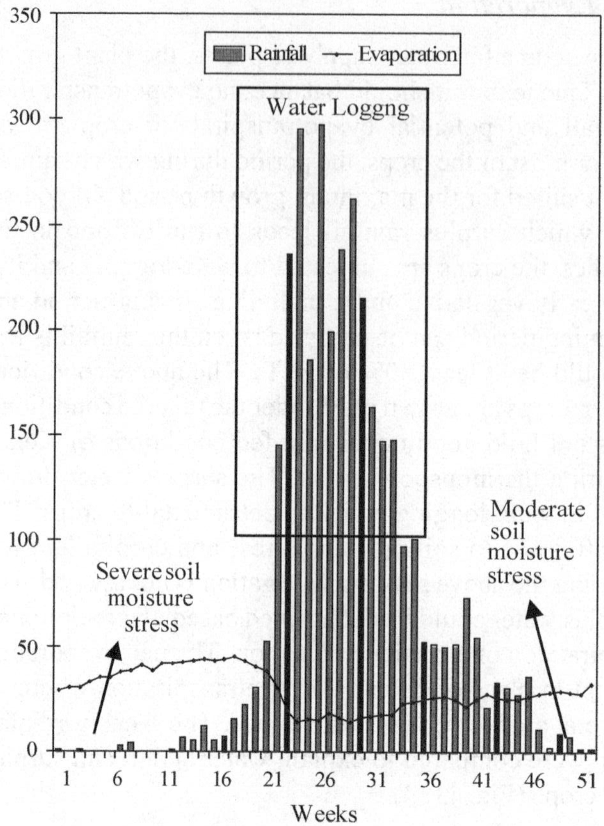

Fig. 11. 4: Weekly rainfall versus evaporation at RARS, Pilicode

11.1.4 Assessment of Drought

Based on ETo and precipitation, irrigation requirement of different crops can be computed using CROPWAT, which needs information on crop management and crop coefficients during different crop stages. Other classical examples are IBSNAT-DSSAT-CERES models, developed for decision making on agrotechnology transfer across different agroclimatic zones of the world. The accumulation of moisture in soils starts with the onset of pre-monsoon showers/monsoon rains, reaches its field capacity with the progress of rains.

In the absence of rains during off-season, the accumulated soil moisture is used for plant growth. The choice of crop and successful crop harvest depend on the availability of soil moisture. The availability of moisture varies between soils depending upon its water holding capacity. The water holding capacity is the difference between field capacity and permanent wilting point. It depends mostly on the mechanical composition of soils. Considering the above, Thorthwaite (1948) developed a book-keeping water balance procedure based on precipitation and potential evapotranspiration. In similar lines, Frere and Popov (1979) developed a cumulative water balance based on decade/weekly precipitation and potential evapotranspiration taking into account crop coefficient factor, which varies from crop to crop, and within the crop from stage to stage. They also introduced the water requirement satisfaction index. It was correlated well with the yield. Using water balance parameters, several indices such as index of moisture adequacy, humidity index, aridity index, moisture index could be worked out. The above climatic indices based on moisture regime are widely used for delineating climatic zones, defining climatic droughts as well for explaining crop yields based on the type of climatic indices used. The thermal use efficiency is also used for climatic zonation based on thermal regime (potential evapotranspiration). A number of thermal indices viz., Stress Degree Days (SDD), Canopy Temperature Variation (CTV), Temperature Stress Days (TSD), Crop Water Stress Index (CWSI) have been developed outside India based on Infrared thermometry for irrigation management. However, such studies are scanty in India. Of course, they are more crop-soil-climatic specific and complicated in case of plantation crops. Most of the plantation crops suffer due to soil moisture stress during summer. In case of coconut, the nut yield is affected in the following year up to 30 to 50 per cent depending upon the variety and crop management due to severe soil moisture stress during summer. The following criterion is developed for assessing agricultural droughts with reference to coconut based on the index of moisture adequacy (Table 11.2).

Table 11.2 Index of moisture adequacy (Ima) and intensity of drought

Intensity of drought	**Mean Ima (%) from December to April**
No drought	Greater than 30
Moderate	25–30
Large	20–25
Severe	15–20
Disastrous	<15

11.1.5 Scarce Water Resources Management

It is understood that the plantation crops that are grown in the low-and- mid-lands of Kasaragod and Kannur Districts experience severe soil moisture stress almost in half of the years. It was probably one of the reasons, why, the fluctuations in plantation crop production was noticed very often. In the recent past, the occurrence of droughts in summer adversely affected the plantation crop production to a great extent. Therefore, Irrigation schedule during the summer months under scarce water resources management as a part of weather risk management is one of the tools to mitigate the ill effects of summer drought through various ways under scarce water resources. If the irrigation requirement is worked out using reference evapotranspiration and crop coefficient, the irrigation water requirement is comparatively less and soil moisture status can be improved to certain extent and in turn plantation crop yields can be increased. To quote an example, the irrigation requirement of coconut was worked out at RARS, Pilicode using the reference evapotranspiration (modified Penman method has been generally accepted as a reliable method to determine evapotranspiration from climatic data) and crop coefficient. The irrigation requirement varied from 1106 litres/palm/month in December to 1488 litres/palm/month in April. The total irrigation requirement from December to May was worked out as 7807 litres (Table 11.3).

Table 11.3 Irrigation requirement of coconut

Month	ET_0 (mm)	Water requirement (mm) (ET_0 x 0.75)	Irrigation requirement (l) (pr^2h)
December	3.79	2.84	1,106.0
January	3.95	2.96	1,154.0
February	4.56	3.42	1,204.0
March	5.01	3.76	1,464.0
April	5.26	3.95	1,488.0
May	4.76	3.57	1,391.0
Total			**7,807.0**

ETo-Reference evapotranspiration

r- Radius of coconut basin- 2 metres

0.75 – Crop coefficient of coconut

The coconut growers can very well adopt the irrigation quantities during summer as recommended above wherever water is available which will improve coconut yield significantly as the yield per palm is well below the normal under the rainfed conditions due to prolonged dryspell from November to May. In another experiment, a field experiment was taken up at RARS, Pilicode, Kasaragode where the palms were provided basin irrigation at the rate of 450 litres/palm/ week till the water is available during summer and leaving the palms without watering further till the onset of monsoon. The yield was better in the above treatments when compared to that of pre-treatments'yield (Table 11.4). The quantity was arrived at 450 litres/palm/week, taking the evapotranspiration of coconut as 5 mm per day and the coconut basin area as radius 2m.

Table 11.4 Duration of soil moisture stress on coconut yield of WCT

Tretheatment	Pre-treatment yield (1976–89) A)	Post-treatment yield (1991–97) (B)	Difference in yield (A – B)	% increase over pre-treatment yield	Whether significant over pre-treatment yield
T1	103	122	19	18.4	S
T2	103	112	9	8.7	S
T3	94	107	13	13.8	S
T4	87	120	33	37.9	S
T5	92	115	23	25.0	S
T7	91	95	04	4.4	S
T6	60	82	22	36.7	S

T1- Irrigating the palms @450l/palm/week during December and January T2- Irrigating the palms @ 450l/palm/week from December to February T3 – Irrigating the palms @450l/palm/week from December to March

T4 – Irrigating the palms @ 450l/palm/week from December to April T5 – Irrigating the palms @ 450l/palm/week from December to May T6 – Irrigating the palms as per climatic water balance procedure

(150 l/palm/week in December, 200 l/palm/week in January, 300 l/palm/week in February, 350 l/palm/week in March, 400 l/palm/week in April, 450 l/palm/week in May)

11.1.6 Agro-climatic Zonation

Based on weather variables such as precipitation and potential evapotranspiration, the agro climatic zonation can be delineated using the water balance techniques. Such agro-climatic zonation if delineated on crop wise, climatic risks can be mitigated to a considerable extent. In the case of cardamom across the Western Ghats, Zone I & II are superior when compared to that of Zone III, where weather risk is high in terms of high temperature, prolonged dry spells and less length of crop growing season (Fig. 11.5). Similarly, cashew can be extended from northern districts of Kerala to the South of Maharastra along the West Coast and North of Tamil Nadu to Orissa along the East Coast and inland areas away from the Coast. However, tea mosquito bug incidence along the West Coast and cyclones along the East Coast are the constraints for obtaining better cashew production. Similarly, simulation models can very well be used to simulate production potential of various crops in a given watershed area through which the climate risk can be minimized with proper crop management practices. In addition, agroadvisory service based on weather forecasting will go a long way in sustenance of crop production. Another multidisciplinary project launched by the Govt. of India, that is FASAL, is a classical example to help in GIS based watershed planning in Agriculture as a part of weather risk management.

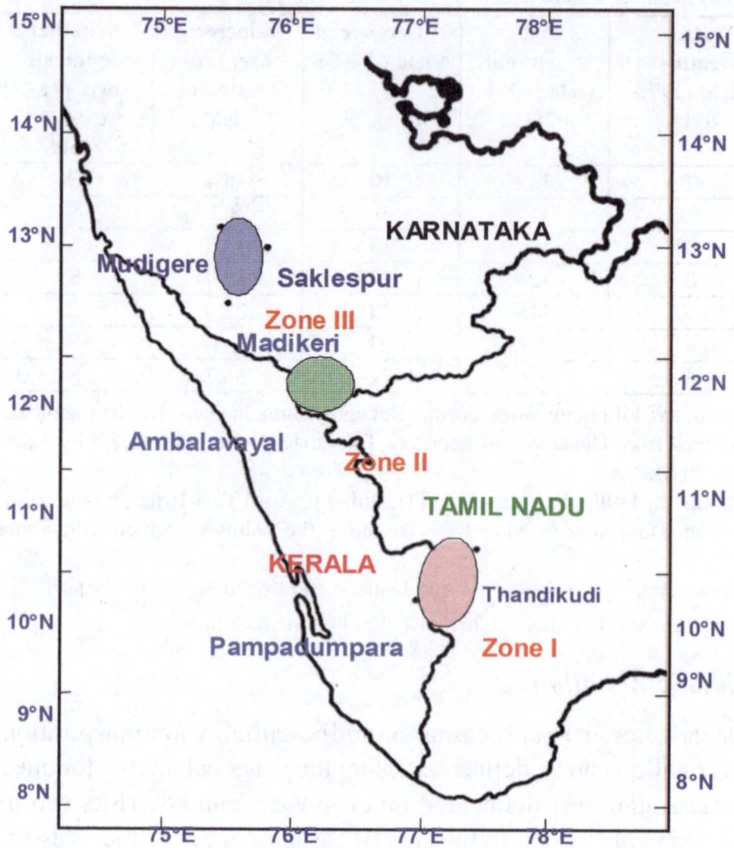

Fig. 11.5: Agroclimatic zones of cardamom across the Western Ghats

The agroclimatic analysis, as indicated above, is a useful tool in not only drought management but also in water/nutritional management as the nutritional uptake is maximum when optimum soil moisture is available to the plant. It is one of the reasons, why, a particular fertilizer element responds positively when optimum soil moisture is available while negatively when soil moisture is not sufficient. Similarly, it is yet to be understood on uptake of nutrients when evapotranspiration is limited - or below optimum due to near zero vapour pressure deficit in the atmosphere, which is not uncommon during rainy season extending three to six months in the Humid Tropics. In such situations, the agroclimatic analysis based on location specific will be of immense use for sustenance of plantation crops' production as part of weather risk management.

11.1.7 Rainwater Harvesting /Groundwater Recharging

Efforts are needed for effective rainwater harvesting in farm lands. Wherever water is running we should make it walk. This is done by terracing, contour bunding, and increasing the soil vegetative cover. All walking water should be lied down. Rain water harvesting is unique and traditional system of recharging of the under-groundwater bodies for sustained water availability. There were several techniques in rainwater harvesting which can be practiced in house hold (roof top rainwater harvesting) as well as fields. Traditional sources of water like very shallow, small diameter wells known as 'keni' in Wayanad district , horizontal wells/tunnels called 'surangams' in Kasaragod district, natural wells, ponds, tanks, springs, lakes and wind breaks and shelter belts should form an integral part of the watershed development programme. Groundwater utilization has to be based on location specific investigations and over exploitation has to be avoided.

11.1.8 Soil Health

The indiscriminate use of fertilizers and plant protection chemical and weedicides are polluting the soil, air and water beyond the tolerable limit. This is true in case of certain areas of Kerala where the practice is creating problem. There was drastic decline in the use of organic manure for the past years due to several reasons. In the process, organic content of the soils is very poor and water-holding capacity of the soils became very weak, resulting in frequent droughts. Therefore, it is essential to improve the organic content of the soils to retain the original status of the soils. Awareness has been created on the use of organic manures among the farmers with reasonable use of chemicals without affecting productivity. There are certain crops and areas where organic farming can be practiced in Kerala. Organic farming also plays an important role as an alternative food production process. It is a crop production system that avoids or largely excludes the use of synthetically produced fertilizers, pesticides, growth regulators and livestock feed additives. As far as possible it relies on crop rotations, animal manures, legumes, green manures, organic wastes, bio-fertilizers and aspects of biological pest control to maintain soil productivity and tillage to supply plant nutrients and to control insects, weeds and other pests. Input availability, productivity, premium price and marketing aspects have to be considered. In other areas, good agricultural practice with the minimum dependence will definitely reduce the pollution of soil, air and water and reduce its influence in greenhouse gas emissions.

11.2 Weather Forecasting

The India Meteorological Department, popularly known as Mausam Bhavan, Government of India is one of the oldest departments in the Country, located in New Delhi. It was established in 1875. It has a network of approximately 6000 rain gauge stations and more than 550 surface meteorological observatories plus about 300 stations which include hydro-agro-meteorological, radiation, air pollution and pilot-balloon stations spread over the entire Country. The whole Country has been divided into 36 meteorological subdivisions based on homogeneity in its rainfall distribution. The aim of the department is to forecast daily weather on meteorological subdivision-wise and disseminate the same on time to the users. All the State Capitals in India also have State Meteorological Centres (SMCs) to serve the weather forecast needs of the States and these centres issue Farmers' Weather Bulletins (FWBs).The bulletins contain a forecast on weather elements valid for 36 hours and an outlook for the subsequent two days. The daily weather forecast is prepared for the entire Country from the weather forecasting division, IMD Pune and communicated to the Mausam Bhavan, New Delhi for dissemination across the Country. There is a vast improvement in precise short range forecasting with the advent of supercomputers along with the radar and satellite information. Cyclone forecast with a network of weather radars along the East Coast of India is a classical example for success of short range weather forecasting. One of the prime objectives of providing weather forecast is to disseminate information to the farmers and users on real time basis so that they can utilize the beneficial aspects of weather information to increase crop and livestock production or to minimise the damage, which may be caused directly or indirectly by unfavourable weather. Incidence of crop pest and diseases and animal diseases could also be forewarned based on weather forecast if information on relationship between weather and specific pest and disease incidence is known in advance. The output of agriculture (includes animal agriculture) could be increased and recurring losses minimized if reliable forecast on incidence of pest and diseases is given timely based on weather variables. Hence, weather forecast plays a key role in sustaining agricultural/livestock production of the Country. This could be possible if information is provided to the farmers, in advance, on the type of weather situations likely to be encountered and method to be adopted for efficient management of inputs. Probably, this is one of the best tools to mitigate the ill effects of climate change/climate variability in plantation crops' production.

The types of weather forecasting generated by the institutes in India, its validity and the present forecast skill are schematically presented in (Fig. 11.6). All the five types of weather forecasting viz., Now Casting, Short Range, Medium Range, Long Range and Climatic/Agro-climatic forecast are one way or other useful from animal agricultural point of view. However, reliability of weather forecasting decreases with increase in duration of its validity and decreasing area as the forecast models are not location specific and the forecast skill is poor if it is for three days and beyond for smaller regions.

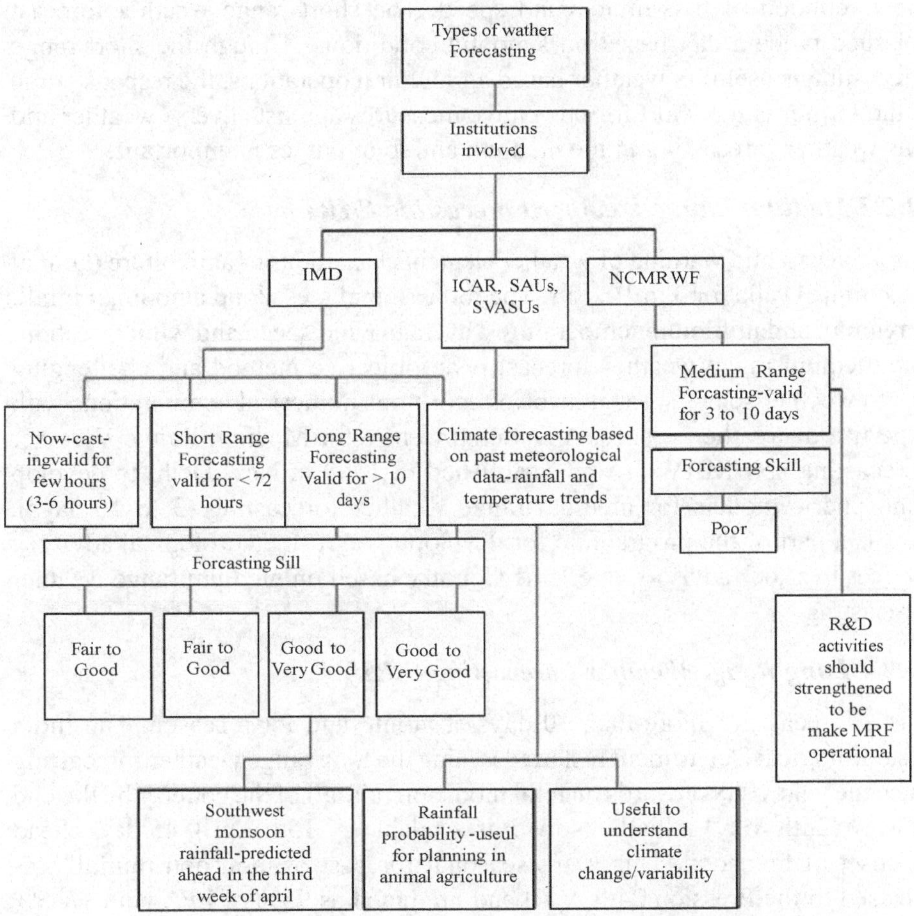

Fig. 11.6: Current weather forecasting skills in India

11.2.1 Nowcasting

It is a type of weather forecast given locally for a time period of three to six hours. Local showers, thunderstorms and cloud bursts can be classified under this type of forecast. This is quite useful for open air functions.

11.2.2 Short Range Weather Forecasting (SRF)

It is a forecast and warning of weather elements hazardous to agriculture (animal agriculture) valid for 36 hours and an outlook for the subsequent two days. The short range forecast includes cloud spread, rainfall distribution, heavy rainfall warnings, maximum and minimum temperatures, heat and cold waves, low pressure areas, cyclone warning, hail/thunderstorm, dust storm, snow, frost and likelihood of maximum wind speed. The short range weather forecast is issued twice a day based on synoptic conditions. Though the short range forecasting is useful in weather based agricultural operations, the response time to the farmer is too short for preventive measures against adverse weather and thus weather forecasting in the medium and long ranges is important.

11.2.3 Medium Range Weather Forecasting (MRF)

It is a forecast and warning of weather elements hazardous to agriculture (animal agriculture) valid for 3 to 10 days. The forecast includes cloud amount, rainfall, maximum and minimum temperatures, average wind speed and wind direction. The medium range weather forecast is an objective method and challenging one to weather scientists as it involves enormous numerical computations with expertise in weather science. A National Centre for Medium Range Weather Forecasting (NCMRWF) was established in 1988 in New Delhi to develop atmospheric models for medium range weather forecasting (3 to 10 days). This is a part of the programme for developing agro-meteorological advisory services/livestock advisory across the Country based on medium range weather forecasting.

11.2.4 Long Range Weather Forecasting (LRF)

It is a forecast for more than 10 days, a month and for a season. The India Meteorological Department restarted issuing the long range weather forecasting since the year 1988 onwards on total monsoon rainfall of the country by the end of May (25th May) initially. Now, it is available by 15th April, 45 days ahead of onset of monsoon. This year's seasonal forecast of monsoon rainfall was released to the Press on 24th April and predicted as 95% of LPA with +/- 5% model error. The onset of monsoon can also be predicted using the predictive models. The predicted and actual long period average (LPA) of monsoon (June-September) rainfall of the Country were broadly in agreement except significant deviations in 1994, 1997, 1999, 2002, 2004, 2007, 2009, 2012, 2013 and 2014 (Fig. 11.7). Though the total monsoon rainfall tallies with predicted in majority of the years, the distribution of rainfall within the season is a very complex one to predict and it is the one which decides agriculture production and to some extent animal agriculture too. Therefore, the forecast need to be improved in many angles including zonal scale. The long range weather forecast can be used for predicting likely trends in foodgrains production of India before beginning of the kharif season as the foodgrains production depends mostly on the distribution and amount of monsoon rains across the Country under rainfed conditions. These forecasts can hold the foodgrain prices in check through buffer stock operations. In good monsoon years when prices fall, the Government may step in and buy, and in bad years when prices tend to rise, it may unload a part of what it had purchased in the past. This will help in price stability, less financial losses, better distribution of the available foodgrains and its import and export policies. With this in view, the crop weather watch groups (CWWGs) in the Agriculture Ministry, Government of India closely follows the forecast and its impact on foodgrains production during the kharif season. The operational Long Range Forecasting (LRF) system in India is based on the statistical methods. IMD issues operational long range forecasts for rainfall/ precipitation during winter (Jan-March), southwest monsoon (Jun to Sept) and northeast monsoon (October-December) seasons. Among these, forecast for monsoon season is most important as the rainfall received during this season accounts for 70–90% of the annual rainfall over most parts of the country and due strong positive association of monsoon season rainfall with both Kharif and rabi crop production in the country. India Meteorological Department (IMD) issues various monthly and seasonal operational forecasts for the south-west monsoon season using models based on latest statistical techniques with useful skill. Operational models are reviewed regularly and improved through in house research activities. Details of the present operational forecasting system used for long range forecasting for southwest monsoon rainfall can be from Pai et al. (2011). Operational forecast for southwest monsoon season rainfall over the country as a whole is issued in two stages; the first forecast in April and update for the April forecast in June. In June, along with the update forecast for the season rainfall, forecasts for the monthly rainfall (for the months of July and August) over the country as a whole and that for season rainfall over the four geographical regions of the country are also issued. In addition, forecast outlook for the rainfall during second half (August-September) of the monsoon season (issued in July) and that for the September rainfall (issued in August) over the country as a whole are also issued. Separate models are used for preparing each of these forecasts. While preparing, the operational forecasts, the experimental and operational forecasts from various research institutions from India and abroad were also taken into account. Along with the operational forecasts, experimental forecasts of monthly and seasonal rainfall over Indian region based on Seasonal Forecasting Model (SFM) are also prepared. In addition to the above long range forecasts, an operational forecast for the monsoon onset over Kerala is also issued in May.

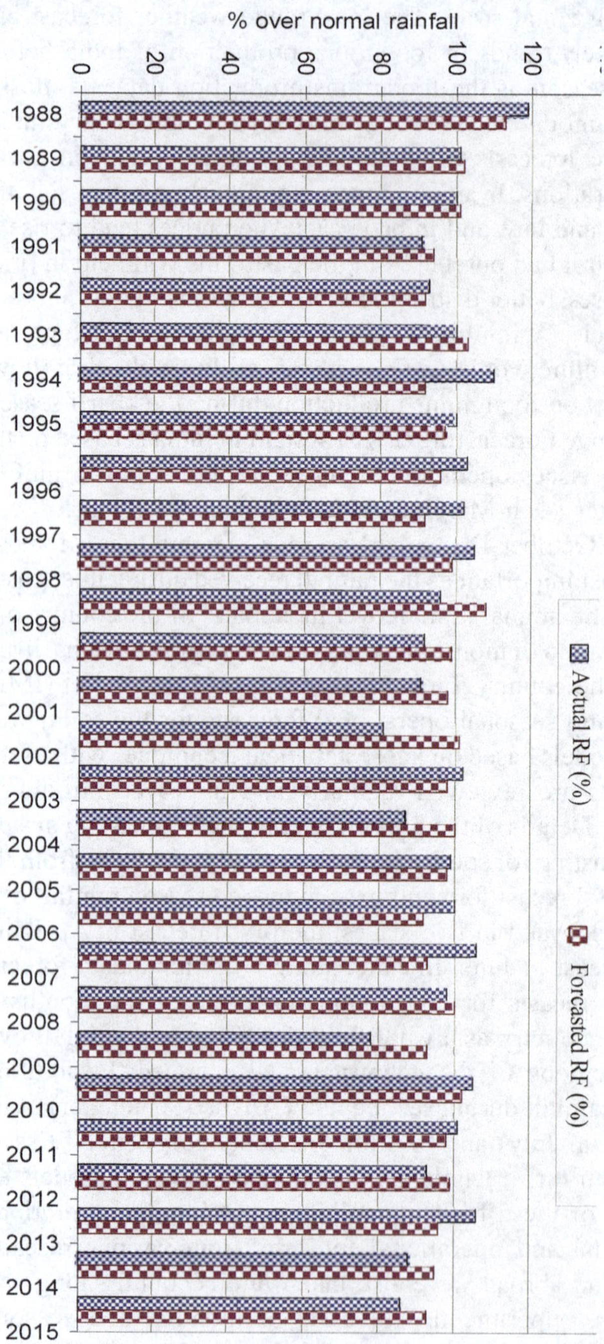

Fig. 11.7: Actual and predicted rainfall (%) from 1988 to 2015

11.2.5 Climatic/Agroclimatic Forecasting/Extended Forecasting

It requires past meteorological data for a good number of years (say 30–50 years). The trends in rainfall and its variability, probability on distribution of rainfall over a season can be determined weekly using the past data on rainfall for a given location. This information is useful to crop planners and farmers as crop growth periods can be adjusted under rainfed conditions depending upon rainfall probabilities. The climatic trends also will help in understanding the impact of climatic variability on animal agricultural production over a period of time. The periods of water surplus and deficit could be worked out for a particular location based on past meteorological data (rainfall and evaporation) through the water balance procedure. This information can be used for selection of crops and adjusting crop growing periods, irrigation scheduling during dryspells and for providing drainage in excess of water surplus. The crop weather relationships could also be well understood based on past phenoogical and meteorological data, which in turn will help in mitigating the ill effects of weather aberrations through better agronomic/livestock practices. In this direction, the research programmes should be strengthened in the field of agricultural/livestock meteorology at the ICAR, SAUs and SVASUs levels with technical personnel having the subject background.

11.2.6 Farmers' Weather Bulletins (FWBs)

A reliable system of short range and medium range weather forecasts is needed for effective on farm management practices such as cultivation of land, preparation of seed bed, planting, choice of crops in case of prolonged monsoon breaks, fertilizer management, harvesting of crops, post-harvest storage and transportation to markets. India Meteorological Department started issuing Farmers' Weather Bulletins (FWBs) on a regular basis since 1945 and later it was entrusted to Regional Meteorological Centres (New Delhi, Mumbai, Calcutta, Chennai and Nagpur), which contain a forecast of expected weather for the next 36 hours and an outlook for another two days. In due course, meteorological centres at State headquarters started to issue FWBs. All India Radio used to broadcast FWBs every evening in regional languages to reach the most remote corner. However, they are found to fall short of the requirements of farmers in terms of duration of forecast validity and capability for small areas, likely impact on agronomic operations. The crop weather calendars prepared for main crops lost their significance with fast changing varieties. As they contain limited information, there was a need to frame them in relation to crops and to specific areas. Thus came the concept of Agro-meteorological Advisory Services (AAS) and it is in operation since 1977. However, the utility of the same is very much limited. Realising the defects and make the AAS more meaningful, National Centre for Medium Range Weather Forecasting (NCMRWF) took up the task in 1988 based on medium range weather forecasting.

11.2.7 Agro-met advisory Services (AAS)

The agro-advisory service (AAS) based on medium range weather forecasting (3 to 10 days) has been made operational by National Centre for Medium Range Weather Forecasting (NCMRWF, IMD). The ICAR and the State Agricultural Universities (including Veterinary and Animal Sciences Universities) co-operate in this task. NCMRWF has established 132 agro-meteorological field units (AMFUs) in all the agro-climatic zones, which cover the entire Country. The agro-climatic zones were delineated by the ICAR under the National Agricultural Research Project (NARP). The AMFUs prepare AAS bulletins. These contain three parts. The first part of AAS consists of weather events occurred during the past week and weather forecast for three days ahead. These forecasts are given to acquaint the farmer with the expected weather on cloud amount, rainfall, average wind speed, wind direction, maximum and minimum temperatures. The second part contains the factual information on the varieties of crops, their state and stage, ongoing agricultural operations, and insect pest and diseases of crops. The livestock component is also included wherever technical service is provided under the scheme. The final and third part of the AAS bulletin provides information on agro-advisory/livestock advisory. The bulletins are prepared by the experts, consisting of scientists belong to various disciplines and agricultural/livestock officers of the locality. Such AAS/livestock bulletins are being prepared and issued in English as well in local languages. The AMFUs functioning at several agro-climatic zones in state agricultural/veterinary universities have made the weekly/biweekly AAS operational effectively. The NCMRWF provides information on MRF every Tuesday and Friday.

The IMD has also established Agro-met Advisory Service Units (AASUs) at the State Meteorological Centres (SMCs). The AASUs issue biweekly agro - advisories to the States on Monday and Thursday based on short-range weather forecasting. These describe the weather over the state during the preceding week and the rainfall forecast valid for the next 48 hours and an outlook for two subsequent days. Information on crops/livestock is obtained from the State Agriculture Department.

11.2.8 Dissemination of Agromet Advisory Services

The agro-advisory bulletins prepared weekly/biweekly by the AMF Us could be communicated on the same day/next day to the selected farmers through a special messenger. The AIR, DD and the other communication facilities can be effectively used to disseminate AAS. It is being done in almost all parts of the Country. The agro-dvisories prepared by the AASUs of the state meteorological centres are sent to the AIR stations and broadcasting in the Farm Radio Programme. The advisories are also faxed to the agromet directorate of IMD, Pune on the same day, where all the advisories sent by the various AASUs are assimilated and then a consolidated brief report is prepared. This report is faxed from Pune to IMD, New Delhi on the subsequent day and is used for Ministerial/ Secretarial briefing.

The agro-advisory bulletins are prepared at the AMF Us by the panel of scientists based on the "Package of practices recommendations" developed by the state agricultural universities for the respective zones. In the absence of a separate package based on R&D in operational agricultural meteorology, it is difficult to improve the quality of AAS. The essence of agro-advisory is to make weather forecasting awareness among the farmers for day -to - day farm operations for sustenance of agricultural production. This is possible if the interaction between weather variables and crop growth and its development is well understood for which crop growth simulation models have to be revalidated/developed for major crops. In similar lines, an interdisciplinary team should work to develop models on pest and disease dynamics. It needs strengthening of research and development activities of the AMF Us in every agro-climatic zone. In addition to that, the NCMRWF in collaboration with the ICAR and SAUs should try to reorient the various research programmes ongoing in the field of agricultural meteorology to tune to AAS. Simultaneously, the skills of medium range weather forecasting must be improved further by the NCMRWF as of now the forecast skill is very poor. The utility of agro-advisory in the field of agriculture based on medium range weather forecasting will be fruitful only if the forecast skill is improved for at least one week ahead along with improvement in quality of AAS. The crop weather watch groups (CWWGs) should be established at the district level on similar lines what we have today at the Centre to integrate the various activities related to weather and agriculture. The CWWGs should be made operational to minimize the crop losses under projected climate change scenario.

11.2.9 Economic Impact of AAS

A reliable and clearly disseminated weather forecast is a very important tool for forewarning crop managers of potential weather hazards. The India Meteorological Department is constantly working to improve forecast skill and help disseminate the forecast in a suitable form to aid farm level decisions. The economic impact of a weekly AAS based on medium range weather forecasting has been assessed for different crops and regions (Table 11.5). This showed that the percentage increase in yield varied from 6.4 –19% depending upon the crop in the case of AAS farmers compared to the non-AAS farmers. Furthermore, it indicated that seasonal crops need intensive advisory, followed by less intensive for biennials and perennials. This is one of the effective tools to mitigate the ill effects of climate change/variability on crops.

Reliability of weather forecast is very important as it declines with increase in time and decrease in space, for which the India Meteorological Department is working to improve the forecast skill. Even the forecast skill is better, the dissemination at farm level is very important. There should be a mechanism to disseminate information effectively through Information Communication Technology such that every farmer and needy people should get weather service at his/her doorstep. The Village Resource Centres under ISRO programme must be linked with online network to disseminate agro advisory based on weather forecasting. It is very pertinent during weather calamities to minimize crop losses. In an event of unusual summer rain, Kuttanad and Thrissur paddy farmers suffered to a large extent. If advance weather service is extended at community level, crop loss would have been minimized to some extent by covering the paddy grains as per the farmers' view. Another way of helping the farmers in such weather calamities is to implement weather insurance. As a long term measure, drainage facilities should be improved at micro-level so that flood damage is minimized in addition to development of flood tolerance/saline tolerant paddy varieties. As the paddy/wetlands act as water sink during the rainy season and water source during summer, such lands should be protected as the area under wetlands is declining vary fast in addition to deforestation reduction across the Western Ghats so as to reduce the run-off and degradation of soils.

Table 11.5: Impact of AAS on crop yields from 2003–04 to 2006–07

Season	Crop	Yield (t/ha)		% increase over in yield over non-AAS farmers	% increase in net return over non-AAS farmers
Rabi 2003–04	Paddy	3.7	3.0	19.0	30.0
	Banana	31.8	27.9	12.2	11.5
	Coconut*	13460	11025	18.1	13.2
Kharif 2004	Paddy	2.8	2.6	7.1	31.6
	Banana	22.8	20.5	10.1	7.9
Rabi 2004–05	Paddy	3.0	2.7	7.5	34.1
	Banana	31.3	29.3	6.4	11.9

Kharif 2005	Paddy	2.7	2.5	6.5	38.5
	Banana	25.3	22.4	11.2	7.1
Rabi 2005–06	Paddy	3.3	2.9	13.6	36.0
	Banana	27.8	24.5	11.8	12.3
Kharif 2006	Paddy	2.8	2.5	9.0	29.4
	Banana	24.5	21.6	11.9	10.0

11.3 Weather Insurance

Weather extremes like flood, drought, landslides, lightening, cloud bursts and hail storms, heat and cold waves are not uncommon now-in-days under the projected global warming and climate change scenario. Both crop and farmer are to be safeguarded against these calamities. Often, the farmer who takes loan and invests in farming cannot repay it due to crop failures in the context of weather extremes. It leads to hardship and rural livelihood is in threat. Farmer suicides are frequent in India and such incidents were noticed in Maharastra and Andhra Pradesh. It was also reported in Wayanad District of Kerala in 2004 due to unprecedented crop loss because of summer drought. Weather insurance is a different one when compared to other crop insurance scheme. If crop loss is noticed due to weather extremes like unusual summer rains in case of paddy, coffee and cashew, monsoon breaks in case of black pepper, occurrence of floods and droughts and heat and cold waves, the farmers need to be paid compensation. Farm level expected income should be assured in an event of weather calamity so as to sustain rural livelihoods. Therefore, insuring farmers as well as crop becomes necessary against weather abnormalities. In this context, precise and accurate information on weather change and impacts are necessary. Short, medium and long range forecasting will definitely help in weather tune farming so as to minimize the adverse effect of weather abnormalities taking suitable precautionary measures to lessen the harmful effects. The meteorological net work of observatories has to be strengthened to disseminate agro-advisory service based on integrated weather forecasting. Pro-active measures against weather aberrations will go a long way in minimizing the crop losses against the weather and climate disasters.

11.4 Weather Risk Managers

Climate change awareness programmes should be organized at various levels, commencing from farmers to planners. These programmes should focus on climate change impacts in agriculture and site specific adaptation and mitigation strategies. The farmers who are trained in all the relevant aspects of climate change and its impacts in agriculture are known as "Weather Risk Managers". They are supposed to disseminate the information and knowledge gained by them through frequent trainings to the neighbouring farmers so as to sustain agriculture against the ill effects of climate change with their wisdom.

Though ill-effects of abnormal weather events are well understood in plantation sector at the field level, it is difficult to quantify the same at the zonal/state level based on secondary data. Hence, a realistic scientific method is to be evolved as such no technique is available to project future crop trends in plantation crops based on climate variability/change. Warming with decline in rainfall and increase in temperature range is projected at some of the locations across the high ranges along the West Coast, where cardamom, coffee, tea and black pepper are predominantly grown under the influence of forest-agro-ecosystems. Deforestation influences the crop boundaries to a considerable extent. Crop growth simulation models indicate that the cropped area and yields are likely to be reduced due to increase in temperature in the case of coffee, tea and cocoa as well in other crops. Though the models indicate reduced crop yields due to long term climate changes (increase in temperature), it is in fact not seen such impacts as on today, but at the same time the adverse impacts of climate variability (occurrence of heat and cold waves and floods and droughts) on plantation crop yields is significant. The GCM models indicate that the frequency of weather abnormalities is likely to increase in ensuing decades under the projected climate change scenario. Therefore, the adverse impact of climate change on plantation crops is seen only through climate variability rather than the direct long term effects of climate change in terms of rise in temperature. Therefore, short term crop management strategies need to be worked out as part of climate resilience plantation crops to combat the ill effects of climate variability/climate change for sustenance of plantation crop production in a changed climate scenario.

The answer is 'Yes' when indirect effect of climate change on plantation crops is considered since the climate change models indicate that the frequency of occurrence of weather extremes such as heat and cold waves and drought and floods is likely to increase in ensuring years/decades under the projected climate change scenario. As mentioned in the preceding sections, the ill effects of cold and heat waves and droughts and floods on plantation crops are well understood and documented across the Country/World. Therefore, the ill effects of climate change are certain through weather extremes and are to be tackled with proactive contingent crop management practices as a part of short term strategy for sustenance of plantation crops' production and rural livelihoods.

The answer is 'No' when direct effects of climate change on plantation crops is considered since the word 'Climate Change' is coined based on long term deviations in weather variables such as temperature and precipitations from the period of 1961–90 for comparison. The GCM outputs indicate long term weather deviations for the periods 2020, 2050, 2080 and so on. All the climate models predict increase in surface atmospheric temperature in varied magnitudes. The negative effects of long term increase in temperature/global warming on plantation crops cannot be singled out much ahead based on the model output since crop growth, development and production is a complex one and depends on many factors viz., the farmer, seed/planting material, soil health, water, ability to spend money for timely farm operations, technology, access to direct marketing and price support to the farm produce. All of them are dynamic and need to be examined to quantify the effect of global warming and climate change on plantation crops for which as of now, there is no mechanism to work out. However, crop growth simulation models highlight reduction in crop yields as well as cropped area if temperature increases by 2°C in the case of coffee and tea. Nevertheless, such impact is unlikely at the farm level as per the present trends noticed by the farming community. Of course, soil health (very poor due to various reasons) and water play a dominant role directly to influence the farm yields on long run. At the same time, change in cropping pattern is noticed in parts of the world depending upon the suitability and profitability, but not due to global warming and climate change.

Climate change adaptation strategies and awareness are particularly important for managing the risks posed by climate variability and change, not only on plantation crops but also across all the society linked sectors that are sensitive to weather and climate. Various agroclimatic techniques have been used in the Humid Tropics to effectively manage some of the risks posed by climate change to plantation crop productivity. Expansion and further development of such techniques will be vital for the continued sustenance of plantation crops' production in the Humid Tropics. As pointed out by Prof. M.S. Swaminathan "India's strength lies in its ability to manage monsoons" instead of saying" Indian agriculture is a gamble of the monsoon".

Climate smart agriculture, climate resilience agriculture, precision farming, hi-tech farming, conservation agriculture, traditional agriculture, protected cultivation, mixed farming and integrated farming are some of the technologies/tools are coined to minimize the ill effects of climate change on crops. These technologies may be sound in the case of field crops/seasonal crops, but not in the case of plantation crops since the plantation crops are in the field round the life time, having economic yield age of 20 to 60 years depending upon the type of crop. In the case of coconut depending upon the type of cultivar or hybrid, the first flowering commences between fourth and eighth year after planting under better management conditions and coconut yield is stabilized by its 20th year and yields economically up to eighty and hundred years. Therefore, integrated farming as done traditionally or crop combinations in the case of coffee and black pepper or multi-tier cropping systems may be strategic approach to withstand against the ill effects of climate change/variability rather than looking for mono-crops. The farm-level weather insurance may be another tool which should be seriously examined for the benefit of the farmers in the event of crop losses due to weather extremes.

The State Action Plan on Climate Change (SAPCC) indicated that the temperature rise is likely around 2oC by 2050 and the rate of increase is likely to be more across the high ranges 2–4.5°C. If sea level rises by one metre, 169 sq.km surrounded Kochi to be inundated. Alappuzha, Palakkad, Wayanad and Idukki are the ‘Climate Change Hot spots’ in Kerala while Alappuzha and Palakkad are the districts most vulnerable to climate change. Idukki, Wayanad, Thiruvananthapuram and Kannur are highly vulnerable to Climate Change. Climate Change is likely to increase the frequency of occurrence of weather extremes. Deforestation, increase in temperature and decline in rainfall along the high ranges may be a threat to plantation crops like tea, cardamom, coffee and black pepper. Unlike seasonal crops, the effects of climate change may be standing many more subsequent years depending upon the phenology of the crop. It is also observed that crop mix/integrated farming may withstand relatively better against the ill effects of climate variability/change as seen in the case of coffee and black pepper. When black pepper performs better coffee suffers and vice-versa. Recent findings indicate that the climate change effects not only crop output but also commodity quality and price. The current hike in coconut price could be attributed to last year’s summer drought in addition to the crop suffered in neighbouring coconut growing states.

Summary

It is high time that the government agencies and policy makers to be proactive with short term and long term strategies to mitigate the ill effects of weather related disasters with the people’s participation and minimize the losses against the weather related disasters. Need of the hour is expect the unexpected guests in the form of weather related disasters under the projected climate change scenario and tune to the system to minimize the losses with well preparedness at various levels. Keeping the above in view, the National Action Plan on Climate Change (NAPCC) identified ‘Sustainable Agriculture’ is one of the thrust areas to tackle the climate change related issues. The mission intends making Agriculture more resistant to climate change by identifying and developing new varieties of crops that are thermal resistance and capable of withstanding extreme weather. These issues are being tackled as a part of

National Initiative on Climate Resilient Agriculture (NICRA) under ICAR. When it comes to the State of Kerala such initiatives need to be taken up on priority basis in developing technologies against the weather extremes such as saline/flood resistance and drought resistance/tolerance rather than thermal resistance, for which a definite State Action Plan on Climate Change (SAPCC) in Agriculture Sector need to be highlighted and tackled with concerted efforts. To tackle the climate change related issues in Agriculture, a definite policy framework must be chalked out and related institutes should look into. To me it appears, we need have an action plan similar to that of NICRA at the State level with technical people in the form of 'State Initiative on Climate Resilient Agriculture'. It can be known as SICRA at the State level similar to that of NICRA at the national level. It may be related to plantation crops or horticulture or medicinal plants or seasonal or as a whole integrated/homestead faming (Fig. 11.8). Such system should be made operational and in vogue at the earliest possible time since climate change/variability is a big threat in plantation sector.

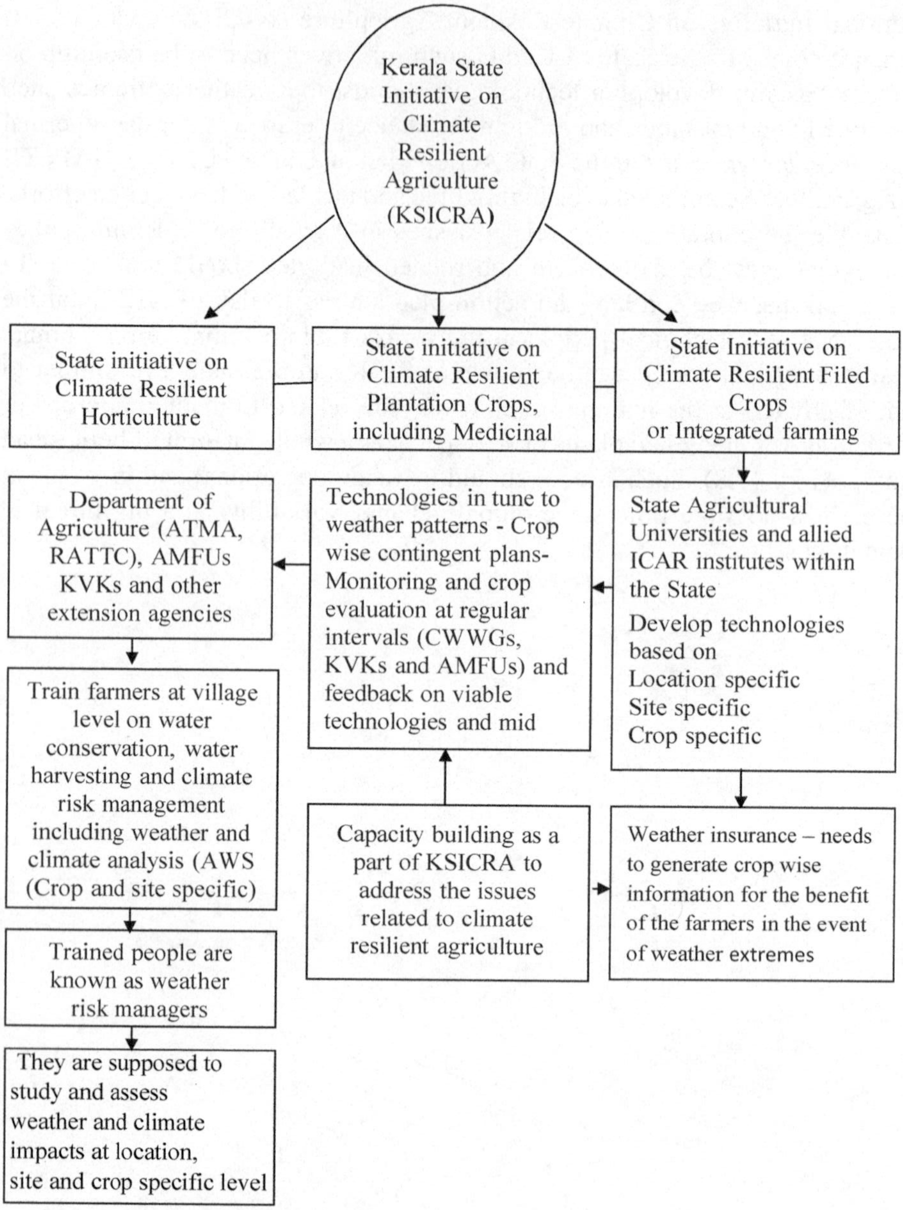

Fig. 11.8: Concept of SICRA – a Model

Suggested Readings

Abdul Salam, M and Mammen, M.K. 1990. Soil and climate based irrigation scheduling for adult coconut. Indian Coconut Journal Vol. 21(3):2–6.

Bhaskaran, U.P and Leela, K. 1978. Response of coconut to irrigation in relation to production status of palms and soil types. In. Proceedings PLACROSYM I, Indian Society of Plantation Crops, Kasaragod. pp 200–206.

Joshi, O.P., Gopalasundaram, P and Muhammed Yusuf. 1988. Moisture depletion pattern in coconut plantation under rainfed and irrigated condition. In: Agrometeorology of Plantation Crops, KAU, Thrissur. pp 22–26.

Kerala State Land Use Board 2001. Project for Agro-ecological zonation in Kerala.

Martinez, R., Hemming, D., Malone, L., Bermudez, N., Cockfield, G., Mushtaq, S., Diongue, A. Hansen, J., Hildebrand, A., Ingram, K., Jakeman, G., Kadi, M., McGregor, G.R. Rao, P., Pulwarty, R., Ndiaye, O., Srinivasan, G., Eh.R., Seck, White, Nand Zougmore, R. 2012. Improving Climate Risk Management at Local Level – Techniques, Case Studies, Good Practices and Guidelines for World Meteorological Organization Members-Chapter 21 (477–532): In Risk Management-Current Issues and Chanllenges (Ed: Nerija Banaitiena),584p.

Muhammed Yusuf and Varadan, K.M. 1993. Water management studies on coconut in India. Advances in coconut research and development. Proceedings of the International Symposium on Coconut Research and Development, CPCRI, Kasaragod. 26–29 November 1991. pp. 337–346.

Nair, R.R, Pillai, G.R. and Rajagoplan, A. 1988. Water management in coconut. In: Six decades of coconut research, Kerala Agricultural University, Thrissur. pp 94–102.

Nelliat, E.V. and Padmaja, E.K. 1978. Irrigation requirement of coconut and response to levels of fertilizer under irrigated condition during the early bearing stage. In: Proc. PLACROSYM – I, Indian Society of Plantation Crops. Kasaragod. pp 196–199.

Pilot study in Kollam District for evolving the methodology. State Planning Board. Govt. of Kerala. Thiruvananthapuram. 105p.

Rajagopal, V. Shivashankar, S and Jacob Mathew. 1996. Impact of dryspells on the ontogeny of coconut fruits and its relation to yield. Agro Physiology. 251–255.

Ramakrishna, Y.S., Rao, G.G.S.N., Srinivasa Rao, G. and Vijayakumar, P. 2006. Impact of climate change in agriculture. Environment and Agriculture (eds. Chadha, K.L. and Swaminathan, M.S.). Malhotra Publishing House, New Delhi, pp. 1–30.

Rao, G.S.L.H.V.P., Gopakumar, C.S., Krishnakumar, K.N. 2013. Chapter 2:Impacts of Climate Change on Horticulture Across India (13–20pp). In: Climate-Resilient Horticulture: Adaptation and Mitigation Strategies (H.P Singh, et.al). Springer India, New Delhi, 302p.

Rao, G.S.L.H.V.P., Rao, G.G.S.N. and Rao, V.U.M. 2010. Climate Change and Agriculture over India. PHI Learning Private Limited, New Delhi. 352p.

Rao, G.S.L.H.V.P., Sudheesh, M.V., Manikandan, N. and Krishnakumar, K.N. 2006.

Rao, G.S.L.H.V.P., V.U.M Rao and G.G.S.N, Rao. 2010. Climate Change and Agriculture over India, PHI Learning India Pvt Ltd.,328p.

Rao, G.S.L.H.V.P. 1991. Agricultural droughts with reference to coconut. Journal of Plantation Crops. 18 (supplement): 166–170.

Rao, G.S.L.H. V. P. 1994. Drought management in coconut. Indian Coco. J. 24(1): 6–8.

Rao, G.S.L.H. V.P. 1994. Drought management in coconut. Indian Coconut J. Vol. XXIV(I 1): pp 6–8.

Rao, G.S.L.H.V.P. 1994. Water management in coconut. Proceedings of the National Seminar on water management for plantation crops. CWRDM, Kozhikode. PP: 65–70.

Rao, G.S.L.H.V.P. 2005. Agricultural Meteorology. Second edition. Kerala Agricultural University, Thrisur, Kerala, P. 326.

Rao, G.S.L.H.V.P. 2008 (Reprint 2014). Agricultural Meteorology. Prentice–Hall of India Private Limited, New Delhi. 366p.

Rao, G.S.L.H.V.P. 2011. Climate change adaptation strategies in agriculture and allied sectors, Scientific Publishers (India), 336p.

Rao, G.S.L.H.V.P. 2011. Climate Change Adaptation Strategies in Agriculture and Allied Sectors. Scientific Publishers, Jodhpur (India). 330p.

Rao, G.S.L.H.V.P. and Gopakumar, C.S.2011. Climate Change Impacts on Monsoon and Indian Foodgrains Production.ACCER, Kerala Agricultural University, Thrissur-680 656, 92p.

Rathore, L.S. and Parvinder Maine. 2008. Economic Impact Assessment of Agrometeorological Advisory Service of NCMRWF. Report NO. NMRF/ PR/1/2008. NCMRWF, MoES, Govt. of India, New Delhi.160p.

Rethinam, P. 1987. Management of drought situation in coconut plantations. Indian Coco. J. 18 (10): 3–5.

Seminar on water management for plantation crops. CWRDM, Kozhikode. pp: 65–70.

Sharangi, A.B., Rao, G.S.L.H.V.P., Das, S., Krishnamurthy, K.S.Upadhyay, T.K., Gopakumar, C.S. and Acharya. S.K. 2024. Brunt of Climate Change and Spice Crops: Scenario, Response, and Resilience. 755-812pp. Eds: Ravindran, P.V., Sivaraman, K.,Devasahayam,S., Nirmal Babu, K. Handbook of Spices in India: 75 Years of Research and Development. Springer Nature Singapore Pte. Ltd.535p

Shukla, P.R., Sharma, S.K. and Venkata Ramana, P. (Eds.) 2002. Climate change and India – Issues, concerns and Opportunities. Tata-McGraw-Hill Publishing Company Limited, New Delhi, p. 317.

Verification of the model on water discharge of the Peechi Dam of Kerala. Proceedings of the International Conference on Hydrology and Watershed Management. 5–8 December 2006, Jawaharlal Nehru Technological University, Hyderabad. Rao, G.S.L.H.V.P. 1994. Water management in coconut. Proceedings of the National

Questions

I. Elucidate on the Following

1. Summer drought management in plantation crops
2. CROPWAT
3. Agroclimatic zones of cardamom
4. Weather forecasting
5. Agromet Advisory Services
6. Climate Risk Managers

II. Write Short notes on

1. Rainfall and pan evaporation curves

2. Economic impact of AAS
3. Weather Insurance
4. Concept of SICRA
5. Scarce water resources management
6. Climate Risk Managers

III. Expand

SDD CTV TSD CWSI IMA

IV. Expand

NCMRWF AMFU AASU CWWG ISRO

V. Expand

NAPCC SAPCC UNFCCC NICRA SICRA

VI. Fill in the Blanks

1. -----------, --------- and --------- are part of agricultural drought
2. ------------- drought related to water logging
3. Temperature are likely to increase between ------------ and -----------^{0}C by 2100AD
4. ---------------- is the ETo in mm annually as per the Penman Method
5. ------------is the relationship between pan evaporation and reference evapotranspiration
6. The values of Kc varies from ---------- to ------------ for many crops
7. ------------ is the season during which the difference between pan
8. evaporation and reference evapotranspiration is large at Vellanikkara.
9. -------------- is the weather element plays a major role for the large difference between pan
10. evaporation and reference evapotranspiration
11. Nowcasting is the weather forecast for a period of --------------
12. Medium Range weather forecasting for a period of -------------
13. Long Range Forecasting is revived in ------------- year
14. Long Range Forecasting is related to ----------------
15. Deforestation influences ---------------- boundaries to a considerable extent
16. Farmers' weather bulletins commenced in ------------- year
17. NCMRWF was established in ----------- year

VII. State true or False

1. Nowcasting is the weather forecast for three days ahead
2. Long range Weather Forecast is related to annual rainfall
3. Zone II is better for cardamom production
4. It is detrimental to coconut productivity if summer irrigation is provided
5. All India drought is defined by IMD in 2002
6. Meteorological drought is defined when rainfall deficiency is more than 20 per cent against the normal rainfall
7. 2014 and 2015 were the consecutive drought years for the first time in India
8. Irrigation scheduling period can be identified based on weekly rainfall and pan evaporation
9. Soil moisture holding capacity of soils depends on organic content of the soils
10. Multi-tier cropping system may be strategic approach to mitigate the ill effects
11. against climate variability/climate change

VIII. Match the Following

Treatment	% increase in coconut yield
T1	37.9
T2	4.4
T3	36.7
T4	18.4
T5	8.7
T6	13.8
T7	25.0

IX. Match the Following

	ETo
December	5.01
January	4.76
February	3.79
March	3.95
April	4.56
May	5.26

X. Match the Following

	Ima in %
No drought	15-20
Moderate	< 15
Large	25-30
Severe	20-25
Disastrous	> 30

XI. Distinguish Between

1. Nowcasting and short range weather forecasting
2. Medium range weather forecasting and long range weather forecasting
3. Zone I and Zone II across the cardamom tract
4. NICRA and SICRA
5. Weather Insurance and Crop Insurance

12

Agrometeorology of the Natural Rubber Plantations in India**

The Government of India, during 1980s, recognized the necessity to improve the latex yield productivity in the traditional rubber growing regions and also expand these plantations to non-traditional regions to meet the ever increasing demands of natural rubber in India (Sethuraj et al., 1989). Agrometeorological studies were initiated during early 1986 with the appointment of a Scientist (Agrometeorology). To begin with, scientific issues were identified as: (a) delineate the potential non-traditional regions for expansion of rubber plantations, (b) estimate the impact of climatic conditions, especially abiotic stress conditions on productivity, (c) quantify influence of meteorological factors on the physiological responses towards plant growth and yield attributes, (d) estimate the irrigation requirements to mitigate the drought conditions, and (e) develop weather based predictive models for `Abnormal leaf fall' disease of rubber plantations. An attempt is made to present synthesis of accomplishments during 1986 to 1992 period of my association with the Rubber Research Institute of India (RRII), Kottayam.

12.1 Agroclimatic Characterization

The prevailing agrometeorological conditions would influence the survival, growth and yields of the rubber (*Hevea brasiliensis)* plantations, whereas the climate (viz. mean monthly rainfall, temperature, sunshine and wind speeds) of the region impacts the overall productivity of the plantations. The optimum climatic requirements of Hevea are: (a) A rainfall of 2000 mm or more, evenly distributed without any marked dry season and with 125-150 rainy days per year, (b) A maximum temperature (Tx) of about 29-34°C, minimum (Tn) of about 20°C or more with a monthly mean of 25-28°C, (c) High atmospheric humidity

P. Sanjeeva Rao**
Scientist G (Rtd.), Department of Science and Technology Govt. of India, New Delhi
**Present Address: H.No.7-4, Gandigunta, Vuyyuru-521165, Krishna, Andhra Pradesh
Email: psanjeevarao@yahoo.com

of the order of 80% with moderate wind and (d) Bright sunshine (S) at the rate of six hours per day in all months. Only a few rubber-growing regions in the world qualify to fall within such a climatic profile. Areas of higher rubber production are characterized by adequate rainfall and less fluctuations in temperature conditions throughout the economic lifespan of about 30 years (Sanjeeva Rao and Vijayakumar, 1992). Any deviation from these optimum conditions may influence the overall performance of the plantation. Further, rainfed rubber cultivation depends on the availability of adequate soil moisture and prolonged dry conditions would adversely influence the establishment of young plants as well as yield of mature plants (Vijayakumar et al., 1988). Established few agrometeorological observatories to assess the prevailing climate in rubber growing areas in India. Meteorological database was developed by collating all the available mean monthly meteorological parameters from the observatories established in the rubber growing regions of the world and all locations from India. This database was utilized to assess the hydrothermal conditions suitable for rubber cultivation and climatological water balance conditions. A simple climatic index (CI) was developed, incorporating the distribution of mean monthly rainfall of 125 mm and temperature (minimum and maximum) fluctuations assigning different ratings as per the deviation from the optimum climatic requirements. The climatological characteristics of the rubber growing regions were categorized into six zones (Sanjeeva Rao et al., 1993). Areas with ideal temperature and rainfall distribution for growth and latex production were grouped under highly suitable region (zone-1) and those areas with the least favorable regime are under not favorable region (zone-6). The highest CI of 0.97 was noted for Senai (03° 08'N, 104°E, 10m) where rainfall occurs throughout the year and temperature was between 21.4°C to 32.6°C, which are very close to the climate requirement. Comparison of CI with already established commercial rubber growing regions elsewhere indicates that most of the locations in India, falls under suitable to moderate zones (Sanjeeva Rao et al., 1993). Most of the traditional rubber-growing regions were categorized as climatically suitable (zone-2). It was found that temperature conditions along West Coast, parts of East Coast and a few locations in Northeast India are more favorable for rubber cultivation than rainfall distribution. Further, none of the locations in India has highly suitable hydrothermal conditions for rubber growth and productivity. However, extension of rubber plantation to the moderate to marginal (zone-3 and zone-4) regions adoption of irrigation and soil moisture conservation techniques would be necessary for successful rubber cultivation.

12.2 Climatic Water Balance and Rubber Yields

Rainfall is the principal source of soil moisture to meet the water requirements of rubber plantations. Soil moisture deficits experienced by the plantation affect the growth, yield and yield components. Variations in the length of moisture availability and the intensity of dry period may regulate growth and productivity in different agroclimatic regions. An understanding of the water balance will guide for planning suitable agronomic and economically viable management strategies. Climatically, evapotranspiration deficit (ETd) or the moisture stress being experienced by the plants is the difference between the water requirement of the plant (ETc) and the actual evapotranspiration (ETa) at a given period and location. In order to assess the severity of drought conditions in different rubber growing areas, climatological water balance model was adopted. In the equatorial regions, annual ETd was below 80 mm whereas the traditional rubber growing areas of south India experience between 52 to 308 mm. The moisture deficit experienced at Agartala (NE India) is comparable to southern parts of Kerala. Most parts of Orissa, drought intensity was more than that in traditional areas and at some locations it was comparable to that of Palghat. Plantations raised at Dapchari experienced severe water deficits of the order of 1100 mm under rainfed conditions (Sanjeeva Rao, 1993). The variations in district rubber yields reported in Kerala are negatively correlated with mean annual water deficit. Ratio of ETa to ETc represents availability of moisture and explains about 93% of variations in regional rubber yields (Sanjeeva Rao et al., 1990). Comparatively low yields in south India than Malaysia may be attributed to the climatic conditions, rather than the soil and management conditions. These investigations suggest that any new areas identified for rubber plantation should have positive water balance so that the excess water during the rainy season can be conserved and reused to reduce the impact of stress.

12.3 Soil-Plant-Atmosphere Interactions

Availability of soil moisture is one of the hypothesized critical factors influencing the latex yields through its effect on plant water relations. In order to investigate the influence of soil moisture status during dry and wet periods on yield, yield components and water relations, initiated collaborative field observational studies, in four young Hevea clones (RRII 105, Gl1, RRII 118 and Tjir 1) in the traditional rubber growing areas (Gururaja Rao et al., 1989; Devakumar et al., 1988). Hourly meteorological observations (between 0600 hrs to 1800 hrs IST) and changes in components of water relations (leaf water potential, latex osmotic potential, stomatal resistance, transpiration rates, xylem sap speeds, latex vessel turgor), daily yield and yield components (initial flow rate, rubber content and plugging index) were recorded during 1987 and 1988 (Devakumar et al., 1988; Gururaja Rao et al., 1989, 1990). Analysis of diurnal observations in two seasons and at two locations indicates clonal variations in yield and yield components and associated physiological parameters in response to soil moisture status and meteorological factors (Gururaja Rao et al., 1990). During the summer months, dry weather prevails with very high evaporative demand resulting not only atmospheric drought but also soil drought. Evapotranspiration vales were around 5.4 mm/d during dry season and 4.0 mm/d in the wet season. Transpiration rates were higher in the afternoon hours when the vapor pressure deficit (VPD) is high, even though the stomatal openings were noted in the morning. Variations in meteorological parameters, viz., irradiance, temperature and VPD exceeding the limits of 300 Wm^{-2}, 35°C and 35 mbar, respectively, might result in stress in Hevea. The transpiration coefficient, defined as the ratio of actual transpiration to potential evapotranspiration, has showed clonal and seasonal variations (ex. RRII 105 (0.104 to 0.544), RRII 118 (0.242 to 1.039)). Transpiration coefficient can be used as a tool to quantify available soil moisture as in tree crops such quantifications are often difficult using soil moisture data. Absence of seasonal variation in minimum stomatal resistance, significant differences in transpiration rates in dry and wet periods, maintenance of higher plant water status were associated with higher stomatal resistance and higher xylem sap speeds (Gururaja Rao et al., 1990; Devakumar et al., 1988). Thus, prevailing meteorological parameters play an important role in maintenance of plant water status through their effects on stomatal resistance and transpiration, thereby influencing latex vessel turgor and latex outputs.

12.4 Irrigation Requirement

Non-traditional rubber growing regions in India, where rainfall distribution is not favorable, the young plants experience prolonged moisture stress and thereby affect the growth during the immaturity period. A field experiment was laid out at Dapchari with an objective of quantifying the requirement of irrigation water with basin and drip irrigation methods. Young budded plants of clone RRII 105 was raised in a randomized block design, 25 plants per plot at 4.9 m x 4.9 m spacing, with three replications. Considering evapotranspiration depends on microclimate and leaf area of the plant canopy, rubber crop evapotranspiration (ETc) at different stages was estimated based on the prevailing meteorological conditions and crop coefficient (0.2 to 1.25). The mean estimated irrigation requirement for the dry season in the traditional rubber growing regions was estimated to be around 10000 l for a mature tree after canopy closure (Vijayakumar et al., 1988). However, estimated water requirement at Dapchari during December to May is around 33,500 l per tree. Methodology followed for imposing irrigation treatments the quantity of irrigation water per plant per irrigation and scheduling of irrigation are described by Vijayakumar et al. (1998). Seven treatments (Control (rainfed), three levels each of basin and drip methods) were imposed. Initial results of these experiments indicated that sufficient quantity of irrigation during stress period can increase the initial establishment of plants, reduce immaturity period and increase latex production (Bhaskar et al., 1990; Mohankrishna et al., 1991). In the experimental site, the drip irrigation was found inferior to that of basin irrigation. Higher rates of photosynthesis were observed in the early hours in plants under adequate irrigation. Unlike the traditional regions, the rainfed plants showed high stomatal resistance even at the morning hours (Vijayakumar et al., 1988, 1998). Water use efficiency in terms of dry matter produced per unit of water applied showed negative relation with quantity of water applied. Results were also indicated that marked and comparable growth could be achieved by providing basin irrigation at 50% of the estimated requirement, thereby overcome the long immaturity periods. Growth rates indicated that once water stress is relieved, conditions are more favorable for overall growth of trees. The biomass of the irrigated trees would increase by 2.8 times to control plants. Even though irrigation water requirement is very high, irrigation is necessary for good adaptation of Hevea plants to water deficits, absence of any direct injury due to atmospheric drought and tolerance to high temperatures experienced in the non-traditional regions was demonstrated (Vijayakumar et al., 1998).

12.5 Crop Weather Relations

Growth and latex production are two distinct physiological phenomena in rubber trees. In order to quantify the relationship between latex yield and prevailing agrometeorological conditions, daily dry rubber yields of 30 trees of RRII 105, planted during 1982 at Kottayam, under tapping system of half-spiral cut in the bark of the tree with alternate daily harvest, were collected during 11 July 1989 to December 1990. Daily meteorological data viz., rainfall (R), maximum (Tx) and minimum (Tn) temperature, sunshine duration (S), forenoon and afternoon relative humidity and pan evaporation (Ep) were recorded. Observed latex yields of RRII 105 are highly variable from 19.8 to 90.5 g per tree per tapping. Analyzed the correlations between yield and meteorological data at different lag periods (1 day, 7 days, 30 days to 180 days) to establish the yield – weather relationships in rubber under humid tropical conditions (Sanjeeva Rao et al., 1998). The weekly meteorological conditions associated with optimum yields were identified as Tx of 30.4°C, Tn of 22.8°C, 5.9 h of sunshine and 72 mm of rainfall. Any deviation from these meteorological conditions up to six months duration may influence latex production. The impact of antecedent weather conditions of Tx, S, VPD and Ep on the latex yield on the day of tapping, showed significant negative correlation with yield, irrespective of the time periods considered. The Tx experienced during 1 day, 7 days and 30 days prior to the tapping explains about 32%, 42% and 64%, respectively, of the day-to-day variability of latex yields. However, significant positive association was observed between rainfall beyond 30 days and latex yields, as latex contains 65-70% water, rainfall through soil moisture influenced the yield (Sanjeeva Rao et al., 1998). These results clearly indicate the harvestable latex yields are influenced by antecedent weather and climatic conditions experienced by the trees beyond a season.

12.6 Abnormal Leaf Fall Disease

Heavy rainfall and low intensity of radiation leads to a number of economically important diseases, like abnormal leaf fall caused by *Phytophthora Palmivora*, require costly protection measures. Prophylactic areal spraying against abnormal leaf fall in the traditional rubber growing regions was undertaken by the growers in routine to protect the plantations. However, in some years, protective measures have not been effective and there was serious concern to provide scientific services to plant protection (Jayarathnam, et al., 1987b). Assessment of daily meteorological conditions at Kottayam associated with abnormal leaf fall disease indicates that `a spell of continuous rainfall of five days accompanied by high humidity and moderate temperature and one overcast day (S <0.1 h)' are essential to trigger the disease, when the disease spores are present, the disease outbreak can be expected within 9 to 15 days after the overcast day ((Jayarathnam, et al., 1987a). Any spraying after noticing these conditions could be of any value in protecting the trees. This disease outbreak forecast model was tested in various regions and found to be satisfactory in providing guidance to planters.

Conclusion

The importance of agrometeorological investigations with respect to rubber plantations was recognized in time and appointed a qualified scientist to initiate the research activity. It was a unique opportunity to conceive many innovative ideas and their timely implementation with the cooperation of many individual scientists, irrespective of their domain expertise. Even though the Agrometeorology unit was part of the Pathology Division of RRII, active scientific collaboration was developed with the Physiologists, Pathologists, Botanists and Agronomists in addressing various issues of contemporary importance. These collaborations could enable to address the burning issues and brought out many scientific publications in peer reviewed high impact scientific journals. Also, technical deliberations at various national and international discussions could able to outreach these results to the rubber planters around the world. Finally, demonstrated the mutual benefits of the culture of `Work Together' within the organization to address the issues of importance to the Society.

Acknowledgement

I take this opportunity to express my sincere gratitude and thanks to all my colleagues and administrators for the technical support, scientific guidance and whole hearted cooperation extended during my association with RRII, Kottayam.

References

Bhaskar, C.V.S., T. Mohan Krishna, T.R. Chandrashekar, P. Sanjeeva Rao, K.R. Vijayakumar and M.R. Sethuraj (1990). Effect of irrigation on growth and other physiological parameters of Hevea brasiliensis in the Konkan region of India. Proceedings of IRRDB Physiology and Exploitation of 'Hevea brasiliensis', Kunning, China, pp.130-135.

Devakumar, A.S., G. Gururaja Rao, R. Ragagopal, P. Sanjeeva Rao, M.J. George, K.R. Vijayakumar and M.R. Sethuraj (1988). Studies on Soil-Plant-Atmosphere System in 'Hevea':II. Seasonal effects on water relations and yield. Indian Journal of Natural Rubber Research,1(2):45-60.

Gururaja Rao, G., R. Rajagopal, A.S. Devakumar, P. Sanjeeva Rao, M.J. George, K.R. Vijayakumar and M.R. Sethuraj (1989). Study on yield and yield components of 'Hevea' clones during water stress in 1987. Proceedings International Congress of Plant Physiology, New Delhi, pp.499-503.

Gururaja Rao, G., P. Sanjeeva Rao, R. Rajagopal, A.S. Devakumar, K.R. Vijayakumar and M.R. Sethuraj (1990). Influence of Soil, Plant and Meteorological factors on water relations and Yield in Hevea brasiliensis. International Journal of Biometeorology, 34:175-180.

Jayarathnam, K., P. Sanjeeva Rao and V. Krishnankutty (1987a). A study on the weather associated with the abnormal leaf fall disease of rubber (Hevea brasiliensis). In: Agrometeorology of Plantation Crops (ed. G.S.L.H.V. Prasada Rao and R.R. Nair). Kerala Agricultural University, pp.93-97.

Jayarathnam, K., P. Sanjeeva Rao, C.K. Jacob and T.T. Edathil (1987b). Prophylactic spraying against abnormal leaf fall disease: Essential or not? Rubber Board Bull., 23(2):24-28.

Mohan Krishna, T., C.V.S. Bhaskar, P. Sanjeeva Rao, T.R. Chandrashekar, M.R. Sethuraj and K.R. Vijayakumar (1991). Effect of irrigation on physiological performance of immature Hevea brasiliensis in Konkan region. Indian Journal of Natural Rubber Research, 4:36-45.

Sanjeeva Rao, P., (1993). Water balance in the rubber growing regions. Plant Physiology and Biochemistry, 20:56-61.

Sanjeeva Rao, P., and K.R. Vijayakumar (1992). Climatic Requirements. In: Natural Rubber: Biology, cultivation and technology (ed. M.R. Sethuraj and N.M. Mathew). Developments in Crop Science 23, Elsevier Science Publishers, B.V., Amsterdam, pp.200-219.

Sanjeeva Rao, P., K. Jayarathnam and M.R. Sethuraj (1990). Water balance studies of the Rubber growing regions of South India. Journal of Applied Hydrology, 3(2):23-30.

Sanjeeva Rao, P., K. Jayarathnam and M.R. Sethuraj (1993). An index to assess areas hydro-thermally suitable for rubber cultivation. Indian Journal of Natural Rubber Research, 6:80-91.

Sanjeeva Rao, P., C.K. Saraswathyamma and M.R. Sethuraj (1998). Studies on the Relationship between Yield and Meteorological Parameters of Para Rubber Tree (Hevea brasiliensis). Agricultural and Forest Meteorology, 90:235-245.

Sethuraj, M.R., S.N. Potty, K.R. Vijayakumar, A.K. Krishnakumar, P. Sanjeeva Rao, A.P. Thapaliyal, T. Mohan Krishna, G. Gururaja Rao, D. Choudhury, M.J. George, T.A. Soman and J.R. Meenattor (1989). Growth performance of Rubber in the non-traditional regions of India'. Proceedings of the R R I M Rubber Growers' Conference, August 1989. pp.212-227.

Vijayakumar, K.R., G. Gururaja Rao, P. Sanjeeva Rao, A.S. Devakumar, R. Ragagopal, M.J. George and M.R. Sethuraj (1988). Physiology of Drought tolerance of 'Hevea'. C.R. du Colloquem Exploitation Physiolgie et Amelioration de l'***Hevea*** France, pp.269-281.

Vijayakumar, K.R., S.K. Dey, T.R. Chandrasekhar, A.S. Devakumar, T. Mohankrishna, P. Sanjeeva Rao, M.R. Sethuraj.(1998). Irrigation requirement of rubber trees Hevea brasiliensis in the sub-humid tropics. Int. J. of Agricultural Water Management, 35:245-259.

Questions

I. Elucidate on

Agromet research of rubber at RRII

II. Write Short Notes on

1. Agroclimatic characterization of rubber
2. Climatic water balance and rubber yields
3. Soil-Plant-Atmosphere relations in rubber
4. Abnormal leaf fall disease of rubber
5. Irrigation requirement of rubber
6. Crop weather relationships of rubber

Appendices

Vertical Profiles of Air Temperature and Vapour Pressure in Coconut Gardens and Open Space[1]

G.S.L.H.V. Prasada Rao and R.R. Nair

Regional Agricultural Research Station Kerala Agricultural University Pilicode-670353, Kasaragod District, Kerala

Abstract

The mean monthly vertical profile of air temperature in December 1986 showed that the air temperature decreased with height in the open whereas no definite pattern was seen inside the coconut garden due to its canopy structure. The difference in temperature between the open and inside the coconut garden was high (1.0°C) at 30 cm above the ground level and low (0.1°C) at 7m height. The differences were high in the afternoon when compared to the morning hours. The vapour pressure was high inside the crop and tended to increase in the afternoon. Higher amplitudes in air temperature were noticed at the lower levels when compared to the canopy level inside the coconut garden. The vertical profiles of air temperature at different time intervals during the day time on 5-2-1987 (clear day) were also presented.

Introduction

The success of any crop association in coconut gardens depends upon the phyto-climate produced by coconut palms due to its structure and stand. The canopy of coconut palm acts as a screen to the incoming solar radiation and develops its own climate. The India Meteorological Department, Government of India has studied the variations in microclimate within the annual crops and in the open at different Agricultural Research Stations in India. A strong inversion with great stability of the air layers is developed during the day inside the sugarcane field in contrast to the turbulent or

[1]*G.S.L.H.V. Prasada Rao and R.R.Nair 1988 (Eds.): Agrometeorology of plantation crops. Proceedings of the National Seminar, held at RARS, Pilicode during 12-13 March, 1987. Kerala Agricultural University, Mannuthy, Thrissur, Kerala, India.187p*

unstable conditions in the open (Anon 1957). No such studies are seen in plantation crops· like coconut. For understanding the phytoclimate in coconut, an attempt ·has been made to ·study the vertical profiles of air temperature and vapour pressure in coconut gardens.

Materials and Methods

A field experiment was taken up at the Regional Agricultural Research Station, Pilicode in order to study the variations in air temperature and vapour pressure inside the coconut garden and in the open. The crop stand had an average height of 7 m and the palms were 12 year-old. The plant density was 180 per ha (7.5 ·m x 7.5 m). The Assmann Psychrometer was used to record dry and wet bulb temperatures at different heights inside the crop and in the open for comparison. The observations were taken daily twice in the morning (7.25 a.m) as well as in the afternoon (2.25 pm) during December, 1986. Semi-diurnal observations on air temperature were also taken on 5-2-1987 to understand the vertical profiles at different time intervals during the day time.

Results and Discussion

The vertical profiles of air temperature and vapour pressure in the coconut garden and in the open are given in Tables 1, 2 and 3 during December 1986. The mean monthly air temperature decreased (27.6-26.8^{0}C) with height in the open while no definite pattern was seen inside the coconut garden due to its canopy structure and texture. The difference in air temperature between the open and inside the coconut garden was the highest at the ground level than at the crown level. The differences were marked in the afternoon (2.25 pm). The same vertical profile in air temperature was maintained (decreasing with height) in the afternoon but differed in the morning in the open space. Interestingly, air temperature in coconut garden was high in the morning.

Table 1: Vertical profile of mean air temperature and vapour pressure in coconut garden and open during December, 1986

Height (cm)	Temperature (^{0}C)			Vapour pressure (mm)		
	Open	Coconut	Difference	Open	Coconut	Difference
30	27.6	26.6	1.0	19.16	19.62	0.46
60	27.3	26.6	1·0	18.98	19.49	0.51
120	27.1	26.7	0.4	‘18.82	19.31	0.49
240	26.9	26.7	0.2	18.75	19.10	0.35
360	26.9	26.7	0.2	18.59	19.10	0.51
540	26.8	26.6	0.2	18.83	19.24	0.41
660	26.8	26.7	0.1	18.94	19.28	0.34

Table 2: Vertical profile of air temperature and vapour pressure in coconut garden and open at 7.25 a.m. during December, 1986

Height (cm)	Temperature (^0C)			Vapour pressure (mm)		
	Open	Coconut	Difference	Open	Coconut	Difference
30	23.4	23.4	0.0	18.57	18.78	0.21
60	23.0	·23.3	0.3	19.54	18.88	0.34
120	23.0	23.4	0.4	18.54	18.57	0.03
240	23.0	23.5	0.5	18. 3	18.68	0.35
360	23.1	23.5	0.4	18.23	18.68	0.45
540	'23.1	23.6	0.5	18.44	18.68	0.24
660	23.4	23.7	0.3	18.57	18.68	0.11

Table 3: Vertical profile of air temperature and vapour pressure in coconut gardens and open at 2.25 PM during December, 1986

Height (cm)	Temperature (^0C)			Vapour pressure (mm)		
	Open	Coconut	Difference	Open	Coconut	Difference
30	31.8	29.8	2.0	19.75	20.46	0.71
60	31.5	29.9	1.6	19.42	20.10	0.6
120	31.2	30.0	1.2	19.09	20.05	0.96
240	30.8	29,8	1.0	19.16	16.51	0.35
360	30.6	29,8	0.8	18.95	19.51	0.56
542	30.4	29.5	0.9	19.22	19.80	0.58
660	30.3	29.7	0.6	19.10	19.87	0.77

Though the vertical profile of vapour pressure decreased with height inside the coconut garden and the open, a second maxima was noticed after 120 cm height from the surface in both the cases. The second maxima of vapour pressure in the open may be due to the influence of surrounding coconut gardens. The vapour pressure was high inside the crop and tended to increase in the afternoon. This is probably due to high rate of transpiration in the afternoon hours.

As these studies pinpoint the changes in phytoclimatic conditions with time, more and more studies of this nature must be taken up in all plantation crops.

Acknowledgement

The authors are grateful to the Director of Research, Kerala gricultural University for providing necessary facilities to take up this study.

Discussion

S. T. Nagaraj (UAS, Bangalore)
Were the coconut palms in the garden of the same height, same age, or of different heights and ages?

G. S. L. H. V. Prasada Rao
The palms selected for the study were of the same age and height.

P. S. Sreenivasan (Jyothinagar, Palghat)
Please mention the location inside the coconut garden where the observations on temperature profile were made.

G. S. L. H. V. Prasada Rao
The observations were recorded in the middle of the coconut garden in N VI Block of R.A.R.S., Pilicode.

R. C. Dubey (IMD, Pune)
While studying temperature and vapor pressure (T & V.P.), did you consider wind profiles?

G. S. L. H. V. Prasada Rao
No, wind profiles were not taken into consideration.

R. C. Mandal (NRCC, Puthur)
1. More data on temperature profiles should be collected across different periods.

2. The studies should be extended to multi-storeyed cropping systems and include cover crops such as Mimosa and Calapogonium to gather additional insights.

G. S. L. H. V. Prasada Rao
I fully agree with your suggestions.

K. Subba Rao (A.U., Waltair)
Combining your two papers, is there an ideal condition in the atmosphere and soil that ensures maximum yield?

G. S. L. H. V. Prasada Rao
The authors have not studied this aspect.

Effect of Weather Parameters on the Productivity of Coffee and Pepper in Wayanad[2]

K. Kannan, V.S. Devadas and C. George Thomas

Regional Agricultural Research Station, Kerala Agricultural University Ambalavayal, Kerala

Abstract

The intensity and distribution of summer showers in any year determined to a considerable extent the crop prospects of both coffee and pepper in that particular year. While coffee required adequate quantities of rainfall at the two critical periods of blossoming (February-March) and fruit set (March-April), a dry spell from February to April should prevail for satisfactory spiking in pepper. In an year of poor or no summer showers, coffee yield was considerably reduced whereas pepper yield was satisfactory (1981 and 1984). On the other hand, fairly high summer rains resulted in high yield of coffee and low yield of pepper (1982, 1983 and 1985). Satisfactory yield could be expected from both the crops in the event of receipt of moderate blossom and backing showers with a fairly long dry spell. Mixed cropping of coffee and pepper is, therefore, more suited to areas like Wayanad instead of mono cropping with coffee or pepper to ensure reasonable net returns from a unit area.

Introduction

Mixed cropping of coffee and pepper is widely prevalent in the sub mountainous region of Wayanad in Kerala. Both these crops are grown under rainfed conditions. Rainfall seems to be the most significant among the various weather factors influencing flowering and yield of these crops. Blossom showers of 20 to 40 mm during February-March and backing showers of 50 to 75 mm during March-April are considered to be critical for flowering and fruit set in Robusta coffee. Failure of any one of these two showers results in poor crop yield (Coffee Guide, 1985). Scanty blossom

[2]G.S.L.H.V. Prasada Rao and R.R.Nair 1988 (Eds.): Agrometeorology of plantation crops. Proceedings of the National Seminar, held at RARS, Pilicode during 12-13 March, 1987. Kerala Agricultural University, Mannuthy, Thrissur, Kerala, India.187p

showers may give rise to 'Padding' or shanking of spikes, resulting in crop loss according to Krishnamoorthy (1986).The beneficial effect of summer rains on the yield of coconut and oil palm have also been reported by Patel (1938), Gardner *et al.* (1952) and Prasad Rao (1984). On the contrary, pepper requires a short spell of drought during summer months preceding to· flowering in June-July (Pillai et al., 1985). Ridley (1912) reports of profuse blooming and fruiting of pepper vines in Sumatra after a continuous drought period of eight months. With these in view, an attempt has been made on the distribution pattern of summer rains and its influence on the yield of coffee and pepper.

Materials and Methods

Yield data of a 37-year-old (planted in 1949) Robusta coffee garden for ten years and a 14-year-old (planted in 1972) Panniyur-1 pepper garden for7 years recorded at the Regional Agricultural Research Station, Ambalavayal in the sub mountainous region of Wayanad, Kerala were analyzed in relation to rainfall with particular reference to summer showers.

Results and Discussion

The monthly rainfall and number of rainy days from 1976 to 1985 are presented in Table 1. No blossom showers were received in 1976 and 1983 which resulted in low per hectare yield of 172 kg and 315 kg of cured coffee beans during the ensuing cropping season (Table 2). Comparatively higher yields of 684, 753 and 811 kg per ha were obtained in 1977, 1981 and 1984, respectively. In all these years, adequate quantities of blossom and backing showers at critical periods were received. In respect of other years, the rainfall was low and erratic which resulted in moderate yields.

With respect to pepper (Table 3) in 1984. no drought condition was experienced as rainfall was received in all the months from January to June. A total of 244.5 mm rainfall was received in 26 days in March-April. In May also 144.1 mm rainfall was received in 14 days. This resulted in very low per hectare yield of (185 kg) dry pepper in that year. As against this in 1982, there was no rainfall in January-February and the month of March was also practically dry since the rainfall was very scanty (40 mm in 6 days). After this drought period, good rains were obtained from third week of April to August and *this* resulted in the maximum yield of 3599 kg of dry pepper per hectare in· that year. So also, the total rainfall recorded in March-April 1985 was only 168·8 mm in 16 days and the yield recorded was 2966 kg/ha. In 1983 also the rainfall was low in March,

April and the yield was high (2372 kg/ha). As only partial drought conditions prevailed in March-April of 1980, 1981 and 1982 the yield was moderate.

The study revealed that the quantum and distribution of summer rains influenced the crop yield of both coffee and pepper to a considerable extent and their effect was antagonistic. The mean per hectare yield of coffee and pepper from 1979 to 1985 are given in Table 4. The maximum coffee yield was obtained in 1984 (811 kg/ha) whereas pepper yield was very low (185 kg/ha). The pepper yield was comparatively low (956 kg/ha) while coffee yield was high (753 kg/ha) in 1981 since no drought condition was experienced during the summer months. Similarly in 1982, a bumper pepper yield was obtained (3599 kg/ha) while the coffee yield was only 514 kg/ha.

The failure or excess of summer showers in a mono cropping system will have disastrous effect on the economy of the farmer. In a situation where the crop prospects of coffee and pepper depend upon the vagaries of nature, it is advisable to go in for a mixed cropping system wherein a minimum return can be expected per unit area of land.

Table 1: Monthly rainfall (mm) and number of rainy days/month from 1976 to 1985.

Year	Jan.	Feb.	March	April	May	June	July	August	Sept.	Oct.	Nov.	Dec.	Annual total (mm.)
1976	0	0	0	91.2 (19)	229,0 (4)	348.0 (16)	761.0 (28)	131•,3 (24)	331.0 (11)	9'2.4 (11)	57.0 (10)	0.6 (1)	2041.5
1977	0	29.0 (2)	88.4 (3)	51.1(11)	86.1 (16)	145.3 (21)	230.1 (23)	76.7 (18)	76.7 (16)	133.8 (16)	38.5 (11)	0	955.7
1978	0	13.0	24.0	31.6	95.1 (15)	205.3 (28)	217.4 (30)	169.0 (30)	30.6 (5)	47.5 (10)	83.2 (7) I	30.7 (4)	947 4
1979	0	15.6 (5)	4.4 (2)	57.7 (10)	203.2 (11)	458.7 (20)	518.2 (24)	510.2 (17)	212.0 (21)	102.4 (12)	143.6 .(13)	0	2226.0
1980	0	0	65.2 (7)	208.5 (14)	193.9 (18)	634.4 (29)	638.0 (31)	277.4 (31)	157.9 (19)	230.1 (16)	195.8 (11)	0.5 (1)	2601.7
1981	0	3.1 (2)	63.1 (11)	78,6 (10)	181.2 114)	434.9 (27)	408 5 (22)	460.0 (30)	334,2 (28)	200.6 (17)	59.3 (11)	29.7 (2)	2253.2
1982	0	0	40.0 (6)	78 6 (9)	225.5 (15)	285.2 (21)	363.0 (26)	343.4 (25)	54.6 (13)	149.4 (11)	221.8 (14)	0	1761.5
1983	0	0	1.5 (1)	21.5 (3)	118.2 (12)	277.2 (20)	458.7 (20)	435 4 (24)	179.8 (25)	196.4 (13)	39.4 (4)	53.9 (6)	1782.0
1984	44.4 (7)	109.2 (8)	114.5 (11)	130.0 (15)	144.1 {14)	531.0 (26)	613.8 (27)	174 0 (25)	197.6 (20)	270.6 (14)	19.8 (2)	31.6 (1)	2380.6
1985	13.6 (1)	3.4 (,)	75.8 (3)	93.0 (13)	56.0 (10)	574.6 (27)	178.6 (28)	313.2 (27)	114r5 (20)	109.4 (12)	170.2 (6)	26.2 (5)	1728.5

( ) Indicates *no.* of rainy days

Table 2: Amount of blossom and backing showers and mean yield of coffee Blossom showers

Year	Backing showers (February-March) in mm	Backing showers (April-May) in mm	Yield of Beans (kg/ha)
1976	Nil	320	172
1977	117.4	137	684
1978	37.0	126	535
1979	20.0	260	377
1980	65.2	402	368
1981	66.2	259	753
1982	50.0	304	5-85
1.983	1.5	139	315
1984	223.7	274	811
1985	79.2	149	514

Table 3: Rainfall during March-April and mean yield of dry pepper

Year	Rainfall (mm)	Number of rainy days	Yield (kg/ha)
1979	62.1	12	1011
1980	273.7	21	1480
1981	141.7	21	956
1882	118.6	15	3599
1983	23.0	4	2372
1984	244.5	26	185
1985	168.8	16	2936

Table 4: Mean yield of coffee and pepper from 1979 to 1985

Year	Yield of coffee beans (kglha)	Yield of dry pepper (kg/ha)
1979	377	1011
'1980	368	1480
1981	753	• 956
1982	585	3599
1983	315	2375
1984	811	185
1985	514	2966

References

Central Coffee Research Institute 1985. Coffee Guide CCR!; Coffee Research Station, Chikmagalore, Karnataka.

Gardner, V. R., Brad_ford; F. C. apd Hooker, H.D. 1952. The fundamentals of fruit production. Mc. Grow - Hill Book Co. Inc., New York.

Krishnamoorthy, K. K. 1986. Soil-Water-Plant relationships in Plantation i;:::rops The Planters' Chronicle LXXX (8): 298.

Patel, J. -S. 1938. The Coconut Monograph-Govt. Press, Madras.

Pillai, V. S:, Abi Cheeran and S. Sasikumar' 1985. Pepper (Malayalam). Tech Bulletin No. 18-KAU.

Prasad Rao, GSLHV 1984. Rainfal I and Coconut yield in the Pilicode region, North Kerala-Proc. Placrosym V-1982: 388-393.

Ridley, H. N. 1912. Spices. Mac Millan and Co.

Influence of Soil Texture on the Thermal Regime of Coconut Root Zone[3]

G.S.L.H.V. Prasada Rao and R.R. Nair

Regional Agricultural Research Stat ion, Kerala Agricultural University, Pilicode-670 353, Kasaragod District, Kerala

Abstract

The soil temperatures in the coconut root zone in lateritic loam and sandy loam were compared based on the data collected at Pilicode and Nileswar during 1983-84. The mean annual soil temperature in the coconut root zone was higher (30.1-32°C) in sandy loam than in loam (27.9-30.6°C). Though the root zone temperature was high during summer in both the soils, the magnitude of difference between the two soils was high (1.9-5.9°C) in the post-monsoon season and low (0.9-2°C) in the South West monsoon. A marked difference in the thermal regime of coconut root zone was seen during the summer months of 1983 (drought year) and 1984 (normal year) in the same soil. Under mulched and exposed conditions in the loam soil, the difference in coconut root zone temperature was about 1°C in summer and about 0.3°C in the South West monsoon. The coconut root zone temperature in the loam soil was 28.6°C under coconut husk mulch and 29.2°C under no mulch at both 40 and 60 cm depths.

Introduction

The effect of soil temperature on root development has been studied *by* several workers (Shaw and Buchele, 1957; Burrows, 1963; and Mederski and Jones, 1963) in annual crops. However, such studies are lacking in perennial crops like coconut. Hence an attempt has been made in this study to understand the variations in soil temperature in the root zone of coconut grown in lateritic loam and sandy loam soils at the Regional Agricultural Research Station, Pilicode during the years 1983 and 1984.

[3]*G.S.L.H.V. Prasada Rao and R.R.Nair 1988 (Eds.): Agrometeorology of plantation crops. Proceedings of the National Seminar, held at RARS, Pilicode during 12-13 March, 1987. Kerala Agricultural University, Mannuthy, Thrissur, Kerala, India.187p*

Materials and Methods

Soil thermometers were installed in the coconut basins at a distance of 1 m from the bole of coconut palms in loam and sandy loam soils. Another set of soil thermometers were installed in the open in sandy loam soil for comparison. The soil temperatures were recorded at 2.5, 10, 40, 60 and 70 cm depths from the surface. The observations on soil temperature were recorded daily twice in the morning (7:25 a. m.) as well as in the afternoon (2:25 p. m.) for two years 1983 (drought year) and 1984 (normal year).The soil temperature was also recorded in the effective root zone of coconut palm under mulched (coconut husk) conditions for comparison in the loam soil. The data on soil temperature were recorded daily during the study period.

Results and Discussion

The data on the annual soil temperature (^{0}C) in the coconut root zone and in the open in sandy loam soil during 1983-84 are given in Table I. It can be seen that the difference was conspicuous (3.0-3.4^{0}C) in the surface soil while it was not so (2.1-2.2^{0}C) in the deeper layers (30-70 cm). The range in soil temperature was high (32.8-36.1^{0}C) in open and low (30.1-32.7^{0}C) in the coconut root zone.

Table 1: Mean annual soil temperature (^{0}C) in coconut root zone and open in sandy loam soils during 1983-84

Depth (cm)	Open (0C)	Coconut (0C)	Difference (0C)
2.5	36.1	32.7	3.4
10	34.0	31:0	3.0
30	32.3	30.1	2.2
40	32.8	30.6	2.2
60	33.0	30.9	2.1
70	32.8	30.6	2.2

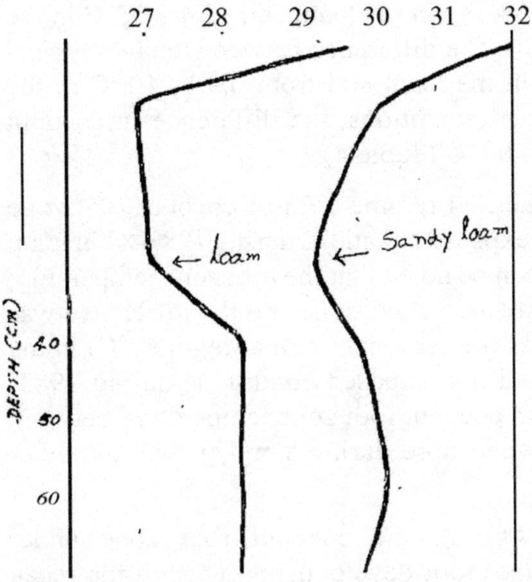

Fig. 1: Profile of soil temperature (⁰C) in two different soils of coconut root zone

The mean annual soil temperatures (⁰C) in the coconut root zone in loam and sandy loam soils during March 1983 to February 1984 are depicted in Fig. I. The_mean annual soil temperature in coconut root zone was higher (30.1-32.7⁰C) in sandy loam than in loam (27.9- 30.6⁰C). The soil temperature inversion was noticed (temperature increases with depth) beyond 10 cm depth in the loam but only beyond 30 cm depth in the sandy loam. This was probably due to the presence of more roots in the loam at 10 cm depth when compared to that of sandy loam.

Table 2 gives the seasonal thermal regime (⁰C) of coconut root zone *in* two different soils. It can be seen that the root zone temperature was high throughout the year under sandy loam. A marked variation in the thermal regime of coconut root zone was seen from one season to the other though the root zone temperature was high during the summer in both the soils. The difference between the two soils was high (1.9-5.9⁰C) in the post-monsoon (October-November) period and low (0.9-2.0⁰C) in the Southwest monsoon. This was probably due to sudden depletion of moisture in the surface soil in the sandy loam immediately after the cessation of monsoonal rains.

The thermal regime (⁰C) of coconut root zone during the summer under two different soils in two typical years, 1983 (drought year) and 1984 (normal year) are presented in Table 3. It is seen that the root zone temperature and its range were higher (30.1-34.1°C in the loam and 32.4-37.3°C in the sandy loam) in 1983 than in 1984. The difference between the two typical years varied from 0.3 to 3.2°C in the loam and from 0.5 to 4.1°C *in* the sandy loam. Even under mulched conditions, the difference in coconut root zone temperature was 0.5 to 0.8°C (Table 4).

The data on the mean annual thermal regime (°C) of coconut root zone under mulch (coconut husk) and exposed situations during 1983-84 in loam soils all furnished in Table 5. I t can be noted that the root zone temperature was comparatively low under mulched conditions and the difference was 0.4 to O 6°C. Table 6 summarizes the seasonal thermal regime (°C) in the coconut root zone under mulched and exposed conditions during 1983-84 in loam soil. The difference in coconut root zone temperature between mulch and no mulch was relatively more during summer than the other seasons.

High soil temperatures (35-38°C) *in* the coconut root zone under unirrigated conditions may hamper root development. Though the mean soil temperature during the summer was around 34°C and 38°C in the loam and sandy loam soils, respectively, the diurnal fluctuations were very high shooting sometimes up to 60°C in the afternoon hours at the surface depths. Cooper (1973) revealed that the root temperature above 33°C drastically decreased the shoot dry weight in maize. Studies of this nature will be useful to understand the effect of soil temperature on the development of coconut roots.

Table 2: Seasonal thermal regime (^0C) of coconut root zone in two differerent soils during 1983-84

Depth (cm)	Summer			Monsoon			Monsoo			Winter		
	SL	L	Diff.	SW	SLL	Diff.	Post	SL L	Dif	SL	L	Diff.
10	32.4	30.1	2.3	29.2	27.2	2.0	32.4	26 5	5.9	31.1	27.7	3.4
30	32.4	30.3	2.1	28.8	27.7	1.1	30.4	26.8	3.6	29.3	27.4	1.9
40	32.9	31.4	1.5	29.5	28.6	0.9	30.4	28.0	2.4	30.6	28.7	1.9
60	32.6	31.1	1.5	30.0	28.8	1.2	30.6	28.0	2.6	30.5	28.7	1.8
70	32.8	30.9	1 9_	29 9	28.9	1.0	30.0	28.1	1.9	30,.0	28.8	1.2

SL-Sandy Loam; L-Loam

Table 3: Thermal regime ("C) of coconut root zone during summer under two different soils in typical years (1983 and 1984)

Depth (cm)	Sandy loam			Loam		
	1983	1984	Diff.	1983	1984	Diff.
2.5	37.3	33.2	4.1	34.1	30.9	3.2
1'0	32.6	32.1	0.5	30.1	29.2	0.9
30	32.4	30.9	1.5	20.3	29.4	0.9
40	32.9	31.5	1.4	31.4	30.7	0.7
60	32.6	31.9	0.7	31.1	30.5	0.6
70	32.8	31.4	1.4	30.9	30.6	0.3

Table 4: Seasonal thermal regime *CC)* in coconut root zone under mulch and no mulch during 1983-84 in loam soil

Depth	Summer			SW monsoon			Post monsoon			Winter		
(cm)	NM	M	Diff.	NM	M	Diff.	NM	M	Diff	NM	M	
10	30.1	29.8	0.3	27.2	27.1	0.1	26.5	26.1	0.4	27.7	26.8	0.9
40	31:4	30.4	1.0	28.6	28.4	0.2	28.0	27.4	0.6	28.7	27.8	0.9
60	31.1	30.3	0.8	28 8	28.5	0.3	28.0	27.5	0.5	2s.1	28.0	0.7

NM-No mulch, M-Mulched with coconut husk.

Table 5: Thermal regime (°C) in coconut root zone during summer (March-May) under mulch (coconut husk) in 1983 and 1984

Depth	1983	1984	Difference (°C)
10	29.8	29.0	0.8
40	30.4	29.8	0.6
60	30.3	29.8	0.5

Table 6: Thermal regime (°C) in coconut root zone under mulched and exposed situations

Depth (cm)	No mulch (°C)	Coconut husk (°C)	Difference (°C}
10	27.9	27.5	0.4
40	29.2	28.6	0.6
60	29.2	28.6	0.6

Acknowledgement

The authors are grateful to the Director of Research, Kerala Agricultural University tor providing necessary facilities during the course of this study.

References

Cooper, A. J. 1973. Root temperature and plant growth. Research Review, 4-Commonwealth Bureau of Horticultural and Plantation Crops, East Melling, Kent.

Burrows, W. C. 1963. Characterisation of soil temperature distribution from various tillage-induced microleafs. Soil Sci. Amer. Proc. 27: 3 50-353,

Mederski, H.J. and Jones, Jr. J.B. 1963. Effect of soil temperature on corn plant development and yield: 1. Studies with a corn hybrid. Soil Sci. Soc. Amer. Proc. 27: 186-187.

Shaw, R.H. and Buchele, W. F. 1957. The effect of the shape of the soil surface profile on soil te.mperature and moisture. Iowa State College. J. Sci. 32: 95-104.

Discussion

Mohamed Kunju (RARS, Kumarakom)

Did you make any observation on the water table in different soil types and correlate it with soil temperature?

In all cases, there is a reduction in the thermal values when the depth increases.

G. S. L. H. V. Prasada Rao

No, the impact of the water table on soil temperature does not arise at the experimental site, as the groundwater table is very low.

If you observe the variations in soil temperature closely, you will not find a conspicuous reduction in temperature up to a depth of 60 cm.

P. S. Sreenivasan (Jyotinagar, Palghat)

Have you tried different types of mulches other than coconut husk?

G. S. L. H. V. Prasada Rao

Yes, soil temperatures were also recorded under tree leaf mulch. However, no significant variation in temperature was observed in the effective root zone when compared to coconut husk mulch.

R.C. Mandal (NRCC, Puttur)

Studies conducted on sandy loam and sandy soils over the past two years have shown only marginal variation in soil moisture.

Weather parameter studies should be correlated with end products such as yield or growth parameters; otherwise, the studies remain incomplete.

G.S.L.H.V. Prasada Rao

Your suggestions are valuable and will be considered in future studies.

Impact of Drought on Stem Bleeding Disease Incidence of Coconut[4]

T.C. Radhakrishnan and G.S.L.H.V. Prasada Rao

Regional Agricultural Research Station. Kerala Agricultural University
Pilicode 670 353, Kasaragod District, Kerala

Abstract

The influence of soil moisture deficiency on stem bleeding disease of coconut was studied utilizing monthly aridity indices during the summer months. The disease incidence was more in 1978 (185 palms), 1979 (189 palms), and 1983 (188 palms) when the mean aridity indices were greater than 95 per cent while it was less in 1980 (37 palms) when the aridity index was 47.65 per cent A significant positive correlation between the mean aridity index during summer and the disease incidence was noticed. A multiple linear regression has been worked out for estimating the disease incidence based on monthly aridity indices during summer (January to May).

Introduction

Stem bleeding of coconut is a disease of unknown etiology. Some workers consider that the disease is caused by the fungus Thelaviopsis paradoxa. However, other fungi like Phomopsis cocoina: Paecilomyces varotii and Schizophyllum commune are also, involved in the disease. The disease is believed to be caused by infection through growth cracks on the stem, which are vulnerable to the fungus. Cracks develop due to sudden changes in moisture content in the soil. Sampson (1923) has described an instance in which lack of soil drainage aggravated a condition of stem bleeding in palms. Briton Jones (1940) opined that to avoid the disease, all attempts should be made to mitigate drastic changes in soil moisture. Soil moisture is a highly complex and dynamic parameter whose depletion from the root zone is controlled by soil, plant and climatic factors (Malik et al., 1986). In order to understand the impact of soil moisture deficiency on the incidence of stem bleeding disease, this study was conducted.

---

[4]*G.S.L.H.V. Prasada Rao and R.R.Nair 1988 (Eds.): Agrometeorology of plantation crops. Proceedings of the National Seminar, held at RARS, Pilicode during 12-13 March, 1987. Kerala Agricultural University, Mannuthy, Thrissur, Kerala, India.187p*

Materials and Methods

The palms in different blocks of RARS, Pilicode constituted the material for the study. The observations on the incidence of stem bleeding disease in the years 1976 to 1983 were taken. The monthly aridity index (la) which is expressed as the ratio of water deficiency (WD} to water need (WN) in percentage (WD/WN x 100) was worked out using Thornthwaite and Mather (1955) book-keeping water balance procedure. Based on the monthly aridity index and .the stem bleeding disease incidence, a multiple regression equation was developed for estimating the diseased incidence.

Results and Discussion

Table 1 gives the mean aridity index (la) and disease incidence during summer (March to May). It can be seen that the disease incidence was more in drought years 1978 (185 palms), 1979 (189 palms) and 1983 (188 palms) during which the mean aridity indices (la) for the summer months (March, April and May) were greater than 95%. The disease incidence in the year 1980 was less (37 palms) when the aridity index was also 47.65%, Also, there was a singificant positive correlation (at 0.01% level) between the aridity index and the disease incidence. The monthly aridity index (la) during summer and the disease incidence are given in Table 2. It can be seen that the disease incidence was high in the years 1978 and 1979 when the aridity indices were high (98.94 in 1978 and 97.29 in 1979) in April and the incidence was low in 1980 when the aridity index was 23.98 though the aridity index till March did not vary much in both the cases. It indicated that the aridity index in April was critical in determining the disease incidence. It was probably due to continuous soil moisture stress over a longer period which caused production of cracks on the stem through which the fungi entered. The following equation is suggested for estimating the disease incidence based on the monthly aridity values

$$Y = 243.0155 + 2.4895\, x1 - 8.8681\, x2 + 3.7445\, x3 + 1.6488\, x\, 4 + 0.2550\, X\, 5$$

Where xl, x2, X3, x4 and x5 are the monthly aridity indices of January, February, March. April and May, respectively. The equation has a multiple correlation coefficient of 0.8995 accounting for 80% of the disease incidence.

Table 1: Mean aridity index (la) during summer(March-May) and disease incidence

Year	Average (Ia)	Disease incidence
1976	72.73	159
1977	70.41	157
1978	97.52	185
1979	96.1	189
1980	47.65	37
1981	72.11	120
1982	79.5	140
1983	97.5	188

Correlation coefficient (r) = 0.9142, Significant at 0.01 level,

Table 2: Monthly aridity index during summer and disease incidence

Year	Jan	Feb	Mar	Apr	May	Disease incidence
1976	79.92	92.45	96.49	86.34	35.35	159
1977	70.63	76.03	84.15	56.67	0	157
1978	69.12	89.67	“96.11	98.94	0	185
1979	34.97	75.52	91.53	97.27	99.5	189
1980	70.63	88.08	95.56	23.98	23.42	37
1981	76.42	89.78	96.77	95.31	24.26	120
1982	73.52	89.55	93.41	98.89	46.20	140
1983	81.54	92.76	97.89	98.89	95.26	180

References

Briton Jones, H. R. 1940.The diseases of coconut palm. Baif liare Tindall & Com London

Malik, R. K., Narda, N. K. and Murty, V. V. N. 1986.- A critical appraisal of Macro scopic scale models of soil moisture dynamics _under cropped conditions for soil moisture estimation. International journal of Tropical Agriculture: 300-309

Sampson, H. C. 1923. The coconut palm. Hohn Balle Sons & Danidson Ltd. London.

Discussion

G. P. Bishnoi (HAU, Hisar)

What was the r value between the aridity index during summer and disease incidence?

T. C. Radhakrishnan

The r value was 0.9142, which is significant at the 0.01 level.

Onset of Monsoon over Kerala (India) from 1870 to 2025

Year	Onset	Year	Onset	year	Onset	Year	Onset
1870	June 3	1909	June 2	1948	June 11	1987	June 2
1871	May 31	1910	June 2	1949	May 23	1988	May 26
1872	June 1	1911	June 6	1950	May 27	1989	June 3
1873	May 23	1912	June 8	1951	May 31	1990	May 19
1874	May 16	1913	June 2	1952	May 20	1991	June 2
1875	June 3	1914	June 4	1953	June 7	1992	June 5
1876	June 8	1915	June 15	1954	May 31	1993	May 28
1877	June 7	1916	June 2	1955	May 29	1994	May 28
1878	June 9	1917	May 31	1956	May 21	1995	June 5
1879	May 17	1918	May 11	1957	June 1	1996	June 3
1880	May 26	1919	June 3	1958	June 14	1997	June 9
1881	June 9	1920	June 3	1959	May 31	1998	June 2
1882	May 31	1921	June 2	1960	May 14	1999	May 25
1883	June 3	1922	May 31	1961	May 18	2000	June 1
1884	June 9	1923	June 11	1962	May 17	2001	May 23
1885	June 4	1924	June 2	1963	May 31	2002	May 29
1886	May 31	1925	May 27	1964	June 6	2003	June 8
1887	June 1	1926	June 6	1965	May 26	2004	May 18
1888	May 28	1927	May 27	1966	May 31	2005	June 5
1889	May 30	1928	June 3	1967	June 9	2006	May 26
1890	May 28	1929	May 29	1968	June 8	2007	May 28
1891	June 2	1930	June 8	1969	May 17	2008	May 31
1892	May 26	1931	June 4	1970	May 26	2009	May 23
1893	May 26	1932	June 2	1971	May 27	2010	May 31
1894	June 4	1933	May 22	1972	June 18	2011	May 31
1895	June 12	1934	June 8	1973	June 4	2012	June 5
1896	June 3	1935	June 12	1974	May 26	2013	June 1
1897	June 6	1936	May 19	1975	May 31	2014	June 6
1898	June 4	1937	June 4	1976	May 31	2015	June 5
1899	May 27	1938	May 26	1977	May 30	2016	June 8
1900	June 10	1939	June 5	1978	May 28	2017	May 30
1901	June 7	1940	June 14	1979	June 13	2018	May 29
1902	June 6	1941	May 23	1980	June 1	2019	June 8
1903	June 12	1942	June 10	1981	May 30	2020	June 1
1904	June 7	1943	May 29	1982	June 1	2021	June 3
1905	June 10	1944	June 3	1983	June 13	2022	May 29
1906	June 14	1945	June 5	1984	May 31	2023	June 8
1907	June 8	1946	May 29	1985	May 28	2024	May 30
1908	June 11	1947	June 3	1986	June 4	2025	May 24

Index